# Formal Logic

# Formal Logic

## Paul A. Gregory

broadview press

BROADVIEW PRESS – www.broadviewpress.com
Peterborough, Ontario, Canada

Founded in 1985, Broadview Press remains a wholly independent publishing house. Broadview's focus is on academic publishing; our titles are accessible to university and college students as well as scholars and general readers. With over 600 titles in print, Broadview has become a leading international publisher in the humanities, with world-wide distribution. Broadview is committed to environmentally responsible publishing and fair business practices.

The interior of this book is printed on 100% recycled paper.

**Library and Archives Canada Cataloguing in Publication**

Gregory, Paul A., author
    Formal logic / Paul A. Gregory.

ISBN 978-1-55481-272-1 (softcover)

    1. Logic, Symbolic and mathematical—Textbooks.
2. Textbooks. I. Title.

BC135.G74 2017          160          C2017-903938-5

*Broadview Press handles its own distribution in North America*
PO Box 1243, Peterborough, Ontario K9J 7H5, Canada
555 Riverwalk Parkway, Tonawanda, NY 14150, USA
Tel: (705) 743-8990; Fax: (705) 743-8353
email: customerservice@broadviewpress.com

Distribution is handled by Eurospan Group in the UK, Europe, Central Asia, Middle East, Africa, India, Southeast Asia, Central America, South America, and the Caribbean. Distribution is handled by Footprint Books in Australia and New Zealand.

Broadview Press acknowledges the financial support of the Government of Canada through the Canada Book Fund for our publishing activities.

# Canadä

Edited by Robert M. Martin
Typeset by Paul A. Gregory
Cover design by Black Eye Design

PRINTED IN CANADA

To Molly, Jamie, John, Sammy, and Ben

# Contents

# Acknowledgements

To say this project has been a labor of love would be a gross understatement on both counts. Fortunately, I have been privileged to share much of that labor—and hopefully some of that love—with a host of people.

Approximately a thousand students have taken my introductory logic courses over the years. Many, especially early on, found errors and warts. I give thanks to all of them for helping me learn how to teach logic—and, hopefully, learn to teach it well. Special thanks go to those who had the keen eyes, confidence, and willingness to bring the errors and warts to my attention.

Sara Sprenkle of the Computer Science Department at Washington and Lee University deserves deep thanks. She and her students developed, debugged, improved, and spruced up the translation practice software that accompanies this text. More detail on the genesis and growth of the application can be found online, but Sara deserves explicit and repeated thanks. Thank you Sara!

My work on this text could not have been completed without the generous summer research support offered by Washington and Lee University. For multiple Glenn Grants early on, I thank Mr. John M. Glenn, and for multiple Lenfest Grants, I thank Mr. H.F. (Gerry) Lenfest.

I am grateful to Aleksander Simonic, creator of WinEdt, the editor in which this whole project has been written. WinEdt has been a constant friend and helper on this journey and with it I am a smarter, better writer.

I thank the editors and designers at Broadview Press. Stephen Latta's kind and encouraging tone in our initial communications was a major factor in my decision to work with Broadview. The book took a bit longer to complete than originally planned, but Stephen's seemingly infinite patience helped me to stay calm and keep working. Bob Martin's careful editing greatly improved the readability of the text. Having lived with the main portions of the book for years, I could not see all the rough spots. Bob zeroed in on them and pressed me to smooth and clarify. Joe Davies proofread with a keen eye, finding more issues than I would have guessed. Thanks to Bob and Joe, no one will know just how many rough spots there were. Those that remain are entirely my fault. Tara Lowes and the design and editing team at Broadview expertly ushered the final product into existence.

Finally, thanks and love go to those closest to me throughout various stages of this project: to TJ for wonderfully blurring the lines between mentor and mentee; to Henry, who, sadly, did not live to see the completion of this project; to his successors, Oreo and Lilly, who may yet; to my parents, who have always supported me; to the Boys: Sammy, John, Jamie, and Ben—my greatest accomplishments. And to Molly—you make it all worth the effort.

# Part I
# Informal Notions

# 1  Informal Introduction

## 1.1  Logic: What, Why, How?

### What Is Logic?

Logic is the study of argument. In particular, logic is the study of criteria for distinguishing successful from unsuccessful arguments and the study of methods for applying those criteria.

By 'argument' I do not mean a shouting match or angry disagreement (though sometimes these accompany the sort of argument that interests us). Rather, an argument is a set of statements, some of which—the *premises*—are supposed to support, or give reasons for, the remaining statement—the *conclusion*. Such statements, and so the argument they constitute, may be written, spoken, thought, or otherwise communicated or privately considered.

Usually the intent behind an argument is to produce understanding or conviction in oneself or another. However, by 'successful argument', I do *not* mean one that succeeds in persuading the reader or listener or thinker. Due to the foibles of human psychology and/or the use of manipulative rhetorical devices, people can be persuaded by arguments even though the premises do not genuinely support the conclusion. Moreover, people can fail to be convinced by even the tightest and clearest reasoning. Strategies for detecting and analyzing manipulative rhetoric and strategies for overcoming natural human obtuseness will not be the focus of this text—those are better placed in a text on general reasoning or critical thinking, rather than in one dedicated to formal logic. Here we shall focus on the core principles that any such critical strategies presuppose: the criteria for successful argument.

The basic idea behind these criteria is that the premises should genuinely support the conclusion—as opposed, for example, to merely inspiring a new, or reinforcing an existing, emotional commitment to the conclusion. To achieve genuine support, the premises should be related to the conclusion in a way that either guarantees or makes probable the preservation of truth from premises to conclusion. In other words, in a successful argument, if the premises are true, then the conclusion is either guaranteed to be true or likely to be true. These rough statements will require much clarification and elaboration. But, before moving on, it is worth considering why one might want to study logic.

### Why Study Logic?

There is a great deal of instrumental value in studying logic—logic is *good for* other desirable things. First, as alluded to above, skill in logical analysis and evaluation is the core of critical thinking. So, increasing your logical abilities will lead to a corresponding increase in your overall critical ability. You will be better able to guard against psychological fallibility and manipulative rhetoric—your own as well as others'. Second, nearly all of the methods presented in this book involve abstracting away from certain details of language in order to analyze the underlying logical form. Thus, you will gain insight into the structure of language and the interconnections among sentences—insight that can increase the clarity and precision of your speaking and writing. Third, the methods here presented will strengthen your abilities to think abstractly and analytically. Such strength is valuable not only in logic, but also in other academic, professional, and creative domains.

It can, however, be easy to lose sight of this broader payoff as we delve deeper and deeper into symbols and abstractions. But consider an analogy with physical exercise: In daily life, or out on the playing field, we are seldom called upon to do reps on a weight machine, run on a treadmill, or flop around on an exercise ball. Yet the repetition of abstracted movements performed in the gym has a clear and definite payoff in terms of your health, strength, flexibility, and precision—both on the street and on the field. Just so, even if you are seldom called upon to perform the symbolic operations learned in this course, you will (if fully committed to the regimen) receive clear and definite payoffs in terms of the quality, strength, flexibility, and precision of your thinking, speaking, and writing—both on the street, and in other classrooms.

Finally, in terms of instrumental value, the formal symbolic approach we pursue here makes direct contact with other fields of study, including philosophy, mathematics, linguistics, computer science, and cognitive science/artificial intelligence. If you continue in one of these fields, you will likely use systems and methods very similar to those presented here.

Of course, beyond instrumental value, there is a great deal of intrinsic value—the study of logic is good *in itself*. First, there is the discovery and learning of new things. Then there is aesthetic and intellectual delight in abstract structures and systems. Less grandiose, perhaps, is the enjoyment of puzzle solving and rising to meet an intellectual challenge. This is not to say that appreciating logic as good in itself is required to do well in a course on logic. Far from it. But it would not be uncommon for you suddenly to find—part way through the term and despite your initial expectations—that you "just like it".

### How Will We Study Logic?

In this chapter I divide criteria for successful arguments into two broad types—*deductive* and *inductive*. In contrast to inductive criteria we will find that the satisfaction of deductive criteria is owed crucially to the *form* of an argument. 'Form' here includes the structure of the statements involved and the manner in which those statements are interrelated as a result of that structure. In all subsequent chapters we will focus exclusively on deductive criteria, exploiting this notion of form in order to be as precise as possible. Our main tools in this effort will be a pair of artificial languages—languages that use a variety of special symbols to stand for logically important parts of statements. These languages will allow us to clearly capture and systematically examine statement and argument form. Thus, we will be able to precisely state logical criteria and develop systematic methods for applying them.

In addition to using these tools, we will occasionally step back and examine the tools themselves, reflecting on strengths, weaknesses, and the trade-offs we have made in their design. At such times we will be engaging in *metalogic*—the study of properties of logical systems—and Part IV of this text is devoted to some of the more advanced portions of logic and some metalogic.

## 1.2  **Arguments, Forms, and Truth Values**

Now it is time to get down to business. To start out, let's officially define some of the terms used in the previous section, as well as some new terms:

**Logic:**
> *Logic* is the study of (i) criteria for distinguishing successful from unsuccessful argument, (ii) methods for applying those criteria, and (iii) related properties of statements such as implication, equivalence, logical truth, consistency, etc.

**Metalogic:**
> *Metalogic* is the study of the properties of logical systems.

Part (iii) of the definition of 'logic' was not discussed above, but the concepts mentioned there will be covered in subsequent chapters. I include them here for the sake of completeness.

**Statement:**
> A *statement* is a declarative sentence; a sentence that attempts to state a fact—as opposed to a question, a command, an exclamation.

**Truth Value:**
> The *truth value* of a statement is just its truth or falsehood. At this point we make the assumption that every statement is either true (has the truth value true) or false (has the truth value false) but not both. The truth value of a given statement is fixed whether or not we *know* what that truth value is.

For variety, throughout most of the text 'statement', 'sentence', and 'claim' will be used interchangeably to refer to declarative sentences. Moreover, our assumption regarding truth values—that all statements are either true or false and not both—is one that may reasonably be questioned. Issues of vagueness, ambiguity, subjectivity, and various sorts of indeterminacy may lead us to think that some statements are neither true nor false, or both true and false, or "somewhere in between". There are logics that attempt to formalize such intuitions by countenancing three or more values (as opposed to our two), and treating compound claims and arguments in correspondingly more complex ways than we will. To genuinely appreciate such logics, however, it is important to have an understanding of classical two-valued (or bi-valent) logic. Such is the subject of this book.

**Argument, Premise, Conclusion:**

> An *argument* is a (finite) set of statements, some of which—the *premises*—are supposed to support, or give reasons for, the remaining statement—the *conclusion*.

This is a standard definition. The restriction to a finite number of premises is mainly a matter of convenience for constructing our formal system. It will allow us to produce truth tables, trees, and derivations, without having to worry about cases of infinite lists of premises. The restriction is also supported by the notion that arguments, as typically conceived, are or could be given by people with finite time, space, and speed of expression. We could choose differently.[1]

When we encounter an argument in the course of reading or during discussion the premises and conclusion may come in any order. Consider the following versions of a classic example:

- Socrates is mortal, for all humans are mortal, and Socrates is human

- Given that Socrates is human, Socrates is mortal; since all humans are mortal

- All humans are mortal; Socrates is human; therefore, Socrates is mortal

Three statements are involved in each of the above examples (two premises and a conclusion), and despite the fact that they appear in different order in each one, all three examples express the same argument. For the sake of clarity we will often transcribe arguments into what is called *standard form*—we list the premises, draw a line, then write the conclusion:

Premise 1
Premise 2
$\vdots$
Premise $n$
Conclusion

All humans are mortal
Socrates is human
Socrates is mortal

Despite the variability of statement order when not expressed in standard form, a good writer usually makes clear which sentences are premises and which is the conclusion. This is usually done through contextual clues, including the indicator words/phrases I used above. Below are two brief lists of conclusion and premise indicators:

**Premise Indicators:**

> as, since, for, because, given that, for the reason that, inasmuch as

**Conclusion Indicators:**

> therefore, hence, thus, so, we may infer, consequently, it follows that

---

1. See, e.g., the discussion of entailment in Section 1.5, where we do allow infinite sets.

So it is usually a relatively simple task to put arguments into standard form. Of course, if the argument is long and complex, with sub-conclusions acting as premises for further conclusions, analysis can get messy.

In addition to standard form, the notions of argument *form* and *instance* will be important in what follows. An *argument form* (sometimes called a *schema*) is the framework of an argument that results when certain portions of the component sentences are replaced by blanks or schematic letters. An *argument instance* is what results when the blanks in a form are appropriately filled in. For example, the argument presented above is an instance of the following form:

> All F are G
> x is F
> ―――――――
> x is G

Here 'F' and 'G' are placeholders for predicate phrases, while 'x' is a placeholder for a name. The form was generated by replacing 'humans' with 'F', 'mortal' with 'G', and 'Socrates' with 'x'. So we have here our first use of symbols as a means to capturing logical form. I trust it is fairly intuitive at this point. With little or no effort you will see that the following is also an instance of the above form:

> All pop-stars are attention-starved
> Mr. Green Jeans is a pop star
> ―――――――――――――――――――――
> Mr. Green Jeans is attention-starved

Here 'pop-star' goes in for 'F', 'attention-starved' for 'G', and 'Mr. Green Jeans' for 'x'. The result is another argument of the same form.

**Argument Form and Instance:**
> An *argument form* (or schema) is the framework of an argument that results when certain portions of the component statements are replaced by blanks, schematic letters, or other symbols. An *argument instance* is what results when the blanks, schematic letters, or other symbols in a form are appropriately filled in.

Now I will divide criteria for evaluating arguments into two basic types. We will have similar but different things to say about what counts as "success" for each type. As mentioned previously, the two types are Deductive and Inductive. These are discussed in Sections 1.3 and 1.4, respectively.

## 1.3 **Deductive Criteria**

Deductive criteria, roughly speaking, require that the premises guarantee the truth of the conclusion. There are two questions we want to ask when applying deductive criteria. One, the question of *validity*, has to do with the connection between the premises and conclusion. The other, the question of *soundness*, has to do in addition with the truth values of the premises. First, validity:

**Deductive Validity, Invalidity:**

> An argument (form) is *deductively valid* if and only if it is NOT possible for ALL the premises to be true AND the conclusion false. An argument (form) is *deductively invalid* if and only if it is not valid.

So an argument is valid if and only if the assumed truth of the premises would guarantee the truth of the conclusion—if you mull it over you'll see that these say more or less the same thing. The statement above, however, is our official, precise definition of validity. (Where no confusion threatens, I will drop the 'deductive' and 'deductively' from my talk of validity.) Note also that the definition of validity applies both to individual arguments and to argument forms. This will be made clear in what follows.

In typical contexts we want more than just validity from our arguments. For, even if they are valid, this means nothing about the truth of the conclusion, unless the premises are true as well—i.e., unless the argument is *sound*.

**Soundness:**

> An argument is *sound* if and only if it is deductively valid AND all its premises are true.

To begin understanding and distinguishing these two criteria, consider the group of arguments in Box 1.1. Upon reflection, you should be able to see that all three arguments of **Form 1** are valid, and that *this has nothing to do with the actual truth values of the component claims*. That is, despite the actual truth value of the component statements in each instance, there is no way (it is NOT possible) that the premises could ALL be true AND the conclusion false. Take **1D**, for example. Despite the fact that all the claims are false, it has to be the case that IF the premises were true, then the conclusion would be true as well—just imagine the premises are true; could the conclusion be false? NO, so it's valid.

As you will have noticed, **1A**, **1B**, and **1D** are all of the same form. Indeed, it is in virtue of having this form that each of the three instances is valid. Any argument in which we consistently substitute predicate phrases for the place-holders F, G, and H will be a valid argument. Thus, **Form 1** is a valid argument form and **1A**, **1B**, and **1D** are valid arguments because they are instances of a valid form. Validity is a question of form. That is why the actual truth values of the component claims are irrelevant. What is relevant is whether the form is such that it is not possible to have all premises true and the conclusion false.

Another way of putting this is that valid arguments are truth preserving. A good metaphor for this is plumbing: if you hook up the pipes correctly (if your argument has a valid form) you know that if you put water in at the top (true premises), you'll get water out at the bottom (true conclusion). But it doesn't actually matter whether you do, indeed, put any water in—the pipes are hooked up correctly (the argument is valid) whether or not you put any water through them (whether or not the premises are true).

Note that of the three arguments of **Form 1**, only **1A** is also sound. This is because, in addition to being valid, it has premises that are actually true.

---

| Box 1.1: Argument Form 1 with Instances | | |
|---|---|---|

|  | All premises True | At least one premise False |  |
|---|---|---|---|
| **Conclusion True** | **1A**<br><br>All whales are mammals<br>All mammals are air-breathers<br><u>All whales are air-breathers</u><br><br>Valid and Sound | **1B**<br><br>All whales are fish<br>All fish are air-breathers<br><u>All whales are air-breathers</u><br><br>Valid but Unsound | **Form 1**<br><br>All F are G<br>All G are H<br>All F are H<br><br>Valid Form |
| **Conclusion False** | **1C**<br><br><br>Not Possible! | **1D**<br><br>All whales are reptiles<br>All reptiles are birds<br><u>All whales are birds</u><br><br>Valid but Unsound |  |

What, then, of **1C**? There can be no such instance of **Form 1**. Since **Form 1** is valid, it is impossible to have an instance with all the premises true and the conclusion false. Compare the arguments of **Form 2**, **2C** in particular (in Box 1.2).

Now look at the four arguments of **Form 2** in Box 1.2. First, note that they are all invalid. That is, despite the actual truth values of the component sentences, in each instance it is possible for all the premises to be true and the conclusion false. **2A**, **2B**, and **2D**, will require some imagination, but you will see that you can consistently imagine all the premises true and the conclusion false in each case.

**2C** takes no imagination at all. Here we have an instance of the form in which it is obvious that the premises are all actually true and the conclusion is actually false. Obviously this particular argument instance is invalid. Since the premises are all true and the conclusion is false, it is *possible* for all the premises to be true and the conclusion false (actuality implies possibility). Moreover, since validity is a matter of form, once we have an instance like **2C** before our eyes, we know that the form is invalid, and so is any instance (such as **2A**, **2B**, or **2D**). Notice that it was precisely the C-type instance (all premises True, conclusion False) that the valid **Form 1** lacked. When we find such a C-type argument we are said to have found a *counterexample* to the argument form. Valid arguments do not have counterexamples.

**Counterexample:**

> A *counterexample* to an argument (form) is an argument instance of exactly the same form having all true premises and a false conclusion. Production of a counterexample shows that the argument form and all instances thereof are invalid. (Failure to produce a counterexample shows nothing, however.)

Stretching the plumbing metaphor a bit, the counterexample, **2C**, illustrates that if your pipes are not correctly connected (the argument form is invalid) there is no guarantee that

---

**Box 1.2: Argument Form 2 with Instances**

| | All premises True | At least one premise False | |
|---|---|---|---|
| **Conclusion True** | **2A**<br><br>Some animals are frogs<br>Some animals are tree-climbers<br>Some frogs are tree-climbers<br><br>Invalid | **2B**<br><br>Some fish are frogs<br>Some fish are tree-climbers<br>Some frogs are tree-climbers<br><br>Invalid | **Form 2**<br><br>Some F are G<br>Some F are H<br>Some G are H<br><br>Invalid Form |
| **Conclusion False** | **2C**<br><br>Some animals are frogs<br>Some animals are birds<br>Some frogs are birds<br><br>Invalid | **2D**<br><br>Some fish are frogs<br>Some fish are birds<br>Some frogs are birds<br><br>Invalid | |

---

putting water in at the top (true premises) will result in water coming out at the bottom (true conclusion). You might get lucky and have water come out at the bottom (you might end up with a true conclusion as in **2A**), but, as shown by **2C**, you might not. And you should never take chances with your plumbing.

It is important to note, however, that the inability to produce a counterexample does not show validity. It may just be that we are (perhaps momentarily) not clever enough to fill in the form so as to have true premises and false conclusion. For determining validity and invalidity, it would be nice to have some method more systematic and reliable than our unschooled imaginations—the formal symbolic approach of subsequent chapters will give us a number of these.

Lastly, because an argument must be valid to be sound, and none of the instances of **Form 2** is valid, none of them is sound—in fact, once we determine that an argument is invalid, we don't bother with the question of soundness. Box 1.3 highlights important points to remember about deductive validity. Box 1.4 presents a flowchart for assessing validity and soundness.

The following are some *valid* forms with example instances. Note that this is by no means an exhaustive list. Technically there are an infinite number of valid argument forms. These are just a few of the more common simple forms you are apt to encounter. (Here 'P' and 'Q' are place-holders for sentences.)

**Disjunctive Syllogism:**

| | |
|---|---|
| Either P or Q | Either Pat is a man or Pat is a woman |
| not-Q | Pat is not a woman |
| P | Pat is a man |

---

> ### Box 1.3: Important Points About Validity
>
> - Validity is a matter of truth preservation, and this is a matter of form, so, except in the case of a counterexample, the actual truth values of the premises and conclusion are irrelevant.
>
> - *All true premises and true conclusion **do not** make a valid argument!* See **1B**, **1D**, and (especially) **2A**.
>
> - If an argument is valid and all its premises are true, then it is sound.
>
> - Soundness *does* have to do with the actual truth value of the premises. Thus, it is not merely a matter of form.
>
> - We can see that a particular argument, and all arguments of the same form, are invalid either by consistently imagining that all the premises are true and the conclusion false, or by finding a counterexample (an instance that actually does have all true premises and a false conclusion).

**Reductio Ad Absurdum:**

|  |  |
|---|---|
| Assume P | Suppose Pat is a mother |
| *deduce a* | All mothers are women |
| *contradiction:* | So, Pat is a woman |
| Q | But, Pat is a man, not a woman |
| not-Q | |
| not-P | So, Pat is not a mother |

**Modus Ponens:**

|  |  |
|---|---|
| If P, then Q | If Pat is a mother, then Pat is a woman |
| P | Pat is a mother |
| Q | So, Pat is a woman |

**Modus Tollens:**

|  |  |
|---|---|
| If P, then Q | If Pat is a mother, then Pat is a woman |
| not-Q | Pat is not a woman |
| not-P | So, Pat is not a mother |

The following two, though tempting, are *invalid*.

**Denying the Antecedent** *(invalid)*:

|  |  |
|---|---|
| If P, then Q | If Pat is a mother, then Pat is a woman |
| not-P | Pat is not a mother |
| not-Q | So, Pat is not a woman |

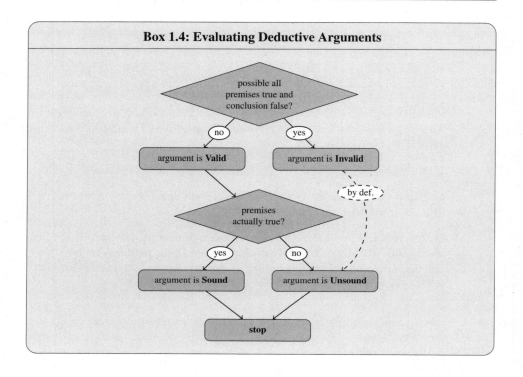

**Box 1.4: Evaluating Deductive Arguments**

**Affirming the Consequent** *(invalid)*:

If P, then Q                    If Pat is a mother, then Pat is a woman
Q                              Pat is a woman
P                              So, Pat is a mother

## 1.3.1  Quirky Cases of Deductive Validity

In discussing deductive validity, we began with the intuitive notion of guaranteed preservation of truth from premises to conclusion. Our technical definition of deductive validity, however, allows two types of cases that might seem counter to that original intuition. Consider the following argument:

Grass is green
Fire is cold
Either Fred is at work or he is not at work

The first thing to note here is that the premises have nothing to do with the conclusion, so one might think that this renders the argument deductively invalid. After all, how could there be guaranteed preservation of truth from premises to conclusion when they are unrelated? On the other hand, note that the conclusion has something special about it. It is structured in such a way that it must be true. No matter where Fred is, one side of the 'Either ... or - - -' must be true, making the whole thing true. In Section 1.5 we will see that this is an example of a logical truth. So the conclusion must be true, it cannot be false. Keeping that in mind, apply the official definition of deductive validity: is it possible for all the premises to be true and the conclusion false? The answer is 'No'. Since it is not possible for the conclusion to be false, it is not possible for the premises to be true and the conclusion false. So, we have a deductively valid argument! Indeed, any time the conclusion is a logical truth the argument will be deductively valid, no matter what the premises are.

Try this one on for size:

Grass is green
Grass is not green
Fred is at work

Here again, the premises have nothing to do with the conclusion. In fact, you might note that the premises contradict one another. Whichever one is true, the other must be false. The premises cannot all be true—they are inconsistent. (Contradictory pairs and inconsistent sets of claims are also discussed in Section 1.5.) So how could this be a valid argument when the premises exhibit what many think of as a logical flaw? Again, we apply the definition: is it possible for all the premises to be true and the conclusion false? The answer is 'No'. Since it is not possible for all the premises to be true, it is not possible for all the premises to be true and the conclusion false. So, once again, we have a deductively valid argument, and any argument with inconsistent premises will be valid, no matter what the conclusion is.

So we have two types of degenerate or trivial cases of validity. Any argument with a logically true conclusion is valid, and any argument with inconsistent premises is valid. Though this may be counterintuitive at first, it is really not problematic. Since the notion of truth preservation was a bit vague, our official definition focused on the impossibility of a counterexample (all premises true with false conclusion). Both types of argument illustrated above pass the test—it is not possible to produce a counterexample.

Moreover, there is no danger we could use these quirky cases of validity to draw any conclusion we should not otherwise draw. In the first type of case where the conclusion is a logical truth, we can identify it as such without the aid of the premises, so we should accept it as true independently of that argument anyway. The second type of case will never lead us astray either. Such arguments, though valid, can never be sound. Because the premises are inconsistent, there will always be at least one premise false, rendering it unsound. If the conclusion happens to be true, then we can likely find some other valid and sound argument to show that it is true. These quirky cases are not common, but we will see examples of them again.

Some students are very bothered by the latter case in which any conclusion validly follows from inconsistent premises. Why would we allow such a strange result? It is really a pragmatic choice. It makes for the simplest, most precise definition of validity, and it simplifies our logic in other ways. There are, however, variant approaches to logic, such as paraconsistent and relevance logics, that avoid these sort of cases—at the cost of added complexity. While not covered in this text, you might find it interesting to look into them.

## 1.4 **Inductive Criteria**

Traditionally, deductive reasoning was said to proceed from the general to the particular, while inductive reasoning was said to proceed from the particular to the general. But this is incorrect. While that contrast in directions does capture one aspect of the difference between some deductive and some inductive arguments, it is not the whole story. Some deductively successful arguments have particular premises and general conclusions. Moreover, some inductively successful arguments invoke general premises and/or arrive at particular conclusions. I, therefore, dispense with this traditional way of making the distinction.

I will distinguish inductive from deductive criteria in terms of the sort of support the premises are required to give the conclusion.

Again, there are two questions we want to ask when applying inductive criteria. One, the question of *strength*, has to do with the connection between the premises and conclusion. The other, the question of *cogency*, has to do with the truth values of the premises. First, strength:

**Inductive Strength:**

> An argument is *inductively strong* to the degree to which the premises provide evidence to make the truth of the conclusion plausible or probable. If an argument is not strong, it is *weak*.

Note the contrast with deductive validity, which requires that premises guarantee the truth of the conclusion. Here, inductive strength is a matter of the degree of plausibility or probability. Also in contrast to the definition of validity, the definition of strength does not apply to argument forms, but only to individual instances. As we shall soon see, this is because strength is not at all a matter of form. Where no confusion threatens, I will omit 'inductive' and 'inductively' when speaking of strength.

**Cogency:**

> An argument is *cogent* if and only if it is inductively strong AND all the premises are true.

With respect to strength, cogency plays a role analogous to that which soundness plays with respect to validity. In both cases we have a question about the connection between premises and conclusion (validity or strength), and then a question about the truth value of the premises.

Consider the following examples:

> This bag has 100 marbles in it
> 80 of them are black
> 20 of them are white
> _____
> The next marble randomly chosen will be black

> It is 5pm on Monday
> But the mail has not come yet
> The mail carrier is almost never late
> _____
> It must be a holiday

In neither of these cases do the premises guarantee the truth of the conclusion. So how successful are these arguments? Well, for the first one we have a pretty good idea: it is quite strong. Barring unforeseen happenings, we would rate the probability of the conclusion at 80%.[2] For the second one, however, it is unclear. It seems pretty strong, but that assessment is vague. And this often is the case with non-statistical assessments of strength. Except where the argument is clearly very weak, often we can only give a vague assessment of its strength.

This is a point to remember about deductive versus inductive criteria: deductive validity is like an on/off switch, an argument is either valid or invalid (and not both); but inductive strength is a matter of degree. Moreover, unlike validity, the strength of an argument is not simply a matter of form. Though form is often relevant to assessing the inductive strength of an argument, we will see below that it is never decisive. With inductive strength, more than form needs to be taken into account.

Consider these two common and simple forms of induction:

**Induction by Enumeration:**

$A_1$ is F
$A_2$ is F

$\vdots$

$A_n$ is F
_____
All As (or the next
    A) will be F

57 trout from Jacob's Creek were all
    infected with the RGH virus
_____
So, all trout (or the next trout found)
    in Jacob's Creek will be infected

**Argument by Analogy:**

A is F, G, H
B is F, G, H, and I
_____
A is I

My car is a 1999 Toyota Camry
Sue's car is a 1999 Toyota Camry and
    it gets over 30 miles per gallon
_____
So, my car will get over 30 mpg

With enumeration, generally speaking, the larger the sample, the stronger the argument. As the number of observed examples exhibiting the target property, F, increases, so does the likelihood of the conclusion (unless the population is infinite). Moreover, the narrower (or more conservative) the conclusion, the stronger the argument. For example, it is a narrower conclusion, and so a safer bet, that the next fish will be infected, than that all fish are (perhaps there are a very small number of resistant fish). With arguments by analogy, strength tends to vary with the number and relevance of shared properties (F, G, and H, three is not a required number). The more the two objects (or groups) have in common, and the more relevant those properties are to the target property, I, the more likely the object in question will also have the target property. In assessing the strength of analogical arguments it is also important to consider relevant dissimilarities. If present, relevant dissimilarities weaken the argument.

_____

2. Of course, if we change the conclusion to 'There is an 80% chance the next marble I pick will be black', then we have a deductive statistical argument.

But inductive arguments are not so simple. Here I will just illustrate a few difficulties. Consider the following, which are of exactly the same forms as the above arguments:

The 13,000 days since my birth have all been days on
    which I did not die
---
So, all days (or the next day) will be a day on which
    I do not die

I like peanuts, am bigger than a breadbox, and have two ears
Bingo the elephant likes peanuts, is bigger than a breadbox,
    has two ears, and has a trunk
---
So, I have a trunk

Note that neither of these arguments is particularly strong. Despite the rather large sample size in the induction, the conclusion that I will live forever is extremely unlikely, and while the conclusion that I will live through the next day seems to be stronger (because it is narrower), it is not particularly comforting. The analogical argument involving the elephant is obviously ludicrous, mainly because the similarities I have cited are largely irrelevant to the question of my having a trunk. Again analogical arguments are stronger when the similarities cited are relevant to the target property, and when there are few relevant dissimilarities.

Determining the relevance of similarities and dissimilarities (as well as the question of the strength of enumerations) depends to a large degree on background knowledge— knowledge which is often left unstated, but which, when made explicit, or inserted as a new premise, may strengthen or weaken the argument. In some cases (say when hypothesizing about the effect on humans of a drug tested on mice) we may not be entirely sure how relevant the similarities and dissimilarities are, and this affects our assessment of the strength of the argument. This issue of background knowledge is part of why strength is not a question of form; more is involved than just the statements that appear in the argument.

Moreover, the emergence of new evidence can radically alter our assessment of inductive strength. If an argument is deductively valid, this status cannot be changed by the introduction of additional evidence in the form of further premises.[3] When applying inductive criteria, however, the introduction of further evidence in the form of additional premises can increase or decrease the strength of the argument. (Imagine pointing out relevant dissimilarities between Bingo and me, or, in the argument about the mail carrier, adding the evidence that the roads are flooded.)

Here are some points to remember:

- Unlike deductive validity, inductive strength is a matter of degree, not an all-or-nothing, on/off switch.

- Unlike deductive validity, inductive strength is *not* a matter of form.

---

3. Even if we introduce a new premise that contradicts one of the old premises! See the discussion of quirky cases of validity, page 12.

- Background knowledge and additional information are relevant to the assessment of strength.

For enumerations and analogies:

- The larger the sample size or comparison base group, the stronger the argument.

- The narrower or more conservative the conclusion, the stronger the argument.

- The greater the number of (relevant) similarities, the stronger the argument.

- The fewer the number of (relevant) dissimilarities, the stronger the argument.

## Abduction

Abduction is a category of reasoning subject to inductive criteria. It deserves special mention because of its ubiquity in daily and scientific reasoning, and because it makes special appeal to the notion of explanation. Criteria for abductive success are essentially inductive criteria. Thus, successful abduction carries no guarantee of truth preservation, strength is a matter of degree rather than a matter of form, and it is subject to reassessment in light of new evidence.

But there is an important difference between abduction and other sorts of inductive attempts. Typical successful induction arrives at a conclusion that is probable when we assume the truth of the premises. In addition to this, abductive reasoning is explicitly aimed at *explaining* the truth of the premises. Abduction tries to answer the question of *why* something is the way it is. Thus, abductive reasoning is often described as *inference to the best explanation*. As a result, assessing the strength of an attempt at abduction will involve assessing whether, and how well, the conclusion explains the premises.

**Abduction:**
> *Abduction* or *abductive reasoning*, also known as *inference to the best explanation*, is a category of reasoning subject to inductive criteria in which the conclusion is supposed to explain the truth of the premises.

The example involving the mail carrier (page 14) is a miniature abductive argument. The conclusion that it is a holiday, would, if true, explain the absence of mail at 5pm on a Monday in a manner consistent with the carrier's past punctuality. The explanatory power of the conclusion comes, in part, from the fact that if it were a holiday, then there would very probably be no mail. Contrast this with the marbles argument. Though that is a strong argument, the truth of the conclusion 'The next marble I pick will be black', does nothing to explain why 80 of the 100 marbles are black and 20 white. So the marbles argument is not an instance of abductive reasoning—no explanation is attempted.

Keep in mind, as well, that not every explanation is the conclusion of an argument. Hence, not every explanation is part of an abduction. If the statements making up the explanation are already accepted or well known, then no argument is being made, merely an explanation, based in accepted claims.

Returning to the mail carrier, the conclusion presented there is not, however, the only hypothesis that would explain the lack of mail. Flooded roads might well explain the absence of mail. Which is the better, or best, explanation? That depends on how much evidence we have and how much more we can gather. Determining what to count as a good explanation is a complex and interesting philosophical problem, well beyond the scope of this text. We can, however, cite a few useful rules of thumb.

Consider a further example:

> I hear scratching in the walls
> I hear the scurrying-clicking sound of little paws at night
> My cereal and rice boxes have holes chewed in them
> ——————————————————————————
> I have mice

My conclusion, together with other relevant information about the behavior of mice, would well explain the data expressed in the premises. Assuming the presence of mice, the observations are to be expected. Hence, we have a pretty good explanation. Of course, I may have an eccentric neighbor who enjoys practical joking. But until I gather some evidence that she is at work (cheese disappears from the unsprung mousetraps; nary a mouse is to be seen; the sounds of mice do not occur on nights when I lock up the house; there are cheesy fingerprints on the window sill) the mice hypothesis seems to be the best.

Clearly, being subject to inductive criteria, abduction shares all the traits of those criteria. Here are some additional points regarding abduction:

- The more known data an explanation can account for in a consistent and coherent manner, the better the explanation.

- The better an explanation coheres with established theory, the better it is (e.g., hypothesizing that the missing cheese has vanished into thin air, does not (among other problems) cohere well with established physical theory).

- Moreover, if an explanation successfully predicts further data not originally observed (I find footprints outside my garage window; and cheesy fingerprints on the window), then that explanation is better than one that cannot do so.

- There are no precise guidelines for abductive reasoning; especially if we focus on the notion of the 'best' explanation. Much depends on plausibility relative to our background knowledge and the quality and quantity of our evidence. Almost always there is more evidence to be had (in principle), and new evidence (as with inductive, but not valid deductive arguments) can radically alter our assessment of the quality of the inference.

Despite the differences between deductive and inductive criteria, the flowchart for assessing strength and cogency, presented in Box 1.5, is parallel to that for validity and soundness (see Box 1.4 on page 12).

It is important to see that the common structure here is the result of the two kinds of question one should ask about any argument. First, the question of the appropriate connection between premises and conclusion (validity or strength): do the premises genuinely

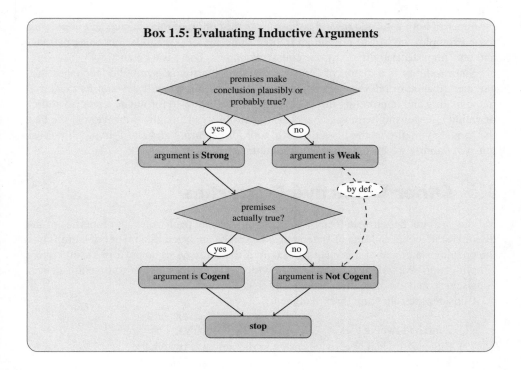

Box 1.5: Evaluating Inductive Arguments

support the conclusion? Second, the truth values of the premises (soundness or cogency): are the premises actually true? These are the two points at which any argument may be criticized.

## Form, Symbols, Deductive Validity

Despite the parallelism cited above, we now narrow our focus exclusively to deductive validity (and some related deductive properties).

Though in logic we define the concepts of soundness and cogency, determining whether the premises of a given argument actually are true is not within the bounds of logic. Rather, determining the truth values of premises often requires some research or investigation (or even further reasoning and argument), and this takes us into the realms of the various sciences, humanities, personal observation, and common knowledge. Logic, instead, is about the first question—the connection between premises and conclusion. Thus, soundness and cogency will be almost entirely absent from the subsequent chapters.

Moreover, as I have stressed, inductive strength is not a matter of form—identical forms can differ greatly in strength, and background knowledge and new evidence are highly relevant to the question of strength. Thus strength will be left behind as well.

What remains? Deductive validity. Recall that an argument is deductively valid if and only if it is not possible for all the premises to be true and the conclusion false. We saw examples of both valid and invalid arguments in the course of this chapter. But in judging these examples we had to rely on an imprecise notion of possibility, our imaginations, and our (imperfect) ability to devise counterexamples. There is a better way.

Since validity is a matter of form we will devise artificial symbolic languages that can clearly capture relevant aspects of form. In addition, we will develop methods for manipulating and interpreting these languages that allow us to formulate a precise notion of validity, dispensing with vague appeals to possibility or imagination. We will see that the formal symbolic method is a powerful tool for capturing and clarifying the intuitive grasp of validity we have been relying on in this chapter.

## 1.5  **Other Deductive Properties**

Validity will not be our sole focus. There are other deductively relevant properties of individual statements and sets of statements that can be analyzed as a matter of form. Thus the opening statement of this chapter, "Logic is the study of argument", is a useful oversimplification. We will also be concerned with such things as logical truth, equivalence, consistency, and entailment, to name just a few.

Consider the following three statements:

(1)  Zeno had twelve toes
(2)  Either Zeno had twelve toes or Zeno did not have twelve toes
(3)  Zeno had twelve toes and Zeno did not have twelve toes

Unless you have unique historical insight, you probably do not know whether (1) is true or false. But what about (2)? Whether or not you even know who Zeno is, you should be able to tell that (2) is true. Indeed, you might be thinking to yourself, 'It has to be true; it just can't be false'. And you would be absolutely correct. (2) is *logically true*. Its logical structure is such that it is not possible for it to be false, it must be true. For, either the guy had twelve toes, in which case (2) is true; or he didn't, but in this case (2) is true again. So in any case the statement is true.

With (2), as with many logical truths, we can see the truth of the statement even if we do not know much about the subject matter. This is because their truth is entirely a matter of logical form. To see this, consider the form of (2), and another instance of the same form (here 'P' is a place holder for statements):

(2′)  Either P or not P
(2″)  Either Zeno was an avid distance runner or Zeno was not an avid distance runner

You may feel that sentences such as (2) and (2″) don't really say that much. They include all the possibilities without really asserting any one of them in particular. You would be in good company—a philosophically important theory of logical truth claims that all logical truths are like this. (2) and (2″) are instances of what are sometimes called tautologies. The word has both the technical use when applied to logical truths, and the original, more common, use when applied to an obvious or redundant statement. But this is just the nature of some logical truths. They may not be very informative or useful for communication, but they are, logically speaking, very interesting statements. Moreover, not all logical truths are as obvious and easy to spot as (2) and (2''); logical truths can be quite complicated and difficult to recognize. But we will develop formal symbolic methods to test for them.

Return now to (3). Here, again, you can tell the truth value without really knowing anything about Zeno. Briefly, because it says that he both did and did not have twelve toes, it must be false. The logical structure of the statement forces it to be false. (3), perhaps unsurprisingly, is what we call *logically false*. Again, it is a matter of form. See if you can give the form of (3) and another instance of the same form (in the style of (2') and (2'')).

Finally, reconsider (1). Very likely it is false—I have no idea—but even if false, it certainly could have been true. Even assuming it is false—for example, assuming that Zeno had exactly ten toes—statement (1) could have been true *if things had gone differently*. Perhaps there was some unusual genetic mutation in his case. In any event, there is nothing about its *logical* form that either requires that it be true or requires that it be false. It could be either. In this case we call the statement logically contingent. Here are some definitions:

**Logically True:**
> A statement is *logically true* if and only if it is not possible for the statement to be false. Such statements are sometimes called tautologies.

**Logically False:**
> A statement is *logically false* if and only if it is not possible for the statement to be true. Such statements are sometimes called self-contradictions.

**Logically Contingent:**
> A statement is *logically contingent* if and only if it is neither logically true nor logically false; i.e., it is both possible for the statement to be true, and possible for the statement to be false.

Now consider some pairs of sentences:

(4)  In addition to having twelve toes, Zeno was an avid distance runner
     Zeno was an avid distance runner, even though he had twelve toes

(5)  Zeno finished some of his races
     Zeno did not finish any of his races

(6)  Zeno had twelve toes
     Zeno never finished a race

The statements in pair (4) do not say *exactly* the same thing. There is a clear difference in rhetorical tone and emphasis, especially with regard to his polydactyly.[4] However, we will not be concerned with such things as tone, emphasis, connotation. From a logical point of view, both statements assert that Zeno had twelve toes and that Zeno was an avid distance runner. Thus, we consider the two sentences *logically equivalent*, meaning that their forms are such that they will always have the same truth value.

---

4. Dodecadactyly!?

Contrast the equivalence of pair (4) with the situation in pair (5). Here, it seems the two statements are saying the exact opposite of one another—the second in the pair contradicts the first (and *vice versa*). We call such a pair *logically contradictory*. Their logical forms are such that they always have opposite truth values.

Pair (6), in contrast to pairs (4) and (5), has nothing very special about it. Unlike pair (4), the statements of (6) could differ in truth value. Unlike pair (5), they do not have to differ in truth value. Hence the pair in (6) is neither logically equivalent, nor logically contradictory.

**Logically Equivalent:**

> A pair of statements is *logically equivalent* if and only if it is not possible for the statements to have different truth values.

**Logically Contradictory:**

> A pair of statements is *logically contradictory* if and only if it is not possible for the statements to have the same truth values.

In addition to looking at individual statements and pairs of statements, we can focus on sets of statements. A set, roughly, is a grouping of any number of objects, considered as a group. A set can contain zero, one, two, three, four,..., an infinite number of members.[5] Below are a couple of sets of statements. Notice that we list the members of the sets (the statements, in this case), separated by commas, with the list enclosed in curly braces.

(7) { Everyone with twelve toes requires custom racing sandals, Zeno had twelve toes, Zeno did not require custom racing sandals }

(8) { Zeno did not require custom racing sandals, Zeno was an avid distance runner, Zeno had twelve toes }

If you think carefully about set (7), you may notice something is not quite right. If the first and second members of the set are true, then Zeno must have required custom racing sandals. But the third member of the set denies this. If the first and third are true, then Zeno must not have had twelve toes; but the second statement claims that he does. Finally, if the second and third are true, then the first must be false. In brief, there is no way for all three of the members of set (7) to be true. The set is *logically inconsistent*. No matter what, at least one of the statements must be false. Possibly more than one is false—we don't really know the truth values. But even without knowing that, we can see that at least one must be false.

Set (8) is a different story. Again, we don't know the truth values of the statements (I suspect only the first is true), but it is clear that they could all be true. That is, it is possible that every member of the set be true all at once. Hence, the set is *logically consistent*.

**Logically Consistent:**

> A set of statements is *logically consistent* if and only if it is possible for all the statements to be true.

---

5. The set with zero objects is called, aptly enough, the empty set or null set. For more on sets and set theory, see **Chapter 8**.

**Logically Inconsistent:**

> A set of statements is *logically inconsistent* if and only if it is not possible for all the statements to be true.

Note that these are complementary properties—a set of statements is logically inconsistent if and only if it is not logically consistent.

The final property we shall discuss here is *entailment*. I'll start with the definition:

**Logically Entails, Logically Follows:**

> A set of statements *logically entails* a target statement if and only if it is NOT possible for every member of the set to be true AND the target statement false. We also say that the target statement *logically follows* from the set.

First, note that this is very close to the definition of deductive validity. Indeed the two concepts are almost identical, with just a few theoretical differences (see below). Here is an example:

(9)  {Everyone with twelve toes requires custom racing sandals, Zeno had twelve toes} *entails* Zeno required custom racing sandals

(10)  {Zeno was an avid distance runner, Zeno had twelve toes} *does not entail* Zeno never finished a race

The claim of entailment in (9) is correct. There is no way for both members of the set to be true and the target statement false. So the target 'Zeno required custom racing sandals' follows from (or is entailed by) the set. The denial of entailment in (10) is likewise correct. It is clearly possible to be a twelve-toed avid runner who finishes at least one race. I.e., it *is* possible for all members of the set to be true and the target false. So the set does not entail the target (the target does not follow from the set).

Again, this notion is very close to that of deductive validity of an argument. If we take the premises of an argument as the set of statements and the conclusion as the target, then we can say that an argument is deductively valid if and only if the premises logically entail the conclusion. There are some differences, however. First, there is the obvious difference in the way in which the question or claim is posed. With validity, we have premises and a conclusion; with entailment, a set and a target. This would be merely cosmetic, if it were not for the second difference. Second, arguments by our definition have a finite number of premises, whereas a set may, in principle, contain an infinite number of statements. This could, for all we know initially, make an important difference. Unexpected things often occur when dealing with infinities.[6] Perhaps there are statements entailed by infinite sets, but that do not follow from any finite set of premises (and so they cannot be the conclusion of any valid argument). As it turns out, a metalogical result called the Compactness Theorem shows that this is not the case for the kind of logic we will pursue here. The two concepts do coincide. This is a minor, though significant, result of metalogic. Finally, the use of the 'entails' and 'follows from' terminology is just as common as the use of 'validity', so it is important for you to be familiar with it.

---

6. See Chapter 8.

In addition to the relationship between entailment and validity, you may have also noticed interrelations between some of the other definitions. For instance, a contradictory pair of statements constitutes a two-member inconsistent set; a logically false statement, or self-contradiction, constitutes a one-member inconsistent set. These and other interrelations will be the focus of Section 3.4.

Note also that the definitions of this section make frequent appeal to possibility, necessity, and impossibility—just as the definition of validity did. Again, I have relied on fairly simple examples and the reader's intuitive sense of possibility and logical correctness. But we need a more systematic way to state and test for these properties. We can do much, much better.

Now, then, is the time to turn to the resources of artificial symbolic languages in order to more clearly analyze form; get a grip on possibility, necessity, and impossibility; clearly define the properties of interest; and develop methods for exploring those properties.

## 1.6  Exercises

> **Note:** Answers to all exercises appear in Appendix A

1. What is a statement?

2. What is an argument?

3. Explain the meaning of truth value.

4. Restate the definition of deductive validity.

5. Restate the definition of soundness.

6. Suppose an argument is valid, what, if anything, does this tell us about the truth values of the premises and conclusion?

7. Suppose an argument is sound, what, if anything, does this tell us about whether it is valid?

8. Suppose an argument is sound, what, if anything, does this tell us about the truth values of the premises and conclusion?

9. Suppose an argument has true premises and a true conclusion, what, if anything, does this tell us about the validity and soundness of the argument?

10. Restate the definition of inductive strength.

11. Explain the differences between deductive validity and inductive strength.

12. Explain what is distinctive about abduction.

13. Compose, in standard form, one example of a valid argument and one example of an invalid argument.

14. Using capital letters as placeholders, give the form of the arguments from your previous answer.

15. Compose, in standard form, two instances of the same inductive argument form. Make one instance strong and the other weak. What does this show about the relationship between form and inductive strength?

16. Using the same premises, give two examples of abductive arguments. Make one example strong (a good explanation) and the other weak (a bad explanation). Explain why the one is a better explanation than the other.

## 1.7 **Chapter Glossary**

> **Note:** A complete glossary appears in Appendix B

**Abduction:**
>    *Abduction* or *abductive reasoning*, also known as *inference to the best explanation*, is a category of reasoning subject to inductive criteria in which the conclusion is supposed to explain the truth of the premises. (17)

**Argument, Premise, Conclusion:**
>    An *argument* is a (finite) set of statements, some of which—the *premises*—are supposed to support, or give reasons for, the remaining statement—the *conclusion*. (6)

**Argument Form and Instance:**
>    An *argument form* (or schema) is the framework of an argument that results when certain portions of the component statements are replaced by blanks, schematic letters, or other symbols. An *argument instance* is what results when the blanks in a form are appropriately filled in. (7)

**Cogency:**
>    An argument is *cogent* if and only if it is inductively strong AND all the premises are true. (14)

**Conclusion:**
>    See **Argument, Premise, Conclusion**. (6)

**Conclusion Indicators:**
>    therefore, hence, thus, so, we may infer, consequently, it follows that (6)

**Counterexample:**
    A *counterexample* to an argument (form) is an argument instance of exactly the same form having all true premises and a false conclusion. Production of a counterexample shows that the argument form and all instances thereof are invalid. (Failure to produce a counterexample shows nothing, however.) (9)

**Deductive Validity, Invalidity:**
    An argument (form) is *deductively valid* if and only if it is NOT possible for ALL the premises to be true AND the conclusion false. An argument (form) is *deductively invalid* if and only if it is not valid. (8)

**Inductive Strength:**
    An argument is *inductively strong* to the degree to which the premises provide evidence to make the truth of the conclusion plausible or probable. If an argument is not strong, it is *weak*. (14)

**Logic:**
    *Logic* is the study of (i) criteria for distinguishing successful from unsuccessful argument, (ii) methods for applying those criteria, and (iii) related properties of statements such as implication, equivalence, logical truth, consistency, etc. (5)

**Logically Consistent:**
    A set of statements is *logically consistent* if and only if it is possible for all the statements to be true. (22)

**Logically Contingent:**
    A statement is *logically contingent* if and only if it is neither logically true nor logically false; i.e., it is both possible for the statement to be true, and possible for the statement to be false. (21)

**Logically Contradictory:**
    A pair of statements is *logically contradictory* if and only if it is not possible for the statements to have the same truth values. (22)

**Logically Entails, Logically Follows:**
    A set of statements *logically entails* a target statement if and only if it is NOT possible for every member of the set to be true AND the target statement false. We also say that the target statement *logically follows* from the set. (23)

**Logically Equivalent:**
    A pair of statements is *logically equivalent* if and only if it is not possible for the statements to have different truth values. (22)

**Logically False:**
    A statement is *logically false* if and only if it is not possible for the statement to be true. Such statements are sometimes called self-contradictions. (21)

**Logically Inconsistent:**
> A set of statements is *logically inconsistent* if and only if it is not possible for all the statements to be true. (23)

**Logically True:**
> A statement is *logically true* if and only if it is not possible for the statement to be false. Such statements are sometimes called tautologies. (21)

**Metalogic:**
> *Metalogic* is the study of the properties of logical systems. (5)

**Premise:**
> See **Argument, Premise, Conclusion**. (6)

**Premise Indicators:**
> as, since, for, because, given that, for the reason that, inasmuch as (6)

**Soundness:**
> An argument is *sound* if and only if it is deductively valid AND all its premises are true. (8)

**Statement:**
> A *statement* is a declarative sentence; a sentence that attempts to state a fact—as opposed to a question, a command, an exclamation. (5)

**Truth Value:**
> The *truth value* of a statement is just its truth or falsehood. At this point we make the assumption that every statement is either true (has the truth value true) or false (has the truth value false) but not both. The truth value of a given statement is fixed whether or not we *know* what that truth value is. (5)

# Part II
# Truth-Functional Logic

# 2 The Language *S*

## 2.1 Introducing *S*

We saw in the previous chapter that deductive validity can be understood as a matter of form. Indeed, a number of other logical properties that interest us are also matters of form. The symbolic languages we develop in this book will allow us to represent in great detail the form of sentences and arguments. Using these languages we are able to explore and reason about logical properties in a highly systematic fashion.

### 2.1.1 Compound Sentences and Truth-Functional Logic

In the course of this book we will be concerned with two aspects of the logical form of sentences. One, which we set aside for a few chapters, has to do with the internal structure of simple sentences and the use of quantity terms such as 'all', 'some', and 'none'. The other, which we take up immediately, has to do with the way *compound sentences* are constructed from *simple sentences*.

**Simple Sentence:**
>A *simple sentence* is a sentence that contains one subject and one predicate.

**Compound Sentence:**
>A *compound sentence* is a sentence that either contains one or more simple sentences and at least one compounding phrase, or contains a compound subject or a compound predicate.

It will help to have a few examples:

>(1) The Earth is at the center of the Universe
>(2) The Sun revolves around the Earth

(1) and (2) are simple sentences. They each have a single subject and a single predicate. We can use these two simple sentences to create some compound sentences:

>(3) *It is not the case that* the Earth is at the center of the Universe
>(4) The Earth is at the center of the Universe, *and* the Sun revolves around the Earth
>(5) *If* the Sun revolves around the Earth *then* the Earth is at the center of the Universe

(6) *If* the Earth is *not* at the center of the Universe, *then* the Sun does *not* revolve around the Earth

(3)–(6) are compound sentences. The subsentences (or component sentences) are in regular type, while the compounding phrases are emphasized. (3) comes from (1) by prefixing 'it is not the case that', while (4) and (5) each use (1) and (2) to make longer, compound sentences. Since 'it is not the case that' modifies a single component, we call it a unary compounding phrase, while the other two compounding phrases, 'and' and 'if-then' are binary, requiring two component sentences. In (6) we have the unary 'not' compounder (a simplification of 'it is not the case that') used together with the binary 'if-then' compounder. As you can see from (6), compound sentences can themselves be further compounded.

Here is a slightly different (and often more natural) way of compounding simple sentences:

(7)  The Earth *and* the Sun are at the center of the Universe

(8)  The Sun is at the center of the Universe *or* revolves around the Earth

(7) and (8) are also compound sentences. Here, however, compounding is achieved through use of a compound subject, as in (7), or a compound predicate, as in (8). The compounding words are emphasized. Often we will want to expand such sentences so they look more like (3)–(6). Thus:

(7′)  The Earth is at the center of the Universe *and* the Sun is at the center of the Universe

(8′)  The Sun is at the center of the Universe *or* the Sun revolves around the Earth

Here (7′) says exactly the same thing as (7), but the expansion allows us to see exactly what the simple components are. Similarly for (8′). We will frequently use this sort of expansion to make clear the structure of sentences we encounter.

Certain sentences (and arguments) have the logical properties they do as a result of the way they (or their premises and conclusion) are compounded from simple sentences. As can be seen above, and in the examples below, there are many ways of compounding sentences, but we shall be interested here only in what are called *truth-functional compounds*.

**Truth-Functional Compound:**
   A sentence is a *truth-functional compound* iff[1] the truth value of the compound sentence is completely and uniquely determined by (is a function of) the truth values of the simple component sentences. Otherwise, the compound sentence is *non-truth-functional*.

For instance, (3) and (4) are each truth-functionally compound sentences. (3) turns out to be true, and it is true precisely because its simple component, (1), is false. Had

---

1. 'iff' is short for 'if and only if'.

(1) been true, (3) would have been false—sentences compounded with 'not' (or 'it is not the case that') have the opposite truth value of their component. (4) is formed with 'and', and its simple components are (1) and (2). (4) happens to be false because at least one (indeed, both) of its components is false. Had both (1) and (2) been true, (4) would have been true as well—sentences compounded with 'and' are true when and only when both components are true. We will consider (5) and (6) also to be truth-functional compounds, though the analysis of the 'if-then' is a bit complicated, and we save it for later (see Section 2.1.5).

Not all compounding phrases generate truth-functional compounds. Sometimes information other than the truth values of the simple components is needed to determine the truth value of the whole compound sentence. For instance:

(9) *Ptolemy believed that* the Earth is at the center of the Universe

(10) *Copernicus doubted that* the Sun revolves around the Earth

(11) *Copernicus argued that* the Sun does *not* revolve around the Earth

(12) *Copernicus believed that Ptolemy believed that* the Sun revolves around the Earth

(13) *Ptolemy believed that* the Earth is at the center of the Universe, *and Copernicus argued that* the Sun does *not* revolve around the Earth

(14) The Sun revolves around the Earth *because* the Earth is at the center of the Universe

(15) The Earth was at the center of the Universe *before* the Sun revolved around the Earth

The 'believes that', 'doubted that', and 'argues that' phrases are each non-truth-functional. We cannot determine the truth values of (9)–(13) simply on the basis of the truth values of the components (1) and (2). Rather, even if we know the truth value of the components (false), we still need further historical information about Ptolemy and Copernicus. The 'because' and 'before' compounders in (14) and (15) are also non-truth-functional. The truth values of the components of such compounds do not give any information as to causal or temporal relationships—again, we need some further scientific or historical information in such cases.

Note that, because (13) is an 'and' sentence, its truth value *is* wholly determined by the truth values of its *immediate* components ((9) and (11)). But those components themselves involve non-truth-functional compounds, hence the truth value of (13) is not wholly determined by its *simple* components, (1) and (2). Hence (13) is non-truth-functional.

### Truth-Functional Logic:

> *Truth-functional logic* is the logic of truth-functional combinations of simple sentences. It investigates the properties that arguments, sentences, and sets of sentences have in virtue of their truth-functional structure.[2]

Truth-functional logic will be the subject of this and the next two chapters. We will study the properties that arguments and sentences have in virtue of their truth-functional

---

2. Truth-functional logic is also frequently called sentential or statement logic. I prefer the former since the truth functions are what generate the interesting logical properties. The name of our language will be '$S$,' to remind us of the basic non-logical unit, statements (statement letters to be precise).

structure. We do not, at this point, delve into the internal logical structure of the simple sentences. Rather, we concern ourselves with simple sentences only insofar as (i) they are bearers of one of the two truth values, and (ii) they can be combined truth-functionally. As a result, in developing our language for truth-functional logic, we will take simple sentences as the basic non-logical units. Our name for this language will be '*S*,' which should remind you that sentences are the basic non-logical unit. We will represent simple sentences with uppercase letters (called *statement letters*), and interpret them as having either the truth value true or the truth value false, but not both. To represent compounding phrases, we will introduce symbols called *truth-functional connectives* as a means of generating compound sentences from simpler components.

Before giving a formal specification of the language *S*, let's take a less technical look. In the remainder of this section I try to use example statements with fairly obvious truth values—in some cases true and in others false—in order to facilitate basic understanding of the truth functions.

## 2.1.2  Negation—It is not the case that ...

The simplest way to generate a truth-functionally compound sentence is to negate (or deny) a sentence.

   (1) Grass is green
   (2) Grass is not green
   (3) It is not the case that grass is green

Choosing 'G' to translate the simple sentence 'Grass is green', and using '¬' (the negation symbol or hook) for 'not' or 'it is not the case that', we can symbolize the above three sentences as:

   (1′) G
   (2′) ¬G
   (3′) ¬G

Translating (1) is just a matter of writing the designated statement letter, as in (1′). Sentences (2) and (3) are only minutely more complex. Note that despite the slight difference in their phrasing, (2) and (3) receive identical treatment when translated into *S* as (2′) and (3′). Moving in the opposite direction, from *S* to English, sentence (3) illustrates the "official" English reading of '¬G'. The somewhat cumbersome phrase 'it is not the case that' is used because it can always be placed at the start of an English sentence to unambiguously negate the whole sentence. Of course, when no ambiguity threatens, we can give '¬G' the more natural reading in (2).

As one would expect, the negation of a sentence has a truth value opposite that of the original sentence. That is, negation is the truth function that maps a truth value onto its opposite. We can communicate this clearly with the *characteristic truth table* for negation:

| $\mathbb{P}$ | $\neg\mathbb{P}$ |
|---|---|
| T | F |
| F | T |

Here 'P' is a placeholder for any statement letter or more complex formula (see Section 2.2.3). The left-most column is a list of all possible assignments of truth values to the relevant component sentences (in this case only one sentence is relevant, so there are only two possibilities). For each row, the resulting truth value of the compound is placed in the column under the hook. The truth value of the compound goes under the connective symbol '¬' to distinguish it from the truth value of the component(s) ('P' in this case).

Let us take (1′) to be true—grass is green. Given the interpretation of the '¬', (2′) and (3′) are false. Negation is pretty simple, but, as we will see throughout the book, it interacts in interesting ways with the other logical operators.

### 2.1.3  Conjunction—Both ... and - - -

Another fairly straightforward way to truth-functionally compound sentences is with 'and'. Here are three new examples:

(1)  Grass is green and the sky is blue
(2)  Grass and snow are green
(3)  Both grass is green and snow is green

Choosing 'B' for 'The sky is blue', 'S' for 'Snow is green', and using '∧' (the conjunction symbol or wedge) for conjunction, we symbolize the above sentences as:

(1′)  G ∧ B
(2′)  G ∧ S
(3′)  G ∧ S

Translating (1) is a simple matter of placing the wedge between the two statement letters, as in (1′). (2) does not, on the face of it, contain two complete sentences. It does, however, have a compound subject, and it is treated as a contraction of the longer (3). Hence, again, despite the slight difference in their phrasing, (2) and (3) receive identical translation into $S$ as (2′) and (3′). Moving from $S$ to English, (3) illustrates the "official" reading of 'G ∧ S'. This official phrasing will sometimes be used to avoid confusion when things get more complicated.

As one would expect, a conjunction is true when and only when both component sentences (called *conjuncts*) are true. As long as one or more of the components is false, the whole conjunction is false. Here is the characteristic truth table:

| P | Q | P ∧ Q |
|---|---|-------|
| T | T | T |
| T | F | F |
| F | T | F |
| F | F | F |

Again, the left-most columns display the possible truth value assignments. We now have a compound with two components, so we use 'Q' as a second placeholder and there are four combinations of the two truth values. The column under the wedge displays the value of the whole conjunction, given the values of the conjuncts listed in the same row. This all means that (1′) is true, while (2′) and (3′) are false.

Note, finally, that conjunction is commutative. Like addition or multiplication in mathematics, changing the order of the components does not change the meaning. Thus, the following are adequate translations of (1)–(3), respectively, and are equivalent to their respective counterparts in (1′)–(3′):

(1″)  B∧G
(2″)  S∧G
(3″)  S∧G

## 2.1.4  Disjunction—Either ... or - - -

The only mildly tricky thing about disjunction is remembering that we choose to use *inclusive* disjunction as opposed to *exclusive* disjunction. The difference is in how we treat the case in which both components (called *disjuncts*) are true. With *inclusive* 'or' the compound is true if and only if one *or both* disjuncts is true (hence it is false only when both disjuncts are false). With *exclusive* 'or' the compound is true if and only if exactly one (*but not both*) of the disjuncts is true. Though perhaps the exclusive predominates, English usage of 'or' varies with the context between inclusive and exclusive. For us it is a matter of convenience which we choose, as we can always define the other via the one, with the help of conjunction and negation. Because it makes certain other definitions and operations more simple, we choose inclusive disjunction.

(1)  Grass is green or the sky is blue
(2)  Grass or snow is green
(3)  Either snow is green or clouds are red

Using 'R' for 'Clouds are red', and '∨' (the disjunction symbol or vee) for disjunction we symbolize as follows:

(1′)  G∨B
(2′)  G∨S
(3′)  S∨R

Again, translating (1) into (1′) should be straightforward. (2) has a compound subject, but we treat it as a contraction of 'Either grass is green or snow is green'. Thus (2) goes over into (2′). (3), of course, translates into (3′), and (3) uses the "official" reading of the vee. Also note that we interpret (1′) as true, but, had we chosen exclusive disjunction, we would have to interpret it as false. (2′) is true as well, and (3′) is false, since both disjuncts are false. Here is the characteristic truth table for disjunction:

$$\begin{array}{cc|c} \mathbb{P} & \mathbb{Q} & \mathbb{P} \vee \mathbb{Q} \\ \hline T & T & T \\ T & F & T \\ F & T & T \\ F & F & F \end{array}$$

Note again that this is the *inclusive* 'or'. That is why the top row, where both disjuncts are true, is itself true; as in (1') above.

Disjunction is also commutative. Thus, the following are adequate translations of (1)–(3), and equivalent to (1')–(3'):

(1″)  B ∨ G
(2″)  S ∨ G
(3″)  R ∨ S

## 2.1.5  Material Conditional—If ..., then - - -

The material conditional is probably the least intuitive of the five basic connectives. This is mainly because in English there are so many different uses of the 'If ..., then - - -' context, most of which are *not* truth-functional uses. We postpone discussion of such non-truth-functional uses (subjunctive, causal, temporal, etc.) until Section 2.1.7. For now, we will explore the material conditional and treat all conditionals as if adequately expressed by the material conditional.

Conditionals consist of two parts that, though their ordering in English is flexible, must be kept in proper order when symbolizing. In other words, conditionals are not commutative! The subsentence that follows the 'if' is the *antecedent*.[3] The remaining subsentence, which sometimes follows a 'then', is the *consequent*. In each case below, 'the Sun is yellow' is the antecedent, while 'grass is green' is the consequent.

(1)  If the Sun is yellow, grass is green
(2)  Grass is green if the Sun is yellow
(3)  If the Sun is yellow, then grass is green

We use the '→' (the material conditional symbol or arrow) with the symbolization of the antecedent on the left of the arrow and the symbolization of the consequent on the right. Using 'Y' for 'the Sun is yellow' we get the following symbolizations:

(1')  Y → G
(2')  Y → G
(3')  Y → G

Note that despite the variations in (1)–(3), their translations all come out the same in (1')–(3'). The antecedent always appears before the arrow, while the consequent appears after

---

3. Assuming there is no 'only if'—more on that below.

it. (3) illustrates the "official" reading of the material conditional. Section 2.4.3 discusses the symbolization of conditionals at length.

If you have no experience with truth-functional logic, then the characteristic truth-table for the material conditional might be a bit confusing.

| $\mathbb{P}$ | $\mathbb{Q}$ | $\mathbb{P} \to \mathbb{Q}$ |
|---|---|---|
| T | T | T |
| T | F | F |
| F | T | T |
| F | F | T |

The first line of the truth table is only slightly problematic. Consider (3) and its translation (3′) from above. Both the antecedent and the consequent are true. So, according to the characteristic truth table, the whole conditional is true. But you are likely to think (3) and its translation (3′) are false. The claim, "If the Sun is yellow, then grass is green", may most naturally be interpreted as a causal claim. But the color of the Sun is at best remotely a causal contributor to the color of grass. So, intuitively, (3) and (3′) are false. But remember that we are setting aside causal, temporal, and other non-truth-functional interpretations of the conditional (Section 2.1.7). So it does not matter that the color of the Sun has little or nothing to do with the color of grass. The material conditional is true *whenever* it has a true antecedent and a true consequent—*even if the antecedent and consequent have nothing to do with each other*. The two components need not be related in any intuitive way. All that matters is their truth values.

The second line of the truth table is probably much more clear. If the antecedent is true but the consequent is false, then the whole conditional sentence clearly does not hold. For, imagine someone says "If I wave my hands, then a rabbit will appear", and she waves her hands, but no rabbit appears—obviously her claim was false. True antecedent and false consequent make the whole conditional false. The rabbit example is intuitively a causal or pseudo-causal claim, and I chose it so that the falsity of the conditional would be clear. But it is still the case that only the truth values of the components matter. So both (4) and its translation (4′) are false

(4) If the Sun is yellow, then clouds are red

(4′) $Y \to R$

Even though (4) seems a bit strange, it and (4′) are false, simply because they each have a true antecedent and a false consequent.

The third and fourth lines of the characteristic truth table are often not at all intuitive. Why should we consider the rabbit claim true if a rabbit appears despite no waving of hands (row three of the table)? Why should it be true if there is no waving of hands and no rabbit (row four)? Again temporal and causal concerns are not in play here.

Why, then, place a T in the final two rows of the table for the '→'? One way of looking at it is that we take the material conditional to be *true in every case except the one that explicitly falsifies it*—true antecedent and false consequent (row two). Thus, whether the antecedent and consequent seem to be relevant to one another or totally unrelated, only one situation renders a material conditional false: true antecedent and false consequent.

Another consideration is that sentences of the form

(5) $(\mathbb{P} \wedge \mathbb{Q}) \rightarrow \mathbb{P}$

ought always to be true. It is the form of a logical truth. Even if one or both $\mathbb{P}$ and $\mathbb{Q}$ are false, it must be the case that *if* both $\mathbb{P}$ and $\mathbb{Q}$ are true, *then* $\mathbb{P}$ is true. Otherwise $\mathbb{P}$ would have to be both true and false, which it cannot be. Another way of putting this is that sentences of form (5) cannot have a true antecedent and a false consequent. Since (5) is true in every other situation, we get the characteristic truth table above—material conditional sentences are true in all cases, except the one in which the antecedent is true and the consequent false.

Thus, each of the following is true:

(6) If clouds are red then grass is green
(6′) $R \rightarrow G$

(7) If clouds are red then snow is green
(7′) $R \rightarrow S$

(6) and (6′) are true since the antecedent is false and the consequent true. (7) and (7′) are true because both components are false.

If the characteristic truth table for the material conditional gets confusing, it helps to remember that material conditionals are true in every case except for true antecedent and false consequent.

## ... only if - - -

The arrow can also be used to symbolize claims of the form '... only if - - -'. In this case— where we have 'only if' without an accompanying 'if'—what follows the 'only if' goes into the consequent of the symbolization, and the remaining part goes into the antecedent. Thus,

(8) Clouds are red only if snow is green
(9) Only if snow is green are clouds red

become,

(8′) $R \rightarrow S$
(9′) $R \rightarrow S$

Despite the difference in the phrase order of (8) and (9), they receive the same treatment in (8′) and (9′). The 'only if', when by itself, indicates the consequent of a material conditional. Hence, in both cases the 'S' for 'Snow is green' goes in consequent position, after the arrow.

Remember: when you have 'if' by itself, use the arrow to symbolize, and what follows the 'if' is the antecedent; when you have 'only if' by itself, use the arrow to symbolize, and what follows the 'only if' is the consequent. The next section discusses when the two appear together as 'if and only if'. Also see Section 2.4.3 for extended discussion of symbolizing conditionals.

## 2.1.6 Material Biconditional— ... if and only if - - -

Given our discussion of the conditional, the material biconditional is pretty straight-forward. As the name suggests it is a bidirectional material conditional. Consider the following sentences.

(1) If the sky is blue, then the Sun is yellow; and if the Sun is yellow, then the sky is blue

(2) The Sun is yellow if the sky is blue, and only if the sky is blue

(3) The Sun is yellow if and only if the sky is blue

As you might expect, we use ↔ (the material biconditional symbol or double arrow), giving us:[4]

(1′) $Y \leftrightarrow B$

(2′) $Y \leftrightarrow B$

(3′) $Y \leftrightarrow B$

Again, despite the variable phrase order in English, the translations are all the same in *S*.

Of course, (1)–(3) could each also be symbolized as conjunctions of material conditionals, yielding

(1″–3″)  $(B \rightarrow Y) \wedge (Y \rightarrow B)$

But it is equivalent, and shorter, to use the '↔' to get (1′)–(3′). Finally, (3) illustrates the "official" reading of '↔' (though we will often shorten 'if and only if' to 'iff').

The truth conditions are simple—the compound is true when the two components have the same truth value, false when they differ in truth value.

| $\mathbb{P}$ $\mathbb{Q}$ | $\mathbb{P} \leftrightarrow \mathbb{Q}$ |
|:---:|:---:|
| T T | T |
| T F | F |
| F T | F |
| F F | T |

So (1′)–(3′) are each true, since their components are both true. For comparison:

(4) The Sun is yellow if and only if snow is green

(4′) $Y \leftrightarrow S$

(5) Clouds are red if and only if snow is green

(5′) $R \leftrightarrow S$

---

4. Note that, while we use a right-pointing arrow '→' and a double arrow '↔', we do *not* have a left-pointing arrow '←' in our symbolic language.

(4′) is false since the components have opposite truth values, while (5′) is true since both components are false.

Finally, like '∧' and '∨', but *unlike* '→', the '↔' is commutative, so (1′)–(3′) are each equivalent to:

(1‴–3‴)  B ↔ Y

Thus, (1‴) is also an adequate translation of each of (1)–(3). Clearly, we could change the order of components in (4′) and (5′) as well.

So far we have looked at extremely simple translations. In Section 2.4 we will deal with more complex and more difficult translations. Before doing so, however, we will discuss non-truth-functional compounds, and then give a technical specification of the language *S*.

## 2.1.7 **Conditionals and Non-Truth-Functionality**

Unlike the material conditional discussed in Section 2.1.5, most everyday uses of the 'If ..., then - - -' construction do not follow any truth-functional pattern. Recall from page 32 that a truth-functional compound is one in which the truth value of the whole is determined by the truth values of the components. Consider the following:

(1) If Fred were in Boston, Massachusetts, then Fred would be in the USA

(2) If Fred were in Boston, Massachusetts, then Fred would be in the United Kingdom

Suppose that Fred is actually in Madrid, Spain. Then (1) and (2) each have a false antecedent and false (though different) consequents. If treated as material conditionals, both sentences would count as true, since a false antecedent and a false consequent result in a true material conditional. But clearly (1) is true while (2) is not. What gives? These two conditionals are what logicians call *counterfactual* (or *subjunctive*) conditionals. The antecedent of each is false, and the claim is about what *would* be the case if the antecedent *were* true. (1) and (2), with the same *component* truth values but different truth values for the *compound*, show that counterfactual conditionals are non-truth-functional. What determines the different truth values of (1) and (2) is not their logical structure, but facts we know about geography. Here is another example:

(3) If John F. Kennedy had not been assassinated, he would have been reelected for a second term

Here it is not clear what, if anything, determines the truth value of (3). There are good historical/political arguments that it is true, good historical/political arguments that it is false, and good philosophical arguments that it is indeterminate—without a truth value. In any case it is not truth-functional.

Another example of non-truth-functionality is the use of 'If ..., then - - -' to express a causal claim.

(4) If I throw this stone, then it will fall back to Earth

(5)  If I throw this stone, then it will burst into flames

Suppose I never even touch the stone. On the material conditional interpretation both (4) and (5) are true. But common sense tells us that, in any typical situation, (4) is true and (5) is false. Their truth values depend on more than the value of the components. Moreover, (4) might be used to express a universal claim like 'Whenever any stone is thrown, that stone falls back to Earth'. In this case, (4) is expressing a claim about an infinite number of actual or possible stone throws, under any and all conditions. On that reading (4) is false, for any number of qualifications or further conditions might intervene—for example, an astronaut might throw a stone while on the surface of the Moon, or Mars.

Two points are important here. First, many uses of 'If ..., then - - -' are non-truth-functional. We, however, will be ignoring that and treating them all as material conditionals whenever we translate into $S$. Second, in doing this we are sacrificing some accuracy in our translation of conditionals. This sacrifice has the benefit of allowing us to treat conditionals in a simple and systematic fashion.[5]

## 2.2  **Some Technical Bits**

Because we will be talking a lot about language, forms, and symbols, and because we want to do so as precisely as possible, it helps to define the technical concepts and vocabulary we will be using. Becoming fluent in the language of the discipline will be of great help as our study advances to more and more difficult topics.

### 2.2.1  Object Language and Metalanguage

At certain points it will be important to distinguish the language we are developing and discussing from the language *in which* we are discussing it. This is the distinction between *object language* and *metalanguage*:

**Object Language:**
> When one is talking about a language, the *object language* is the language being talked about.

**Metalanguage:**
> When one is talking about a language, the *metalanguage* is the language in which one is talking about the object language.

For instance, in a Spanish class for native English speakers, the instructor will have occasion to discuss, in English, Spanish grammar. In such a situation, Spanish is the object language, and English is the metalanguage. For most of this and the next two chapters, the object language will be $S$, while the metalanguage is English (slightly augmented with some symbols).

---

5. Conditionals are by no means the only non-truth-functional contexts. See Section 2.1.1 for just a few others. Certain intensional logics attempt to deal systematically with non-truth-functional contexts. Indeed, there is a vast philosophical and logical literature on conditionals and other non-truth-functional contexts, but that is beyond the scope of this book.

### 2.2.2  Use and Mention

A related distinction is that between *use* and *mention*. Most of the time we simply *use* words or phrases, and they have their normal role in meaningful communication. At certain times, however, we talk about (or *mention*) specific words or phrases themselves, either by placing them between quotation marks or displaying them in a special manner on the page. For instance, the Spanish language instructor might write on the board:

(1)  The word 'comer' is a verb

(1) is a claim about a particular Spanish word, and that word is *mentioned* (talked about) by placing it in single quotation marks (so Spanish is the object language). Moreover, the claim itself is made by *using* five other English words (so English is the metalanguage). As further examples, consider the following:

(2)  'Chicago' has three syllables
(3)  'Chicago' rolls off your tongue
(4)  'Chicago' is my kind of town

In each of the above I am *mentioning* the name of a large city in the USA, and this is appropriate in (2) and (3) but not in (4). Sentence (4) involves a use/mention confusion. Compare:

(5)  Chicago has three syllables
(6)  Chicago rolls off your tongue
(7)  Chicago is my kind of town

In each of these I am *using* the name of a large US city, and this is appropriate only in (7). Sentences (5) and (6) involve a use/mention confusion.[6]

### 2.2.3  Metavariables

In the course of our explorations we will frequently have occasion to use the metalanguage to mention, and thereby talk about, specific expressions of the object language $S$. Indeed, I have already done this in the process of introducing the connectives of $S$, sometimes by quotation and sometimes by displaying example formulas set off from the main text. But, while quotation and display allow us to talk about specific expressions and formulas of $S$, these devices do not allow us to speak more generally of all expressions or formulas of a certain form or type. And, since the logical properties we are interested in are matters of form, we will need to be able to refer more and less generally to expressions of $S$. We will also need to talk generally about expressions and formulas of $S$ when specifying the syntax (or grammar) of the language. And the ability to generalize about the object

---

6. The astute reader will notice that, within the context of this discussion, each of the seven sentences only gets mentioned. Thus, except for in certain special circumstances, it is possible for a language (English, say) to be both the object language and the metalanguage: 'The word "eat" is a verb'.

language will become especially important when we start our metalogical investigation of the language and of the logical system itself.

The key to doing this is through the use of *metavariables*—variables of the metalanguage that range over (take as possible values) expressions of the object language. For this purpose we will use uppercase letters set in the Blackboard Bold typeface (the option of placing a positive integer subscript on the letters gives us an infinitely large supply).

**Metavariables:**

> *Metavariables* are variables of the metalanguage that range over (take as possible values) expressions of the object language. We use uppercase Blackboard Bold:
>
> $$\mathbb{A}, \mathbb{B}, \mathbb{C}, \ldots, \mathbb{Z}, \mathbb{A}_1, \ldots$$

We have already seen this practice in our specification of the characteristic truth tables. For instance, the table for conjunction involves

(1)  $\mathbb{P} \wedge \mathbb{Q}$

This (indeed the whole truth table) is an expression of the metalanguage, and can be understood as referring to any object language sentence that is an instance of a certain form—i.e., those consisting of some sentence of the object language followed by the wedge followed by some sentence of the object language (possibly, but not necessarily, distinct from the first). The characteristic table for the wedge thereby tells us the truth function for *any* conjunction of two formulas, no matter how complex the conjuncts themselves might be. Do not let the simple appearance of the '$\mathbb{P}$' and '$\mathbb{Q}$' distract you; they are metavariables that take as their values any well-formed symbolic sentences of $S$ (no matter how complex). Moreover, '$\mathbb{P}$' and '$\mathbb{Q}$' need not be distinct, each may take the same sentence or formula as its value. See Box 2.1 for an illustration of this. Thus, (1) refers to a certain set of sentences of $S$ by specifying the form of its members.

It is worthwhile to clarify a certain point of usage with respect to object language symbols in the metalanguage. So far, when discussing object language expressions I have used either quotation or display, except in contexts (such as (1)) involving metavariables. Strictly speaking, (1) and other expressions like it are problematic. (1) appears to be hybrid, using signs from both the object language (the wedge) and the metalanguage (the metavariables), but it cannot do both—which is it? If we try to understand (1) as part of the object language $S$, then it is ungrammatical, for metavariables are not part of the object language. If we try to understand it as part of the metalanguage (as we should), we have to explain the presence of the wedge. Either it is being used—in which case this wedge is part of the metalanguage and we need to explain its relation to the object language wedge; or the expression is somehow mentioning the object language wedge—in which case the absence of quotation marks needs to be explained. Actually, both are the case. The wedge symbol is being used in the metalanguage to mention the object language wedge. That is, unless otherwise noted, we adopt the convention that, in contexts involving metavariables, expressions of the object language are used to name themselves. This eliminates the need for distracting quotation marks in contexts such as (1). This convention shall be adhered to except where potential confusion may occur.

---

**Box 2.1: Form and Instances of Formulas**

| instance of form $\mathbb{P} \wedge \mathbb{Q}$ | the $\mathbb{P}$ part | the $\mathbb{Q}$ part |
|:---:|:---:|:---:|
| $A \wedge B$ | $A$ | $B$ |
| $A \wedge A$ | $A$ | $A$ |
| $(F \rightarrow G) \wedge \neg J$ | $(F \rightarrow G)$ | $\neg J$ |
| $(S \vee D) \wedge ((E \leftrightarrow A) \wedge \neg D)$ | $(S \vee D)$ | $((E \leftrightarrow A) \wedge \neg D)$ |

---

### 2.2.4 Syntax and Semantics

The study of *syntax* is the study of the signs of a language with regard only to their formal properties—e.g., which signs, specified in terms of their form or shape, are part of the language and what are the permissible combinations and transformations of them. That is, the study of syntax is carried on without any official recognition of the meaning or interpretation of the elements of the language.

The study of *semantics* is the study of language with regard to meaningful interpretations or valuations of the components. While introducing $S$ we have so far been paying attention to interpretation or meaning in two ways. First, we have discussed how English sentences can be translated into symbolizations of $S$; and, second, we introduced the characteristic truth tables for the connectives, specifying how the truth values of compound sentences depend on those of their constituents.

**Syntax:**
> *Syntax* is the study of the signs of a language with regard only to their formal properties.

**Semantics:**
> *Semantics* is the study of language with regard to meaningful interpretations or valuations of the components.

In the next section we will officially focus only on the syntax of $S$. I say 'officially' because, with a few exceptions, our specifications and discussions of syntax are carried out with an eye to eventual interpretations of the language—that is, with an eye to semantics. Despite this, such discussion can in principle be carried out with explicit reference only to syntactic properties of the language.

## 2.3 The Syntax of $S$

### 2.3.1 Defining the Language

The language $S$ shall consist of the following symbols:

**Symbols of *S*:**

> **Statement Letters:**
> $A, B, C, \ldots, Z, A_1, B_1, C_1, \ldots, Z_1, A_2, \ldots$

> **Truth-Functional Connectives:**
> $\neg \wedge \vee \rightarrow \leftrightarrow$

> **Punctuation Marks:**
> ( )

This completes the lexicon of *S*.

Notice that, as with metavariables in the metalanguage, the subscripting of statement letters gives us an infinite supply. Now that we have catalogued the available symbols and punctuation, we need to specify the set of proper grammatical constructions. Called *well-formed formulas* (or wffs),[7] these are the analogues of sentences in natural languages such as English, Spanish, or isiZulu. Indeed, many logic books speak of sentences rather than wffs, and I will sometimes do so. (When we specify the language *P*, we will have occasion to distinguish sentences as a subset of wffs.)

In order to define 'wff of *S*' we first define:

**Expression of *S*:**

> An *expression of S* is any finite sequence of the symbols of *S*.

Consider the following examples:

| Expressions of *S*: | Not Expressions of *S*: |
|---|---|
| (1) $ABC\neg\wedge)($ | (5) $\mathbb{A}\mathbb{B}\mathbb{C}\neg\wedge)($ |
| (2) $Z_1 \neg\neg\neg\neg\neg\neg$ | (6) $z_1 \neg\neg\neg\neg\neg\neg$ |
| (3) $A_1 \rightarrow B_{22}$ | (7) $A^1 \rightarrow B^{22}$ |
| (4) $((A \rightarrow B) \rightarrow C)$ | (8) $((A \leftarrow B) \leftarrow C)$ |

The sequences of symbols in the left column are all expressions of *S*. Even though (1) and (2) appear to be just jumbles of symbols, they are all symbols of *S*, hence (1) and (2) qualify as expressions of *S*. Expressions need not be grammatical. The same goes for (3) and (4); though these should look a bit more grammatical than (1) and (2). On the other hand, the sequences on the right are not expressions of *S*, because each contains at least one symbol that is not a symbol of *S*. (5) contains metavariables, and metavariables are not symbols of *S* (they are metalanguage symbols, and *S* is the object language). (6) contains 'z', which is not a symbol of *S*, because it is lowercase. (7) contains superscripts, but *S* does not allow superscripts (only subscripts). (8) fails because '$\leftarrow$' is not a symbol of *S*.

Now we can define 'wff of *S*':

---

7. The abbreviation 'wff' is typically pronounced *whiff*, but some also say it *woof* or *wuff*. If these all feel silly, you can just say 'formula' or 'statement'.

**Well-Formed Formula, Wff, of** $S$:
>   Where $\mathbb{P}$ and $\mathbb{Q}$ range over expressions of $S$,

>>   (1)  If $\mathbb{P}$ is a statement letter, then $\mathbb{P}$ is a wff of $S$
>>   (2)  If $\mathbb{P}$ and $\mathbb{Q}$ are wffs of $S$, then

>>>   (a)  $\neg\mathbb{P}$ is a wff of $S$
>>>   (b)  $(\mathbb{P} \wedge \mathbb{Q})$ is a wff of $S$
>>>   (c)  $(\mathbb{P} \vee \mathbb{Q})$ is a wff of $S$
>>>   (d)  $(\mathbb{P} \to \mathbb{Q})$ is a wff of $S$
>>>   (e)  $(\mathbb{P} \leftrightarrow \mathbb{Q})$ is a wff of $S$

>>   (3)  Nothing is a wff of $S$ unless it can be shown so by a finite number of applications of clauses (1) and (2)

The definition of a wff happens to be a *recursive definition*. Recursive definitions consist of three parts. First, a *basis* clause specifies the initial members of the set of things to which the term in question applies. In our case this is clause (1): all statement letters are wffs. Second, a *recursive* portion specifies how to generate (or determine whether objects qualify as) further members of the set of things to which the term applies. Above, this is clause (2): certain concatenations of established wffs, parentheses and connectives are also wffs. This portion of the definition can be repeatedly applied an indefinite, though finite, number of times. (The recursive portion need not be a single clause.) Third, the *extremal* clause states that the term applies to *only* those things specified by the basis and recursive clauses.[8]

### Recursive Definition:
>   *Recursive definitions* consist of three parts: the basis, the recursive clause(s), and the extremal clause.

## Syntax Trees

The definition and construction of a wff is much easier to illustrate than to describe. We will use *syntax trees* to represent the syntactic structure of a wff. Consider the following:

This simple tree illustrates the structure of applying clause (2a) to the statement letter 'B'. 'B' qualifies as a wff in virtue of clause (1), then an application of (2a) results in '¬B'. If we then take 'D' (a wff in virtue of clause (1)), and apply clause (2d), we have

---

8. The extremal clause is sometimes called the Porky Pig clause: "That's All Folks!" This is clause (3) above.

---

**Box 2.2: A Labeled Syntax Tree**

$$
\begin{array}{c}
B \\
2a \mid \\
\neg B \quad D \\
\end{array}
$$

$$
\begin{array}{c}
C \\
2a \mid \\
\neg C \quad D \\
\underbrace{\phantom{xxxx}}_{2e} \\
(\neg C \leftrightarrow D) \qquad (G \wedge (\neg B \rightarrow D)) \\
\underbrace{\phantom{xxxxxxxxxxxxxx}}_{2c} \\
((\neg C \leftrightarrow D) \vee (G \wedge (\neg B \rightarrow D)))
\end{array}
$$

$$
\begin{array}{c}
G \quad (\neg B \rightarrow D) \\
\underbrace{\phantom{xxx}}_{2b} \\
\end{array}
$$

$$
\begin{array}{c}
\neg B \quad D \\
\underbrace{\phantom{xx}}_{2d} \\
(\neg B \rightarrow D)
\end{array}
$$

---

$$
\begin{array}{c}
B \\
\mid \\
\neg B \quad D \\
\underbrace{\phantom{xx}} \\
(\neg B \rightarrow D)
\end{array}
$$

If we now take 'G', and apply clause (2b) to our previous result, we get

$$
\begin{array}{c}
B \\
\mid \\
\neg B \quad D \\
\underbrace{\phantom{xx}} \\
G \quad (\neg B \rightarrow D) \\
\underbrace{\phantom{xxxx}} \\
(G \wedge (\neg B \rightarrow D))
\end{array}
$$

Note that the tree has statement letters at the tips of its branches. Branches involving negation do not fork, since negation is a unary connective. In contrast, branches involving the binary connectives do fork. Note, as well, that the completed wff forms the base of the tree. These trees can be read either top to bottom—which reveals the order of construction (or justification via the definition). Or they can be read bottom to top—which reveals the order of decomposition into sub-wffs. Box 2.2 contains a more complex tree with the branches labeled according to which clause of the definition they represent (normally we will not bother labeling branches).

Here are some further examples of wffs and non-wffs:

| Wffs of $S$: | Not Wffs of $S$: |
|---|---|
| (9) $A_{73}$ | (15) $(A_{73})$ |
| (10) $\neg A_{73}$ | (16) $A_{73}\neg$ |
| (11) $(D \rightarrow E)$ | (17) $D \rightarrow E$ |
| (12) $\neg\neg(D \rightarrow E)$ | (18) $\neg\neg(\mathbb{D} \rightarrow E)$ |
| (13) $((A \rightarrow B) \rightarrow C)$ | (19) $(A \rightarrow B \rightarrow C)$ |
| (14) $((A \vee B) \wedge (\neg B \rightarrow C))$ | (20) $((A \vee \wedge B) \wedge (\neg B \rightarrow C))$ |

Note that (9) is a wff simply in virtue of being a statement letter. Thus, we simply cite clause (1) of the definition of a wff to show that (9) is a wff. (10) qualifies in virtue of clauses (1) and (2a). Two applications of clause (1) (one each for 'D' and 'E') and an application of (2d) show that (11) is a wff as well. Two applications of (2a) turn (11) into (12). The reader should be able to draw a syntax tree for each of these wffs (also see **Exercises 2.3.3**). Expression (15) fails to be a wff due to the illegal parentheses. In (16) the hook is misplaced. (17) lacks the requisite outer parentheses, though we shall say more on this below. (18) contains a metavariable, and so is not even an expression, hence it cannot be a wff (see the preamble of the definition). (19) lacks a pair of necessary inner parentheses, and is therefore ungrammatical. The lack of parentheses fails to distinguish which of the two arrows is the main connective. It should either look like (13), or like this: '$(A \rightarrow (B \rightarrow C))$'. Whenever the subformulas of a binary connective are not sentence letters or negations of sentence letters, it is important that the left and right subformulas be clearly indicated by parentheses. Finally, (20) contains a stray connective.

## 2.3.2 Syntactic Concepts and Conventions

### Parentheses and Brackets

It will be convenient to adopt two conventions concerning the syntax of $S$. First, in most contexts we will drop the outermost parentheses of any wff, as doing so cannot lead to any semantic ambiguity. Indeed, in introducing the connectives I was already following this convention. Under this convention, (17) above would qualify as a wff. We could also remove the outermost parentheses of (13) and (14) as well, but not the parentheses in (12). Second, though not required, we reserve the option of using square brackets, '[' and ']', in the place of some pairs of parentheses in order to facilitate visual chunking of formulas. For instance:

$$\neg(((A \vee B) \wedge C) \rightarrow ((D \leftrightarrow E) \wedge F))$$

may become:

$$\neg([(A \vee B) \wedge C] \rightarrow [(D \leftrightarrow E) \wedge F])$$

### Main Connective, Well-Formed Components, and Scope

It is worthwhile defining certain syntactic notions for later use. The '$\neg$' is a *unary* connective (since it attaches to a single formula). The other four connectives are *binary*.

Further,

**Atomic Formula, Molecular Formula:**
>    Any wff that qualifies simply in virtue of clause (1) of the definition of a wff (that is, any wff that just is some statement letter), is called an *atomic* formula, wff, or sentence. By analogy, all other wffs are *molecular*.

For example, (9) is an atomic wff, while (10)–(14) are molecular.

**Main Connective, Well-Formed Components:**
>    Atomic wffs have no main connective. The *main connective* of a molecular wff $\mathbb{R}$ is the connective appearing in the clause of the definition of a wff cited last in showing $\mathbb{R}$ to be a wff. The *immediate well-formed components* of a molecular wff are the values of $\mathbb{P}$ and $\mathbb{Q}$ (in the case of clause (2a) simply $\mathbb{P}$) in the last-cited clause of the definition of a wff. The *well-formed components* of a wff are the wff itself, its immediate well-formed components, and the well-formed components of its immediate well-formed components. The *atomic components* of a wff are the well-formed components that are atomic wffs.

Thus, (9) has no main connective. The main connective of (10) is the hook. The main connective of (11) is the arrow. In (12) it is the leftmost hook. In (13) the main connective is the rightmost arrow; the immediate well-formed components are '$(A \rightarrow B)$' and 'C'; the well-formed components are (13) itself, the two previously mentioned components, and 'A', 'B', and 'C'; these latter three are obviously atomic components. Recall that the problem with (19) is that the lack of parentheses made it impossible to determine which arrow was the main connective.

As you will notice, this chopping up of a wff to find its main connective and various levels of its components is basically the reverse process of using the definition of a wff to construct (or to show an existing expression to be) a wff. See Box 2.3 for a detailed example.

**Scope:**
>    The *scope* of a connective is that portion of the wff containing its immediate well-formed component(s).

In (10), for instance, the scope of the hook is the '$A_{73}$'. In (12) the scope of the leftmost hook is '$\neg(D \rightarrow E)$', while the scope of the rightmost hook is '$(D \rightarrow E)$'. The scope of the arrow in (12) is the portion enclosed within parentheses.

See Box 2.4 for a graphical representation of scope. You can also think of the scope as the tree component(s) that are connected immediately above a well-formed component.

## 2.3.3 **Exercises**

> **Note:** Answers to all exercises appear in Appendix A

**Box 2.3: Syntax Tree with Well-Formed Components Indicated**

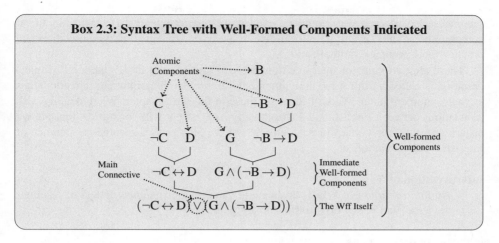

**Box 2.4: Scope of Connectives**

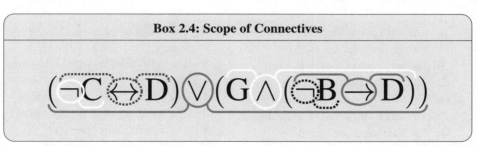

For each of the following, determine whether the sequence of signs is an expression of *S*; if it is not, write 'Not an expression'; if it is, then determine whether the expression is a wff of *S*; if it is not, write 'Not a wff'; if it is a wff, then draw its syntax tree—you do not need to label the tree. Assume the conventions concerning outer parentheses and optional brackets are in effect.

1. $\neg b \to \mathbb{F}$
2. $\neg(B\neg F)$
3. $\neg(A \land B) \to B$
4. $((A \land B) \land C) \land D$
5. $(F \lor (K \land (\neg G \to J))) \leftrightarrow (F \to K)$
6. $(S \to (R \to (Q \to (P \lor O)))) \land (M \land N)$

## 2.4 **Symbolizations**

So far we have dealt with only very simple translations from English to *S*. But, as is obvious from the discussion of syntax, wffs can be of arbitrary complexity. Moreover, English sentences also become very complex, even if we restrict ourselves to truth-functional compounding. Thus, it will help to discuss some basic strategies of symbolization and practice more complex symbolizations.

There are two kinds of interpretations we can give to wffs of *S*. One kind of interpretation is called a truth value assignment, and is simply an assignment of truth values to each statement letter. We will focus on these in the next chapter. When dealing with translations between English and *S* a truth value assignment will not suffice. Instead, we need a translation key that, for each atomic wff we plan to use, specifies a sentence of English for it to symbolize.

**Interpretation of *S*:**
> An *interpretation of S* assigns semantic value to the statement letters of *S*, either via a translation key or a truth value assignment.

In this section we shall use the following key:

K: Khumalo plays for the Falcons
N: Nakata plays for the Falcons
S: Santacruz plays for the Falcons
C: Santacruz captains the Falcons
O: Khumalo plays for the Mustangs
Z: Santacruz plays for the Mustangs
G: The Falcons score a goal
H: The Mustangs score a goal

W: The Falcons win their semifinal match
V: The Mustangs win their semifinal match
L: The Falcons win the final match
T: The Mustangs win the final match
F: The Falcons win the Cup
M: The Mustangs win the Cup

## 2.4.1 **Negation, Conjunction, Disjunction**

First, note that when compounding sentences in English, we almost always find a way to shorten the sentence. English allows compound subjects and predicates, but *S* does not. Thus, for purposes of symbolizing in *S*:

(1) Khumalo and Nakata play for the Falcons
(2) Santacruz plays for the Falcons or the Mustangs

should be understood as containing two complete subsentences that are then compounded:

(1*) Both Khumalo plays for the Falcons and Nakata plays for the Falcons
(2*) Either Santacruz plays for the Falcons or Santacruz plays for the Mustangs

Thus, the symbolization should be as follows:

(1′) K ∧ N
(2′) S ∨ Z

Sentences (1*) and (2*) illustrate an intermediate step in translating between English and *S*, in which we paraphrase in a sort of official logico-English. I encourage you to perform this step in your head, or, if the sentence in question is particularly complex, to write out the paraphrase before symbolizing. You will find, I am sure, that for most symbolizations you will automatically paraphrase in your head, or be able to jump right to the symbols. In most cases I will not display the paraphrase, but simply symbolize the sentences in question.

It is not always the case that a compound expression indicates two separate sentences compounded. For example, "They served macaroni and cheese" should not usually be read as, "They served macaroni, and they served cheese". Though perhaps not strictly false, the latter sentence suggests that two separate dishes were served, and in most contexts "macaroni and cheese" is the name of a single dish. English, like all natural languages, is much richer than our formal language, especially since the meaning of natural language words and phrases is often highly context-dependent. When translating, we sometimes have to make decisions about how to understand the English, and logicians might differ on how much context to include in interpreting a natural language sentence. My preference is to be as strictly logical as possible, leaving out contextual considerations.

Connectives can, of course, be used in combination:

(3) Nakata played for the Falcons and Santacruz did not
(4) Neither Khumalo nor Santacruz played for the Falcons
(5) Khumalo and Santacruz do not both play for the Falcons
(6) Khumalo or Nakata plays for the Falcons, but Santacruz does not
(7) Khumalo plays for the Falcons, or Nakata does and Santacruz does not

become:

(3') $N \wedge \neg S$
(4') $\neg(K \vee S)$
(5') $\neg(K \wedge S)$
(6') $(K \vee N) \wedge \neg S$
(7') $K \vee (N \wedge \neg S)$

(3') should be straightforward—we have a conjunction, one of whose conjuncts is negated. (4') is a negated disjunction, but (5') is a negated conjunction. Note that (3)–(4) are in the past tense. Our symbolizations do not take tense into account at all. Thus, so we can use the same statement letter, 'N', for 'Nakata plays...', 'Nakata played...', 'Nakata will play...', and so on.

If you think through (6') and (7'), you will see that the placement of parentheses is crucial and is indicated by the compound subject in (6) and placement of the comma in each (6) and (7). If 'K' and 'S' are both true, then (6') will be false while (7') will be true, so, again, the placement of parentheses in our symbolization makes an important difference. Note that while no 'and' appears in (6), we still consider it a conjunction. The difference between 'and' and 'but' is mostly rhetorical and not relevant to symbolization in *S*.[9]

---

9. This goes for other English conjunctions as well: 'moreover', 'in addition', etc.

---

**Box 2.5: 'Not', 'And', 'Or'**

---

| Not Both $\mathbb{P}$ and $\mathbb{Q}$ | Both Not-$\mathbb{P}$ and Not-$\mathbb{Q}$ |
|:---:|:---:|
| $\neg(\mathbb{P} \wedge \mathbb{Q})$ | $\neg\mathbb{P} \wedge \neg\mathbb{Q}$ |
| $\Updownarrow$ | $\Updownarrow$ |
| Either Not-$\mathbb{P}$ or Not-$\mathbb{Q}$ | Neither $\mathbb{P}$ nor $\mathbb{Q}$ |
| $\neg\mathbb{P} \vee \neg\mathbb{Q}$ | $\neg(\mathbb{P} \vee \mathbb{Q})$ |

---

The 'neither nor' (as in (4)), 'not both' (as in (5)), 'both not', 'either not or not' locutions can be confusing. It will be helpful to have a general schematic grasp of these.[10]

(8) Not Both $\mathbb{P}$ and $\mathbb{Q}$        (10) Both Not-$\mathbb{P}$ and Not-$\mathbb{Q}$

(8′) $\neg(\mathbb{P} \wedge \mathbb{Q})$             (10′) $\neg\mathbb{P} \wedge \neg\mathbb{Q}$

(9) Either Not-$\mathbb{P}$ or Not-$\mathbb{Q}$      (11) Neither $\mathbb{P}$ nor $\mathbb{Q}$

(9′) $\neg\mathbb{P} \vee \neg\mathbb{Q}$              (11′) $\neg(\mathbb{P} \vee \mathbb{Q})$

It is important to note that (8) and (9) are truth-functionally equivalent, as are their translations, (8′) and (9′). Similarly, (10) and (11) are truth-functionally equivalent, and so are their translations, (10′) and (11′).[11] Since, in translating, we are most concerned with capturing the truth-functional content of the English, (9′) would be an acceptable (though perhaps less natural) translation of (8), as (8′) would be of (9). Similar remarks apply to the other pair of equivalent wffs. Also note the crucial differences in the placement of the parentheses, and the way in which this affects the scope of the hook. These schemas are highlighted in Box 2.5, with equivalences indicated by '$\Updownarrow$'.

Let's look at some concrete examples of these equivalences:

(12) Not both Khumalo and Santacruz play for the Falcons

(12′) $\neg(K \wedge S)$

(13) Either Khumalo does not play for the Falcons or Santacruz does not

(13′) $\neg K \vee \neg S$

In (12) we have the 'not' preceding, or outside of, the 'both ... and - - -. This is the denial that both of them play for the Falcons. So the sentence allows three possibilities: Khumalo does not, Santacruz does not, or they both do not play for the Falcons. This is symbolized in (12′) by taking the simple conjunction 'K $\wedge$ S', enclosing it in parentheses, and placing a '$\neg$' to the left in order to negate or deny the conjunction as a whole. In (13) we have the

---

10. Note that in the English instances here I am allowing the metavariables to range, not over expressions of *S*, but over English sentences. The reasons should be clear, and no confusion threatens, so I will continue to engage in this variant usage where convenient.

11. These equivalences are known as DeMorgan's Laws.

'either ... or - - -' connecting two 'not's. This assertion allows the same three possibilities as (12): Khumalo does not, Santacruz does not, or they both do not play for the Falcons. (Remember that we are reading 'either ... or - - -' as inclusive!) So (12) and (13) are truth-functionally equivalent. To symbolize (13), we simply take '¬K' and '¬S' and make a disjunction of them with '∨', yielding (13′). Of course, since (12) and (13) are equivalent, so are (12′) and (13′). So each could be used as a correct translation of the other. The only real difference is how closely the translation follows the original English.

Now consider these:

(14)  Both Khumalo does not play for the Falcons and Santacruz does not
(14′)  ¬K ∧ ¬S

(15)  Neither Khumalo nor Santacruz play for the Falcons
(15′)  ¬(K ∨ S)

First, it is worth pointing out explicitly that 'neither ... nor - - -' is equivalent to 'not either ... or - - -', with the 'not' preceding, or outside of the 'either ... or - - -'. Next, note that (14) and (15) are equivalent. Each allows only one possibility: that Khumalo does not play for the Falcons and Santacruz also does not play for the Falcons. Neither of them does! So (14′) and (15′) are each adequate translations of (14) and (15). The only difference is that (14′) more closely mirrors the structure of (14), with the '∧' connecting two '¬'s, while (15′) more closely mirrors the structure of (15), with the '¬' extending over the '∨'.

Finally, while (12) and (13) are equivalent to each other, and (14) and (15) are equivalent to each other, (12)/(13) are not equivalent to (14)/(15). If, for instance, Khumalo plays for the Falcons and Santacruz does not, then (12)/(13) will be true and (14)/(15) will be false. The same can be said of their respective translations into *S*. The schemas in Box 2.5 present a visual summary of this discussion.

The placement of the 'not' in English can sometimes introduce ambiguity. Consider the following sentence:

(16)  Both Khumalo and Santacruz do not play for the Falcons

Which of the following is the best interpretation?

(16*)  It is not the case that both Khumalo plays for the Falcons and Santacruz plays for the Falcons
(16*′)  ¬(K ∧ S)

(16**)  Both it is not the case that Khumalo plays for the Falcons and it is not the case that Santacruz plays for the Falcons
(16**′)  ¬K ∧ ¬S
(16**″)  ¬(K ∨ S)

In (16*), and its translation (16*′), we are taking the 'not' in (16) to have greater scope than the 'both-and'. This leaves open the possibility that one of them does play for the

Falcons. We are reading it as 'not both'. One might plausibly argue that this is the correct interpretation of (16)—especially if we imagine it spoken in the right context with special emphasis on the 'not':

>       Fred:  Do both Khumalo and Santacruz play for the Falcons?
>       Kate:  No. Both Khumalo and Santacruz do *not* play for the Falcons.

Here, given the context, Kate plausibly means that not both of them play for the Falcons, suggesting that (at least possibly) one of them *does* play for the Falcons. This is suggested by her opening "No" and her emphasis on '*not*'. But I think Kate is not being clear. If she wanted to deny what Fred said, she should have said, "No. Not both of them. Only Khumalo." In the absence of any context—or if we are being strictly logical and eliminating context where possible—we should interpret (16) as in (16\*\*) and translate it as (16\*\*′) or the equivalent (16\*\*″). We take the English word order—in particular the placement of the 'both' prior to the 'not'—to indicate that the scope of the 'both-and' is greater than the scope of the 'not'. This yields (16\*\*) and (16\*\*′). Even so, had Kate meant (16\*\*), she would have done better to say, "No. Neither of them." This is (16\*\*″). In general, especially in writing, but even in speaking, we should try to avoid potential ambiguity, and use (16) only when we mean (16\*\*). Better yet, say neither-nor.[12]

Finally, consider the following two sentences.

(17)  Either Khumalo or Santacruz plays for the Falcons

(18)  Either Khumalo or Santacruz plays for the Falcons, but not both

Recall that the vee, '$\lor$', represents *inclusive* disjunction. It is true when one, the other, or *both* sides are true. Even though a simple 'or' in natural language is frequently meant exclusively, we will always translate 'or' as inclusive *unless explicitly stated otherwise*. Of course, when needed, we can translate an exclusive 'or' with the help of conjunction and negation. So (17) and (18) translate to:

(17′)  $K \lor S$

(18′)  $(K \lor S) \land \neg (K \land S)$

For (18), we simply add "and not both" to the original 'either-or'. See (12) above.[13]

## 2.4.2  Exercises

Using the key given on page 52, symbolize each of the following as wffs of $S$:

**1.** The Falcons do not score a goal

---

12. Similar considerations apply to the relative scopes of 'all' and 'not' and will be discussed in Chapter 5.
13. We can also reproduce exclusive 'or' with negation and the biconditional. Thus, (18′) is equivalent to each of these:

   (18″)  $K \leftrightarrow \neg S$
   (18‴)  $\neg K \leftrightarrow S$

2. The Falcons and the Mustangs win their semifinal matches
3. Either the Falcons or the Mustangs win the Cup
4. Santacruz plays for the Falcons but Khumalo does not
5. Khumalo either does or does not play for the Falcons
6. Neither the Falcons nor the Mustangs win the cup
7. Both the Falcons and the Mustangs don't win the cup
8. Not both the Falcons and the Mustangs win the Cup
9. Either the Falcons or the Mustangs do not win the cup
10. Santacruz plays for the Falcons and captains the Falcons, but does not play for the Mustangs
11. Either Nakata and Khumalo both play for the Falcons, or Khumalo plays for the Mustangs
12. Nakata plays for the Falcons, and Khumalo plays for either the Falcons or the Mustangs
13. Either Khumalo, Nakata, or Santacruz plays for the Falcons
14. Either both teams score a goal, or neither team wins the Cup
15. Santacruz plays for and captains the Falcons, and either they score a goal or they don't win the Cup

## 2.4.3 Conditionals and Biconditionals

Conditionals can be variously expressed in English.

(1) If Santacruz captains the Falcons, then she plays for the Falcons
(2) Santacruz plays for the Falcons, if she captains them
(3) Assuming that (provided that…, given that…,) Santacruz captains the Falcons, then she plays for them
(4) Only if Santacruz plays for the Falcons, does she captain them
(5) Santacruz captains the Falcons only if she plays for them

The key is to identify the antecedent and consequent; placing the wff symbolizing the antecedent on the left side of the arrow and that symbolizing the consequent on the right side. Remember that when 'if' (or an equivalent expression) appears without an 'only if', then what follows the 'if' is the antecedent. When 'only if' appears without an additional 'if', then what follows the 'only if' is the consequent. Briefly, 'if' introduces the antecedent, 'only if' introduces the consequent. When they appear together we have a biconditional, see below. Thus (1)–(5) are all properly symbolized by:

$(1'-5')$  $C \rightarrow S$

When we have both 'if' and 'only if' we have a biconditional:

(6) The Falcons win the Cup if and only if they win the final match
(7) The Falcons win the Cup if they win the final match, and only if they win the final match

---

**Box 2.6: 'If', 'Only If', and 'If and Only If'**

---

| 'if' alone: | 'only if' alone: |
|---|---|
| If $\mathbb{P}$, then $\mathbb{Q}$ $\left.\vphantom{\begin{matrix}a\\b\end{matrix}}\right\}$ $\mathbb{P} \to \mathbb{Q}$ <br> $\mathbb{Q}$, if $\mathbb{P}$ | $\mathbb{Q}$ only if $\mathbb{P}$ $\left.\vphantom{\begin{matrix}a\\b\end{matrix}}\right\}$ $\mathbb{Q} \to \mathbb{P}$ <br> Only if $\mathbb{P}$, $\mathbb{Q}$ |

---

'if' and 'only if':

$\mathbb{P}$ if and only if $\mathbb{Q}$

$\mathbb{P}$ if $\mathbb{Q}$, and only if $\mathbb{Q}$ $\qquad \mathbb{P} \leftrightarrow \mathbb{Q}$

If $\mathbb{P}$ then $\mathbb{Q}$, and if $\mathbb{Q}$ then $\mathbb{P}$ $\qquad \mathbb{Q} \leftrightarrow \mathbb{P}$

$\mathbb{P}$ iff $\mathbb{Q}$

---

The order of the wffs surrounding the double arrow is irrelevant for purposes of symbolization, so each of the following wffs is a proper symbolization of each of the preceding sentences:

$(6', 7'')$ $\text{F} \leftrightarrow \text{L}$

$(7', 6'')$ $\text{L} \leftrightarrow \text{F}$

Box 2.6 highlights some useful schemas.

### Necessary, Sufficient, Necessary and Sufficient

The arrow and double arrow can also be used to state necessary, sufficient, and necessary and sufficient conditions. A *necessary condition* is one that must hold for some further condition to hold. For example, for the Falcons to win their semifinal match, it is necessary that they score a goal.[14] A *sufficient condition* is a condition that, if it holds, guarantees that some further condition holds. For example, the Falcons's winning the final is sufficient for their having won their semifinal match. This may seem strange, because the winning of the semifinal comes first temporally. But keep in mind that we are capturing only logical (specifically, truth-functional) relationships between sentences—not temporal or causal relationships between the objects or events discussed in those sentences. Assuming the standard single-elimination tournament rules, then a team's winning of the final does guarantee that they won their semifinal match.

Note that conditions may be necessary but not sufficient (being a quadrilateral is necessary but not sufficient for being a square); sufficient but not necessary (being a square is sufficient but not necessary for being a quadrilateral); or both necessary and sufficient (being a regular quadrilateral is necessary and sufficient for being a square). Necessary conditions can be expressed by putting them in the consequent of a conditional;

---

14. Here I include shootout goals.

hence, symbolized on the right side of the arrow. Sufficient conditions can be expressed by putting them in the antecedent of a conditional; hence, on the left side of the arrow. Necessary and sufficient conditions can be expressed in a biconditional; hence, symbolized with the double arrow. Here are some examples:

(8) For the Falcons to win the final match it is necessary that they win their semifinal match

(9) The Falcons win the final only if they win their semifinal

both become:

$(8', 9')$  $L \rightarrow W$

And,

(10) The Falcons's winning the Cup is a sufficient condition for their having scored a goal

(11) If the Falcons win the Cup, then they scored a goal

translate to:

$(10', 11')$  $F \rightarrow G$

Finally,

(12) For the Falcons to win the Cup it is necessary and sufficient for them to win the final

(13) The Falcons win the Cup if and only if they win the final

become:

$(12', 13')$  $N \leftrightarrow E$

Note that these examples illustrate the relation between conditionals, 'if', 'only if' and necessary and sufficient conditions. In particular:

- 'if' by itself introduces a sufficient condition, and sufficient conditions appear in the antecedent of a conditional

- 'only if' by itself introduces a necessary condition, and necessary conditions appear in the consequent of a conditional

- 'if' and 'only if' together state necessary and sufficient conditions, and they may appear in a biconditional

Box 2.7 contains some helpful schema pairs.

People often wrongly use (or wrongly understand) 'only if' to indicate necessary and sufficient conditions, when it really only states necessary conditions.[15] Thus consider a sentence we saw earlier:

---

15. Actually, all conditionals express the consequent as necessary for the antecedent, and the antecedent as sufficient for the consequent.

---

### Box 2.7: 'Necessary' and 'Sufficient'

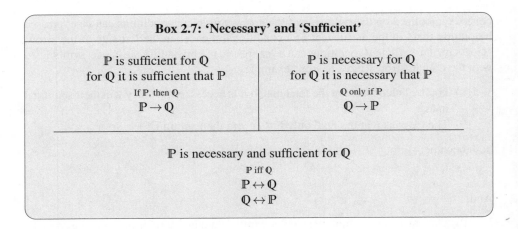

$\mathbb{P}$ is sufficient for $\mathbb{Q}$
for $\mathbb{Q}$ it is sufficient that $\mathbb{P}$

If $\mathbb{P}$, then $\mathbb{Q}$

$$\mathbb{P} \rightarrow \mathbb{Q}$$

$\mathbb{P}$ is necessary for $\mathbb{Q}$
for $\mathbb{Q}$ it is necessary that $\mathbb{P}$

$\mathbb{Q}$ only if $\mathbb{P}$

$$\mathbb{Q} \rightarrow \mathbb{P}$$

$\mathbb{P}$ is necessary and sufficient for $\mathbb{Q}$

$\mathbb{P}$ iff $\mathbb{Q}$

$$\mathbb{P} \leftrightarrow \mathbb{Q}$$
$$\mathbb{Q} \leftrightarrow \mathbb{P}$$

---

(5)  Santacruz captains the Falcons only if she plays for them

This sentence points out that to captain the team Santacruz must play for the team. It is silent on whether playing for the team is enough to be captain of the team (which it is not). That is, it does *not* say that Santacruz captains the team if she plays for them. This example may well be perfectly clear, but suppose a professor says to a student, "You will pass the class only if you pass all major assignments." The student may (wrongly) assume that it is enough (sufficient) to pass all the major assignments. But suppose there are some minor assignments that he fails miserably, he might well fail the class despite passing the major assignments. The professor stated necessary, not sufficient conditions. This sort of misunderstanding is encouraged, perhaps, by the following sort of situation. Suppose it is near the end of the semester, and the professor and student have reviewed all of the student's grades, except for the impending final exam, and the professor says, "You will pass the class only if you ace the final exam." Similar to before, the professor has only *explicitly* stated that acing the final is necessary; but, given that all the other grades are in, and they have just been looking them over, the professor likely would make that pronouncement only if acing the final is also sufficient, for the student will likely assume that the professor wouldn't say it otherwise. There is an issue of conversational norms here that goes beyond the truth-functional content of statements. In symbolizing we ignore all such issues, so both of the professor's statements should be symbolized as expressing only necessary conditions (i.e., with an arrow and not a double arrow). To be fully explicit and as clear as possible the professor might say either "You will pass the class if you ace the final, and that's the only way you'll pass the class," or "There's no way for you to pass the class now."

A related confusion arises over 'unless'. Consider the following:

(14)  The Mustangs win the Cup unless Santacruz plays for the Falcons

This might be misunderstood as claiming that if Santacruz plays for the Falcons, then the Mustangs will not win the Cup. This **INCORRECT** understanding would be symbolized as:

(14$^X$)  S $\rightarrow \neg$M                                **INCORRECT**

As above, this might not be a misunderstanding in certain conversational contexts. But (14) explicitly claims only that Santacruz not playing for the Falcons is sufficient for the Mustangs to win; or that Santacruz playing for the Falcons is necessary for the Mustangs to lose. (14) does not claim what (14$^X$) says—that Santacruz playing for the Falcons is sufficient for the Mustangs to lose. Contrary to (14$^X$), Santacruz might well play for the Falcons, and the Mustangs still win—Santacruz turns out to be necessary, but not sufficient for the Mustangs to lose. This suggests a pair of equivalent readings for (14):

(14*)  If Santacruz does not play for the Falcons, then the Mustangs will win the Cup
(14**)  If the Mustangs do not win the Cup, then Santacruz plays for (must have played for) the Falcons

that translate to:

(14$'$)  $\neg$S $\rightarrow$ M
(14$''$)  $\neg$M $\rightarrow$ S

That is, what follows the 'unless' ('S') may either be taken as a condition whose failure is sufficient for the holding of the other subsentence of the claim (as in (14$'$)), or as a condition that is necessary (though not claimed as sufficient) for the failure to hold of the other subsentence (as in (14$''$)). Note that (14$'$) and (14$''$) are truth-functionally equivalent (as we'll see in detail down the line), and in a loose sense they each say that if one of the subsentences is false, then the other is true, which is to say at least one of them is true. And this suggests a further way of symbolizing (14):

(14$'''$)  M $\vee$ S

Wff (14$'''$), which is truth-functionally equivalent to (14$'$) and (14$''$), helps to show that there is no claim that Santacruz's playing is sufficient for the Mustangs's losing. The wedge symbolizes an inclusive 'or' and so the wff allows for the possibility that both Santacruz plays for the Falcons and the Mustangs win the Cup. Thus, we do well to remind ourselves to read (14) as 'the Mustangs win the Cup unless Santacruz plays for the Falcons (and even then the Mustangs might win)'. The disjunctive symbolization is perhaps the easiest to remember—when symbolizing 'unless', simply treat it as an inclusive 'or', and use the vee, '$\vee$'.

The confusion with 'unless' results from its frequent use in conversational contexts where, either explicitly or implicitly, the inclusive case has been ruled out. I choose to symbolize the barest logical content of the 'unless' as used in a single sentence. This is the inclusive reading. If the inclusive case is explicitly denied, then we can symbolize appropriately:

(15)  If Santacruz plays for the Falcons, then the Mustangs won't win the Cup; and the Mustangs win unless Santacruz plays for the Falcons

---

### Box 2.8: 'Unless'

$$\left.\begin{array}{c} \mathbb{P} \text{ unless } \mathbb{Q} \\ \text{Unless } \mathbb{Q}, \mathbb{P} \end{array}\right\} \quad \begin{array}{c} \neg\mathbb{Q} \to \mathbb{P} \\ \neg\mathbb{P} \to \mathbb{Q} \\ \mathbb{P} \vee \mathbb{Q} \end{array}$$

---

(15*) The Mustangs win the Cup unless Santacruz plays for the Falcons, in which case the Mustangs don't win

(15′) $(S \to \neg M) \wedge (\neg M \to S)$
(15″) $\neg M \leftrightarrow S$
(15‴) $(M \vee S) \wedge \neg(M \wedge S)$

The formulas above are just three equivalent ways that (15) and (15*) can be translated. Box 2.8 presents schemas for 'unless'.

## 2.4.4 Exercises

Using the key given on page 52, symbolize each of the following as wffs of *S*:

1. If Nakata plays for the Falcons, then the Falcons win the Cup
2. The Falcons score a goal, if Khumalo plays for them
3. The Falcons score a goal only if Khumalo plays for them
4. The Falcons score a goal if and only if Khumalo plays for them
5. Santacruz plays for the Falcons, assuming she doesn't play for the Mustangs
6. The Falcons winning their semifinal match is necessary for their winning the Cup
7. The Falcons winning the final match is sufficient for their winning the Cup
8. The Falcons winning the final is necessary and sufficient for winning the Cup
9. The Mustangs win the Cup unless the Falcons score a goal
10. Unless Santacruz doesn't play for the Falcons, the Mustangs will not win the Cup

## 2.4.5 Complex Symbolizations

In theory, truth-functionally compound sentences of English can be of any finite level of complexity. In practice, we will rarely encounter any sentence of English that would require more than, perhaps, seven or eight atomic components. There are no hard and fast rules for translating from English to *S*, and, indeed, the English can often be ambiguous. The examples and exercises in the text avoid this by making clear through word order and punctuation what the correct translation should be. Learning to translate successfully requires practice and some ability to abstract the truth-functional structure from the more complex structure of natural English. The best approach is to proceed with multiple examples.

(1) Assuming Nakata and Khumalo play for the Falcons, the Falcons will win the Cup

(1′) $(N \wedge K) \rightarrow F$

(2) The Falcons will win the Cup only if Santacruz plays for and captains the Falcons

(3) Santacruz must play for and captain the Falcons, if they are to win the Cup

(2′, 3′) $F \rightarrow (S \wedge C)$

(4) The Falcons win the Cup if and only if they win both their semifinal and final matches

(4′) $F \leftrightarrow (W \wedge L)$

(5) If the Falcons and the Mustangs win their semifinal matches, then either the Falcons or the Mustangs win the Cup, but not both

(5′) $(W \wedge V) \rightarrow [(F \vee M) \wedge \neg(F \wedge M)]$

(5″) $(W \wedge V) \rightarrow (\neg F \leftrightarrow M)$

The consequent of (5′) includes '$\wedge \neg(F \wedge M)$' in order to generate the exclusive 'or' indicated by the 'but not both' in (5). The consequent of (5″) achieves this in a different, but equivalent, manner. (5′) better reflects the structure of the English, but (5″) is more economical. Both are acceptable.

(6) If Santacruz and Khumalo don't both play for the Mustangs, then the Mustangs will win the Cup only if Santacruz or Nakata doesn't play for the Falcons

(6′) $\neg(Z \wedge O) \rightarrow [M \rightarrow (\neg S \vee \neg N)]$

(6″) $(\neg Z \vee \neg O) \rightarrow [M \rightarrow \neg(S \wedge N)]$

Note how the placement of the main connective is indicated by the locations of 'If', 'then', and the comma. The antecedents of (6′) and (6″) are equivalent, as are the consequents of their consequents. Again, both are acceptable.

(7) The Mustangs win the Cup or Nakata and Santacruz don't play for them

(7′) $M \vee (\neg N \wedge \neg S)$

(8) Either the Falcons win the Cup or Nakata doesn't play for them; moreover Nakata doesn't play for them

(8′) $(F \vee \neg N) \wedge \neg N$

(9) If Santacruz plays for the Falcons, then if she captains the Falcons then the Falcons will score a goal and win the Cup

(9′) $S \rightarrow [C \rightarrow (G \wedge F)]$

## Translating Quantity Terms in $S$

Strictly speaking, $S$ cannot handle quantity terms such as 'all', 'some', and 'none'. To deal with these very important logical terms properly, in Chapter 5 we will develop a more complex and more robust formal language, $P$. But for now, we can deal with quantity terms in a limited fashion. Consider the following:

(10)  All of them play for the Falcons

If we here understand 'All' to refer to just the three players mentioned in the current interpretation (page 52), then we can easily translate (10) as:

(10′)  $(K \wedge N) \wedge S$

All we have done here is use a conjunction to explicitly state of each player that he or she plays for the Falcons. This works as long as we restrict reference to those we can name in the interpretation. Of course, if we named 20 players in the interpretation, the conjunction for translating (10) would have 20 conjuncts. So this is practical only for smaller numbers, and cannot handle infinite quantities since wffs of $S$ must be finite in length.[16]

Keeping in mind that we are restricting ourselves to just these three players, we can also translate the other common quantity terms:

(11)  Some of them play for the Falcons
(11′)  $(K \vee N) \vee S$

Logicians always treat 'some' as meaning at least one, up to and including all. In conversational contexts, 'some' often suggests 'not all'. We will again leave context aside to get at the purely logical use of 'some'. If 'not all' is made explicit, we can easily include it in our translation (see exercises).[17] Thus using the inclusive 'or' in (11′) gives us a wff which is true so long as one, two, or all three of them play for the Falcons. The only way (11′) could be false is if none of them play for the Falcons, which is exactly what we want for our translation of (11). This also suggests a way to translate 'none'.

(12)  None of them play for the Falcons
(12′)  $\neg[(K \vee N) \vee S]$

Here we just negate the whole of (11′), creating a long 'neither nor' wff. Another way to translate (12) is to explicitly deny of each player that he or she plays for the Falcons. Thus:

(12″)  $(\neg K \wedge \neg N) \wedge \neg S$

There are plenty of variations on this theme, many of which are covered in the exercises. Here is just one more:

(13)  At least two of them play for the Falcons
(13′)  $[(K \wedge N) \vee (K \wedge S)] \vee (N \wedge S)$

As you can see, this gets rather long and unwieldy. But it does express what we want. (13′) asserts that at least one pair of all the possible pairs consists of two players who play for the Falcons. Since the 'or' is inclusive (13) and (13′) also allow that all three of them play for the Falcons.

---

16. We *will* be able to handle infinite quantities with the language $P$, see Chapter 5.
17. Note that this is the same issue with the inclusive 'or' and 'not both' discussed above.

## 2.4.6 **Exercises**

**A.** Using the key given on page 52, symbolize each of the following as wffs of *S*:

1. Santacruz plays for the Falcons or the Mustangs, but she does not captain the Falcons
2. Either Nakata and Khumalo play for the Falcons, or Santacruz plays for the Mustangs
3. Either the Falcons or the Mustangs win the Cup, but not both
4. The Falcons and the Mustangs both win their semifinal match, and the Mustangs or the Falcons don't win the Cup
5. Either Santacruz plays for the Falcons and the Falcons score a goal, or Khumalo plays for the Mustangs and the Mustangs score a goal
6. The Falcons win the Cup if and only if either Nakata or Santacruz play for them
7. If the Mustangs won their semifinal match but did not win the Cup, then they did not win the final match
8. The Falcons will not win the Cup unless Nakata plays for them
9. Unless Santacruz and Khumalo both play for the Falcons, the Mustangs will win the Cup
10. Only if Nakata plays for the Falcons and the Mustangs don't score a goal, will the Falcons win the final match and the Cup
11. For the Falcons to score a goal it is necessary that Santacruz and Khumalo play for them
12. For the Falcons to win the final match and the Cup it is sufficient that Santacruz plays for them and captains them
13. If Khumalo, Nakata, and Santacruz all play for the Falcons, then the Mustangs will not score a goal nor will they win the Cup
14. The Falcons score a goal and win the Cup if Santacruz and Khumalo play for them, and only if Santacruz and Khumalo play for them
15. If Khumalo and Santacruz play for the Mustangs, then if Nakata doesn't play for the Falcons, then the Falcons will win the Cup only if the Mustangs don't score a goal

**B.** Using the key provided, symbolize each of the following as wffs of *S*:

| | |
|---|---|
| A: Ann goes to the fair | N: Carol eats lots of popcorn |
| B: Bob goes to the fair | O: Ann pays |
| C: Carol goes to the fair | P: Bob pays |
| D: Ann drives | Q: Carol pays |
| E: Bob drives | R: Ann rides the roller coaster |
| F: Carol drives | S: Bob rides the roller coaster |
| L: Ann eats lots of popcorn | T: Carol rides the roller coaster |
| M: Bob eats lots of popcorn | X: Ann throws up |

Y:  Bob throws up                    Z:  Carol throws up

1. Provided that Ann drives, Bob will go to the fair
2. If Ann goes to the fair, she will eat lots of popcorn and ride the roller coaster
3. Ann will go to the fair if and only if Carol pays and Bob drives
4. Bob and Carol will go to the fair only if Ann drives
5. Both Ann and Carol will not go to the fair
6. Neither Ann nor Carol go to the fair
7. Either Ann or Carol will not go to the fair
8. Not both Ann and Carol will go to the fair
9. Ann will throw up only if she rides the roller coaster and eats lots of popcorn
10. Ann will throw up unless she eats lots of popcorn
11. If Bob throws up, then Ann will not drive unless Carol and Bob pay
12. Bob will ride the roller coaster and eat lots of popcorn only if Carol or Ann pays; but if Bob rides the roller coaster and eats lots of popcorn, then he'll throw up
13. Ann will ride the roller coaster only if at least one of the others does
14. All three go to the fair
15. At least one of them goes to the fair
16. At least two of them go to the fair
17. At most one of them goes to the fair
18. At most two of them go to the fair
19. Some, but not all, go to the fair
20. Exactly one of them goes to the fair

## 2.5  **Alternate Symbols and Other Choices**

While reading this chapter you likely noticed that there are many choices to be made in developing a formal language for logic. Should we opt for an inclusive or exclusive 'or'? The typical choice of logicians is to introduce a symbol for inclusive 'or' and recover the exclusive 'or' through a slightly more complex symbolization. Further, I choose to translate every natural language 'or' as inclusive, unless explicitly specified as exclusive. How should we treat conditionals? Given that most natural language use of conditionals is non-truth-functional, the standard choice in introductory logic is to use the thoroughly truth-functional material conditional. More advanced logics try to tackle non-truth-functional conditionals with more sophisticated approaches. How should we deal with the natural language contexts surrounding 'only if', 'unless', and 'some'? In these cases, I opt for the barest logical reading of the sentence, disregarding context. This allows us to isolate the logical structure of individual sentences. If we want to account for implicit or explicit context, we can complicate our symbolization.

These choices, to varying degrees, decrease fidelity to the original natural language. This cost is balanced, however, by gains in simplicity and clarity with respect to the logical features that interest us—truth-functional properties of statements and arguments. So it is a price we are willing to pay, especially at the introductory level. Starting in Chapter 5 we will enhance our language so we can deal with quantifiers, predicates, and singular terms. But this machinery still won't capture all there is to language. Natural languages such as English are incredibly rich and expressive. The art of communication involves more than just the logic of sentences. It includes such things as conversational context, communicative conventions, background assumptions, tone of voice, body language, narrative and poetic conventions, and shared culture. Addressing these issues would take us out of the realm of logic and into philosophy of language, linguistics, literature, and anthropology. As interesting as these subjects are, I won't be addressing them here. But you should definitely explore them in other books and classes!

When constructing a formal language we also have to choose the symbols. I have chosen a set that is widely used, easy to write, and—I think—aesthetically pleasing. But, since you may encounter (or have encountered) other logic books or logical systems, it is worth pointing out other common symbols:

| (0) | $\neg P$ | $P \wedge Q$ | $P \vee Q$ | $P \rightarrow Q$ | $P \leftrightarrow Q$ |
|---|---|---|---|---|---|
| (1) | $\sim P$ | $P \,\&\, Q$ | $P \vee Q$ | $P \supset Q$ | $P \equiv Q$ |
| (2) | $-P$ | $P \cdot Q$ | $P \mid Q$ | | |
| (3) | $!P$ | | $P \parallel Q$ | | |
| (4) | $\overline{P}$ | $PQ$ | | | |
| (5) | $Np$ | $Kpq$ | $Apq$ | $Cpq$ | $Epq$ |

Row (0) in this table shows the symbols I have chosen for the truth-functional connectives. Row (1) shows the other most common combination of symbols, the tilde, the ampersand, the vee again (though some books calls this the wedge, our wedge points up!), the horseshoe, and the triple bar. In row (2) we see the hyphen, the centered dot, and the vertical line or pipe. The dot for conjunction has been around for a long time, but seems to be falling out of use. The pipe, or double pipe in row (3), is often used in computer programming languages, as is the exclamation point in row (3). Row (4) shows a very economical (though difficult to work with) notation in which a bar over a wff is a negation, and simple concatenation (putting next to each other) of two formulas makes a conjunction. So, for example:

$$\neg P \wedge Q \;\Leftrightarrow\; \overline{P}Q \qquad\qquad P \rightarrow Q \;\Leftrightarrow\; \overline{P\overline{Q}}$$
$$\neg(P \wedge Q) \;\Leftrightarrow\; \overline{PQ} \qquad\qquad P \rightarrow \neg Q \;\Leftrightarrow\; \overline{PQ}$$

Here is a long one:

$$\neg((P \vee \neg Q) \leftrightarrow \neg(R \rightarrow (S \wedge T))) \;\Leftrightarrow\; \overline{\overline{\overline{P\overline{Q}}\,R\overline{ST}}\,R\overline{R}\,\overline{P}Q}$$

Clearly, this becomes very unwieldy for formulas of even moderate complexity. One might choose such a system for its theoretical simplicity, not for ease of translation. Row

(5) displays the notation developed by Polish logician Jan Łukasiewicz (1878–1956). It is called Polish Notation or Łukasiewicz Notation. The capital letter prefix indicates the truth function, each being the first letter of the name of the function in Polish. From left to right: *negacja, koniunkcja, alternatywa, implikacja, ekwiwalencja*. One supposed asset of Polish notation is that parentheses are totally unnecessary. The scope of negation is the smallest wff to the right and the scope of the binary functions is the pair of wffs to the right. Thus:

$$\neg \mathbb{P} \wedge \mathbb{Q} \Leftrightarrow \text{KNpq} \qquad\qquad \mathbb{P} \rightarrow \mathbb{Q} \Leftrightarrow \text{Cpq}$$
$$\neg (\mathbb{P} \wedge \mathbb{Q}) \Leftrightarrow \text{NKpq} \qquad\qquad \mathbb{P} \rightarrow \neg \mathbb{Q} \Leftrightarrow \text{CpNq}$$

And for a long one:

$$\neg((\mathbb{P} \vee \neg \mathbb{Q}) \leftrightarrow \neg(\mathbb{R} \rightarrow (S \wedge \mathbb{T}))) \Leftrightarrow \text{NEApNqNCrKst}$$

While it is said that Łukasiewicz and his students were very adept at working with this system, it does take some getting used to. It also becomes less useful when quantifiers are added. Except for the use of Reverse Polish Notation (the operator is postfixed instead of prefixed) in some scientific calculators, it has largely fallen out of favor.

Finally, if you have to type logical symbols and do not have easy access to special characters you can simply use the following:

$$\sim P \ or \ \text{-P} \ | \ P \char`\^ Q \ or \ P \ \& \ Q \ | \ P \ v \ Q \ | \ \text{P->Q} \ | \ \text{P<->Q}$$

As you see there are a lot of choices to be made in developing a logical system, and they are all pragmatic choices. The system is a tool and it can be designed in various ways with various simplicities and complexities, strengths and weaknesses. We try to balance simplicity, power, and usability.

## 2.6 **Chapter Glossary**

**Atomic Formula, Molecular Formula:**
> Any wff that qualifies simply in virtue of clause (1) of the definition of a wff (that is, any wff that just is some statement letter), is called an *atomic* wff. By analogy, all other wffs are *molecular*. (50)

**Compound Sentence:**
> A *compound sentence* is a sentence that either contains one or more simple sentences and at least one compounding phrase, or contains a compound subject or a compound predicate. (31)

**Expression of** $S$**:**
> An *expression of* $S$ is any finite sequence of the symbols of $S$. (46)

**Interpretation of** $S$**:**
> An *interpretation of* $S$ assigns semantic value to the statement letters of $S$, either via a translation key or a truth value assignment. (52)

**Main Connective, Well-Formed Components:**
  Atomic wffs have no main connective. The *main connective* of a molecular wff
  $\mathbb{R}$ is the connective appearing in the clause of the definition of a wff cited last in
  showing $\mathbb{R}$ to be a wff. The *immediate well-formed components* of a molecular wff
  are the values of $\mathbb{P}$ and $\mathbb{Q}$ (in the case of clause (2a) simply $\mathbb{P}$) in the last-cited
  clause of the definition of a wff. The *well-formed components* of a wff are the wff
  itself, its immediate well-formed components, and the well-formed components of
  its immediate well-formed components. The *atomic components* of a wff are the
  well-formed components that are atomic wffs. (50)

**Metalanguage:**
  When one is talking about a language, the *metalanguage* is the language in which
  one is talking about the object language. (42)

**Metavariables:**
  *Metavariables* are variables of the metalanguage that range over (take as possible
  values) expressions of the object language. We use uppercase Blackboard Bold:

  $$\mathbb{A}, \mathbb{B}, \mathbb{C}, \ldots, \mathbb{Z}, \mathbb{A}_1, \ldots \qquad (44)$$

**Molecular Formula:**
  See Atomic Formula. (50)

**Object Language:**
  When one is talking about a language, the *object language* is the language being
  talked about. (42)

**Punctuation Marks:**
  $(\ )$     (46)

**Recursive Definition:**
  *Recursive definitions* consist of three parts: the basis, the recursive clause(s), and
  the extremal clause. (47)

**Scope:**
  The *scope* of a connective is that portion of the wff containing its immediate sentential component(s). (50)

**Semantics:**
  *Semantics* is the study of language with regard to meaningful interpretations or
  valuations of the components. (45)

**Statement Letters:**
  $A, B, C, \ldots, Z, A_1, B_1, C_1, \ldots, Z_1, A_2, \ldots$     (46)

**Simple Sentence:**
  A *simple sentence* is a sentence that contains one subject and one predicate. (31)

**Symbols of** S**:**
>   See Statement Letters (46), Truth-Functional Connectives (46), Punctuation Marks
>   (46).

**Syntax:**
>   *Syntax* is the study of the signs of a language with regard only to their formal
>   properties. (45)

**Truth-Functional Compound:**
>   A sentence is a *truth-functional compound* iff the truth value of the compound
>   sentence is completely and uniquely determined by (is a function of) the truth
>   values of the simple component sentences. Otherwise, the compound sentence is
>   *non-truth-functional*. (32)

**Truth-Functional Connectives:**
>   ¬ ∧ ∨ → ↔    (46)

**Truth-Functional Logic:**
>   *Truth-functional logic* is the logic of truth-functional combinations of simple sen-
>   tences. It investigates the properties that arguments, sentences, and sets of sentences
>   have in virtue of their truth-functional structure. (33)

**Well-Formed Components:**
>   See Main Connective, Well-Formed Components. (50)

**Well-Formed Formula, Wff, of** S**:**
>   A *well-formed formula* or *wff* of S is a grammatical sentence of the language S.
>   See full definition in text. (47)

# 3 **Formal Semantics for** $S$

## 3.1 **Truth Value Assignments and Truth Tables**

We noted in the previous chapter that there are two kinds of interpretation we could give to wffs of $S$. One was to assign atomic wffs as translations of English sentences, as we have just been doing. The other sort of interpretation sets aside questions of translation into English and treats the wffs of $S$ as simply having the truth value true or false (and not both). This approach will allow us to investigate certain truth-functional semantic properties such as validity, consistency, equivalence, etc. For such properties, translation into a natural language is irrelevant—we need only consider truth-functional form and the possible truth values of the atomic components.

Two things will be of utmost importance here: the characteristic truth tables that define the truth functions of the connectives, and the notion of an interpretation or truth value assignment. We have seen the characteristic truth tables, Box 3.1 reviews them.

---

**Box 3.1: Characteristic Truth Tables**

| $\mathbb{P}\,\mathbb{Q}$ | $\neg\mathbb{P}$ | $\mathbb{P}\wedge\mathbb{Q}$ | $\mathbb{P}\vee\mathbb{Q}$ | $\mathbb{P}\rightarrow\mathbb{Q}$ | $\mathbb{P}\leftrightarrow\mathbb{Q}$ |
|---|---|---|---|---|---|
| T T | F | T | T | T | T |
| T F |   | F | T | F | F |
| F T | T | F | T | T | F |
| F F |   | F | F | T | T |

---

This tells us the value of any compound wff given the values of its immediate components. If we are given an assignment of values to the relevant atomic wffs (an interpretation), then we can easily compute the value of any compound wff of whatever complexity. Let us define truth value assignment and the related notion of a model before moving on.

**Truth Value Assignment:**

A *truth value assignment* (or tva) is an assignment of the value true (abbreviated by a 'T') or the value false ('F'), but not both, to each of the atomic wffs of $S$. A truth value assignment is a function from the set of atomic wffs into the set $\{T, F\}$. Also called an interpretation.

**Model:**

A tva is a *model* for (or *models*) a wff $\mathbb{P}$ (or set of wffs $\Gamma$) iff the wff $\mathbb{P}$ is (or all wffs in $\Gamma$ are) true on that tva.

Since there are an infinite number of atomic wffs each truth value assignment is infinite. (Each tva must assign values to, among others, '$A_1$', '$A_2$', '$A_3$', ...) Moreover, there are an infinite number of truth value assignments. (One that assigns T to '$A_1$' and F to all others, one that assigns T to '$A_2$' and F to all others, one that assigns T to '$A_1$' and T to '$A_2$' and F to all others, and so on.) Most of the time, however, we will be focused only on that portion of a tva that deals with the atomic wffs of immediate relevance, and since we will only ever occupy ourselves with a finite number of atomic wffs, we can treat the relevant truth value assignments as finite in size and number.

For example, suppose we wish to calculate the value of

(1)  $A \leftrightarrow [B \rightarrow (C \wedge A)]$

where '$A$' is T, '$B$' is T, and '$C$' is F. There are an infinite number of tvas which make that particular assignment to those atomic wffs, but they differ only in what they assign to wffs other than '$A$', '$B$', and '$C$'. Thus, we will ignore those differences and speak loosely of *the* truth value assignment that assigns '$A$' T, '$B$' T, and '$C$' F.

It is a simple matter to calculate the value of (1) for the assignment in question. We simply (i) write the assigned values under the atomic wffs, (ii) calculate (based on the values of their immediate well-formed components) the values of all the smallest well-formed components yet to be calculated, continue doing (ii) until we have a value under the main connective, at which point we (iii) stop and box the column under the main connective. (You should see a parallel between this process and the definition of wff (p. 47), is this just a coincidence? Why not?) Below I show these steps on separate lines, but normally we would proceed on one line only:

$$
\begin{array}{r|c|c c c}
 & A \leftrightarrow & [B \rightarrow & (C \wedge & A)] \\
\hline
\text{step (i)} & \boxed{\phantom{T}}\,T & T & F & T \\
\text{step (ii)} & & & F & \\
\text{repeat (ii)} & & F & & \\
\text{steps (ii) \& (iii)} & \boxed{F} & & &
\end{array}
$$

Thus, the truth value of (1) is F on the truth value assignment in question.

In many cases we will want to consider not just one tva but all truth value assignments to the relevant atomic wffs. That is, we may want to know the value of (1) for each possible assignment of T or F to '$A$', '$B$', and '$C$'. Since we have 3 atomic components and each can take 2 values, the number of possible tvas is $2^3 = 8$. Had we only 1 atomic component there would be only $2^1 = 2$ distinct tvas; had there been 2 atomics we would have $2^2 = 4$ tvas. As you will have already intuited, given $n$ distinct atomic components, there are $2^n$ distinct assignments of truth values to those atomic components. To calculate these values, we generate a table that displays all possible tvas and the values, based on those tvas, for all well-formed components of the wff in question. Here is how it's done:

(I) We start by listing the *n* atomic components in alphabetical order at the top left of the table, and then, separated by a vertical line, we write out the wff in question; all of this gets underlined.

(II) Next we generate the *basis columns*. Starting with the rightmost atomic component we fill in the column under it by alternating Ts and Fs for $2^n$ rows. Then, move one column to the left and fill in its $2^n$ rows by alternating pairs of Ts and Fs. Repeat this process of moving one column to the left and doubling the number of consecutive Ts and Fs, until you have a column of $2^n$ rows under each atomic component, at which point you stop.

(III) Fill in the columns under the wff in question as follows:

   (i) Copy the basis columns to the columns under the corresponding atomic components of the wff.[1]

   (ii) Calculate the values in the columns of all the smallest well-formed components yet to be calculated (based on the values in the columns under their immediate well-formed components). Repeat this process until the column under the main connective is complete.

   (iii) Box the column under the main connective.

This is much easier to do than to explain how to do. Seeing someone do it on the blackboard is the best, but I will illustrate a couple of steps in the process below. Here is the result of steps (I) and (II):

| A B C | A ↔ [B → (C ∧ A)] |
|-------|-------------------|
| T T T | |
| T T F | |
| T F T | |
| T F F | |
| F T T | |
| F T F | |
| F F T | |
| F F F | |

Note that the table has eight rows as a result of having three distinct atomic components. Here is the table after step (III.i) and one iteration of step (III.ii):

---

1. This step may be skipped once you become proficient, but you'll have to take care to consult the proper basis columns for the first iteration of step (III.ii).

```
A B C | A ↔ [B → (C ∧ A)]
T T T | T    T    T T T
T T F | T    T    F F T
T F T | T    F    T T T
T F F | T    F    F F T
F T T | F    T    T F F
F T F | F    T    F F F
F F T | F    F    T F F
F F F | F    F    F F F
```

First, per step (III.i), the basis columns were copied over to the relevant columns. Next, per step (III.ii), the values under the wedge are calculated based on the two immediately adjacent columns, since 'C' and 'A' are the immediate components of '(C ∧ A)'. Here it is after the next iteration of step (III.ii):

```
A B C | A ↔ [B → (C ∧ A)]
T T T | T    T T  T T T
T T F | T    T F  F F T
T F T | T    F T  T T T
T F F | T    F T  F F T
F T T | F    T F  T F F
F T F | F    T F  F F F
F F T | F    F T  T F F
F F F | F    F T  F F F
```

The column under the arrow is calculated based on the values under the 'B' and the wedge, since these columns contain the values for the immediate subcomponents 'B' and '(C ∧ A)' of '[B → (C ∧ A)]'. Here is the completed table:

```
A B C | A ↔ [B → (C ∧ A)]
T T T | T T  T T  T T T
T T F | T F  T F  F F T
T F T | T T  F T  T T T
T F F | T T  F T  F F T
F T T | F T  T F  T F F
F T F | F T  T F  F F F
F F T | F F  F T  T F F
F F F | F F  F T  F F F
```

The column under the main connective (the double arrow) is based on the columns under the 'A' and the arrow since these contain the values for the immediate subcomponents. We are done.

Now the rows of the truth table tell us what value (1) has for each distinct assignment of truth values to its atomic components—i.e., for any tva. Indeed, if we look at the second line of the truth table, we will see the tva that we calculated individually above.

**Truth Table:**

A *truth table* for a wff or set of wffs is a structure that lists the wff or wffs, all relevant truth value assignments, and the truth value of each wff on each truth value assignment.

Here are a few more examples:

(2)

| A B | ¬ | (¬ A ∧ B) |
|-----|---|-----------|
| T T | T | F T F T |
| T F | T | F T F F |
| F T | F | T F T T |
| F F | T | T F F F |

(3)

| C D E | (E ∧ ¬ D) | ↔ | ¬ C |
|-------|-----------|---|-----|
| T T T | T F F T | T | F T |
| T T F | F F F T | T | F T |
| T F T | T T T F | F | F T |
| T F F | F F T F | T | F T |
| F T T | T F F T | F | T F |
| F T F | F F F T | F | T F |
| F F T | T T T F | T | T F |
| F F F | F F T F | F | T F |

# 3.2 **Semantic Properties of Individual Wffs**

Generating truth tables simply for the heck of it is soon dull (if it wasn't to begin with). But there is more that we can do. Certain classes of wffs have interesting properties. For instance, certain wffs are always true, simply in virtue of their truth-functional structure, while others are always false. Truth tables give us a simple and mechanical procedure for deciding whether any given wff has one of these properties. I will define these properties, and describe the method of testing for them.

**Truth-Functionally True Wff:**

A wff $\mathbb{P}$ of $S$ is *truth-functionally true* iff $\mathbb{P}$ is true on every truth value assignment (every tva is a model).[2]

**Truth Table Test:** $\mathbb{P}$ is T-Fly[3] true iff, in its truth table, only Ts appear in the column under the main connective.

**Truth-Functionally False Wff:**

A wff $\mathbb{P}$ of $S$ is *truth-functionally false* iff $\mathbb{P}$ is false on every truth value assignment (no tva is a model).

**Truth Table Test:** $\mathbb{P}$ is T-Fly false iff, in its truth table, only Fs appear in the column under the main connective.

**Truth-Functionally Contingent Wff:**

A wff $\mathbb{P}$ of $S$ is *truth-functionally contingent* iff $\mathbb{P}$ is neither truth-functionally true nor truth-functionally false; i.e., iff it is false on at least one truth value assignment

---

2. Remember, 'iff' is short for 'if and only if'.
3. 'T-Fly', will be used to abbreviate 'truth-functionally'.

and true on at least one truth value assignment (at least one tva is model, but not all are).

**Truth Table Test:** $\mathbb{P}$ is T-Fly contingent iff, in its truth table, at least one F and at least one T appear in the column under the main connective.

### Truth-Functionally Satisfiable Wff:

A wff $\mathbb{P}$ of $S$ is *truth-functionally satisfiable* iff $\mathbb{P}$ is not truth-functionally false; i.e., it is true on at least one truth value assignment. We also say that the wff has a *model* or is *modeled*.

**Truth Table Test:** $\mathbb{P}$ is T-Fly satisfiable (has a model) iff, in its truth table, at least one T appears in the column under the main connective.

### Truth-Functionally Falsifiable Wff:

A wff $\mathbb{P}$ of $S$ is *truth-functionally falsifiable* iff $\mathbb{P}$ is not truth-functionally true; i.e., it is false on at least one truth value assignment (at least one tva is not a model).

**Truth Table Test:** $\mathbb{P}$ is T-Fly falsifiable iff, in its truth table, at least one F appears in the column under the main connective.

In the following examples I refrain from copying out the basis columns. This makes for easier reading.

(1)

| B | C | C | → | ¬B |
|---|---|---|---|----|
| T | T |   | F | F  |
| T | F |   | T | F  |
| F | T |   | T | T  |
| F | F |   | T | T  |

(2)

| A | B | (A ∧ B) | → | B |
|---|---|---------|---|---|
| T | T | T       | T |   |
| T | F | F       | T |   |
| F | T | F       | T |   |
| F | F | F       | T |   |

(3)

| A | A | ∧ | ¬A |
|---|---|---|----|
| T |   | F | F  |
| F |   | F | T  |

According to the definitions above, wff (1) is truth-functionally contingent (there are some Ts and some Fs under the main connective). As a result, (1) is also truth-functionally satisfiable (some Ts under the main connective), and truth-functionally falsifiable (some Fs under the main connective). Wff (2) is truth-functionally true (all Ts) and so truth-functionally satisfiable. Wff (3) is truth-functionally false (all Fs) and so falsifiable.

Box 3.2 illustrates the relationships between the five categories of wffs.

## Possibility, Necessity, and Truth Value Assignments

Intuitively, a truth-functionally true wff is one that, as a result of its truth-functional structure, is guaranteed to be true, necessarily true, or for which it is impossible to be false. A truth-functionally false wff is one that is necessarily false, or for which it is impossible to be true. But what exactly do we mean by possibility, necessity, and impossibility here? Recall that we worried a bit about these notions in Section 1.5.

In fact, we now have a very precise notion of truth-functional possibility—the truth value assignment. The basis columns of a properly constructed truth table list all relevantly distinct tvas—that is, all possible assignments of values to the atomic components of the wff. The column under the main connective lists the value of the complete wff in question

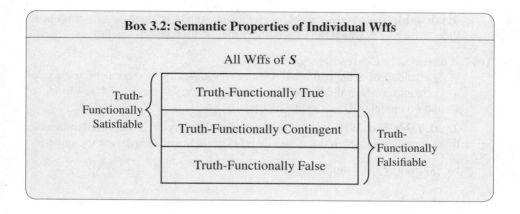

for each possible tva. So we need only scan the complete truth table to see whether or not there is any possibility of the wff being true or false. This is one of the many ways formalization helps us to get a firm grip on the logical properties in which we are interested. At least for truth-functional logical properties—and we will see this across the next few sections as well—with truth tables we render the notions of truth-functional possibility and necessity with full clarity.

Note that the properties discussed in this section apply to *individual* wffs, and an individual wff will have either two or three of these properties.

## 3.2.1 Exercises

For each of the following, construct a truth table to determine which of the five semantic properties the wff has. Write your determination under your table.

1. $A \lor \neg A$
2. $A \land \neg A$
3. $F \rightarrow \neg(G \lor H)$
4. $\neg(J \land K) \leftrightarrow (J \rightarrow \neg K)$
5. $(D \rightarrow (D \rightarrow D)) \leftrightarrow D$

# 3.3 **Semantic Properties of Sets of Wffs**

There are certain interesting semantic properties that apply, not to individual wffs, but to sets of one, two, or more wffs. And, of course, we have methods of testing for these properties with truth tables.

**Truth-Functionally Equivalent Pair of Wffs:**

Wffs $\mathbb{P}$ and $\mathbb{Q}$ of $S$ are *truth-functionally equivalent* iff there is no truth value assignment on which they differ in truth value (every model of $\mathbb{P}$ is a model of $\mathbb{Q}$, and every model of $\mathbb{Q}$ is a model of $\mathbb{P}$).

**Truth Table Test:** Wffs $\mathbb{P}$ and $\mathbb{Q}$ are T-Fly equivalent iff, in their joint truth table, there is no row in which the values under their main connectives differ.

### Truth-Functionally Contradictory Pair of Wffs:

Wffs $\mathbb{P}$ and $\mathbb{Q}$ of *S* are *truth-functionally contradictory* iff there is no truth value assignment on which they have the same truth value (no model of $\mathbb{P}$ is a model of $\mathbb{Q}$, and no model of $\mathbb{Q}$ is a model of $\mathbb{P}$).

**Truth Table Test:** Wffs $\mathbb{P}$ and $\mathbb{Q}$ are T-Fly contradictory iff, in their joint truth table, there is no row in which the values under their main connectives are the same.

Take, for example, the following wffs:

(1) $\neg A \vee \neg B$
(2) $A \rightarrow \neg B$
(3) $A \wedge B$
(4) $\neg A \rightarrow B$

To test for equivalence or contradiction we need to generate a joint truth table. This is simply a table that includes more than one wff, allowing for easy comparison of the value of each wff on a given truth value assignment. As an example, let's compare (1) to (2).

| A B | ¬A ∨ ¬B | | | A → ¬B | |
|-----|---------|---|---|--------|---|
| T T | F | F | F | F | F |
| T F | F | T | T | T | T |
| F T | T | T | F | T | F |
| F F | T | T | T | T | T |

Thus, (1) and (2) are truth-functionally equivalent. As the table shows, on each tva (1) has the same truth value as (2). Now compare (1) to (3).

| A B | ¬A ∨ ¬B | | | A ∧ B |
|-----|---------|---|---|-------|
| T T | F | F | F | T |
| T F | F | T | T | F |
| F T | T | T | F | F |
| F F | T | T | T | F |

As we can see, there is no tva on which (1) and (3) have the same truth value. Hence, they are contradictory. What, given the equivalence of (1) and (2), does this tell us about (2) and (3)? Will (2) and (3) be contradictory or equivalent?

Let's compare (1) and (4):

| A B | ¬A ∨ ¬B | | | ¬A → B | | |
|-----|---------|---|---|--------|---|---|
| T T | F | F | F | F | T | ⇐ |
| T F | F | T | T | F | T | ⇐ |
| F T | T | T | F | T | T | |
| F F | T | T | T | T | F | |

As is shown by the first line of the table, (1) and (4) are not equivalent; as shown by the second line, they are not contradictory. I.e., there is at least one tva on which they have the same value and at least one on which they have different values.

Keep in mind that the properties discussed so far in this section apply to *pairs* of wffs— sets of two—and the possibilities are that the members of the pair are truth-functionally equivalent, truth-functionally contradictory, or neither equivalent nor contradictory.

**Truth-Functionally Consistent Set of Wffs:**

A set $\Gamma$ of wffs of $S$ is *truth-functionally consistent* iff there is at least one truth value assignment on which all members of $\Gamma$ are true. We also say that $\Gamma$ has a *model* or is *modeled*. $\Gamma$ is *truth-functionally inconsistent* iff it is not truth-functionally consistent.

**Truth Table Test:** A set $\Gamma$ of wffs is T-Fly consistent iff, in the set's joint truth table, there is at least one row in which all the members of $\Gamma$ have a T under their main connectives. $\Gamma$ is T-Fly inconsistent iff there is no such row.[4]

Take, for example, the following sets of wffs:

(5) $\{\neg A \lor \neg B, \ \neg A \to B, \ \neg A \land B\}$
(6) $\{\neg A \lor \neg B, \ \neg A \to B, \ A \leftrightarrow B\}$[5]

| A B | ¬A ∨ ¬B | ¬A → B | ¬A ∧ B | |
|-----|---------|--------|--------|---|
| T T | F  F F | F  T | F  F | |
| T F | F  T T | F  T | F  F | |
| F T | T  T F | T  T | T  T | ⟸ |
| F F | T  T T | T  F | T  F | |

As shown by the third line of the table above, all the wffs of set (5) are true on the tva that assigns F to 'A' and T to 'B'. Thus, (5) is a consistent set. Here is the table for (6):

| A B | ¬A ∨ ¬B | ¬A → B | A ↔ B |
|-----|---------|--------|-------|
| T T | F  F F | F  T | T |
| T F | F  T T | F  T | F |
| F T | T  T F | T  T | F |
| F F | T  T T | T  F | T |

Set (6) is inconsistent. There is no line of the truth table, and so no tva, on which all members of the set are true. Note that while each individual wff is truth-functionally satisfiable, the set is not consistent.

---

4. Strictly speaking, while the definition applies to finite or infinite $\Gamma$, the test applies only to finite $\Gamma$.
5. Since I am mentioning wffs in specifying the sets, the wffs should appear in single quotation marks. For simplicity's sake and the avoidance of visual clutter, we allow wffs to name themselves in this context.

**Truth-Functionally Entails:**

> A set $\Gamma$ of wffs of *S* *truth-functionally entails* a wff $\mathbb{P}$ iff there is no truth value assignment on which all the members of $\Gamma$ are true and $\mathbb{P}$ is false. (Every model of $\Gamma$ is a model of $\mathbb{P}$.)

> **Truth Table Test:** A set $\Gamma$ T-Fly entails a wff $\mathbb{P}$ iff there is no row of their joint truth table on which all the members of $\Gamma$ have a T under their main connectives and $\mathbb{P}$ has an F under its main connective.[6]

We use the double turnstile, '$\vDash$', and negated double turnstile, '$\nvDash$', respectively, to assert or deny entailment:

$$\Gamma \vDash \mathbb{P}$$
$$\Gamma \nvDash \mathbb{P}$$

Typically, we will spell out the members of the set, as well as the target; for example:

$$\{A \wedge B, B \leftrightarrow \neg C, D \rightarrow C\} \vDash \neg D \wedge A$$
$$\{A \wedge B, B \leftrightarrow C, D \rightarrow C\} \nvDash \neg D \wedge A$$

Some sentences are entailed by the empty set; that is:

$$\varnothing \vDash \mathbb{P}$$
$$\varnothing \vDash C \leftrightarrow \neg \neg C$$

But, as is the practice, we will drop the empty set symbol and write simply:

$$\vDash \mathbb{P}$$
$$\vDash C \leftrightarrow \neg \neg C$$

Moreover, when the set in question contains only a single wff (is a unit set or singleton) we drop the set brackets. So, for example,

$$\{\mathbb{R}\} \vDash \mathbb{P}$$
$$\{A \wedge B\} \vDash B$$

become,

$$\mathbb{R} \vDash \mathbb{P}$$
$$A \wedge B \vDash B$$

Let's see how to carry out the test for entailment. Take, for example, the following entailment claim:

$$(7) \quad \{A \vee B, \neg A\} \vDash B$$

---

6. Strictly speaking, while the definition applies to finite or infinite $\Gamma$, the test applies only to finite $\Gamma$.

To test the entailment claim made in (7) we list the members of the set first and the wff
that is a candidate for entailment at the far right:

| A B | A ∨ B | ¬A | B |
|-----|-------|----|----|
| T T | T | F | T |
| T F | T | F | F |
| F T | T | T | T |
| F F | F | T | F |

As there is no line in which the members of the set are all true and the candidate wff false,
the entailment claim is true: $\{A \vee B, \neg A\} \vDash B$. Note that while there is a line in which
all the wffs are true, this is not the crucial case, for this may appear even in cases where
entailment fails. The crucial thing is the absence of a line in which all members of the set
are true and the candidate wff is false. Let's look at a case of failure of entailment.

(8) $\{A \vee B, A\} \nvDash B$

If (8) is true, then we should find a row in the relevant joint truth table where the members
of the set are true but the target wff is false.

| A B | A ∨ B | A | B |
|-----|-------|----|----|
| T T | T | T | T |
| T F | T | T | F ⟸ |
| F T | T | F | T |
| F F | F | F | F |

As the second row shows, (8) is true. There is no entailment. Note that, as shown by the
first line, there is a tva on which all the wffs are true. But, while this shows the set of three
wffs is consistent, it is irrelevant to the question of entailment. It is the presence of the
second line that shows that $\{A \vee B, A\} \nvDash B$.

As we saw in the opening chapter, the concepts of entailment and validity are closely
related. So are the truth table tests. First we get explicit about what constitutes an argument.

### Argument of $S$:

An *argument* of $S$ is a finite set of two or more wffs of $S$, one of which is the
conclusion, while the others are premises.

### Truth-Functionally Valid Argument:

An argument of $S$ is *truth-functionally valid* iff there is no truth value assignment
on which all the premises are true and the conclusion is false. An argument of $S$ is
*truth-functionally invalid* iff it is not truth-functionally valid.

**Truth Table Test:** An argument is T-Fly valid iff there is no row of their joint
truth table on which all the premises have a T under their main connectives and the
conclusion has an F under its main connective. If there is such a row, the argument
is T-Fly invalid.

This should sound rather familiar. Indeed, we could define validity in terms of entailment—an argument is valid iff the set consisting of all and only its premises entails the conclusion. Obviously, the truth table tests for entailment and validity are basically the same. Consider the following two arguments that correspond to claims (7) and (8), respectively:

(9)  A ∨ B
    ¬A
    ―――
    B

(10)  A ∨ B
    A
    ―――
    B

The truth table for argument (9) will be identical to that for entailment claim (7):

| A B | A ∨ B | ¬A | B |
|-----|-------|-----|---|
| T T | T | F | T |
| T F | T | F | F |
| F T | T | T | T |
| F F | F | T | F |

The absence of a row in which all its premises are true and its conclusion false shows that argument (9) is valid. Here is the table for argument (10):

| A B | A ∨ B | A | B | |
|-----|-------|---|---|---|
| T T | T | T | T | |
| T F | T | T | F | ⟸ |
| F T | T | F | T | |
| F F | F | F | F | |

Obviously this is the same as the table we produced for the failure of entailment in claim (8). The presence of a row with all the premises true and the conclusion false, shows that (10) is truth-functionally invalid. Note again that the presence of a row with all premises true and the conclusion true is irrelevant as it may appear in the tables of both valid and invalid arguments. Indeed, we have a case of that here. What matters to validity (and entailment) is the presence or absence of a row with all the premises (all members of the set) true and the conclusion (target) false.

    Note as well, the relation of truth-functional validity (of arguments) to deductive validity generally. In Sec. 1.3 we defined a deductively valid argument as one in which it is not possible for all the premises to be true and the conclusion false. As mentioned above, the truth table displays all truth-functionally relevant possibilities. Hence, if there is no row of the table on which all the premises are true and the conclusion false, then that is impossible, and the argument is truth-functionally valid. If there is such a row, then, clearly, all true premises and a false conclusion is possible, and the argument is truth-functionally invalid.[7]

---

7. Keep in mind that truth-functional validity is a more narrow notion than deductive validity. Indeed, in later chapters we will see arguments that are truth-functionally invalid, but deductively valid, in virtue of being quantificationally valid.

Finally, the definition for truth-functional validity (and the related truth table test) fits with and clarifies the quirky cases of validity discussed in Section 1.3.1. There we saw that every argument that has a logically true conclusion is deductively valid, and that every argument with inconsistent premises is deductively valid. The same goes for the truth-functional properties. Every argument with a truth-functionally true conclusion is truth-functionally valid, and every argument with truth-functionally inconsistent premises is truth-functionally valid. Consider the following arguments:

(11)  A → B
     B
    —————
    A ∨ ¬A

(12)  A ∧ B
     ¬A
    ———
     B

Here is the table for argument (11):

| A B | A → B | B | A ∨ ¬A |
|-----|-------|---|--------|
| T T | T | T | T F |
| T F | F | F | T F |
| F T | T | T | T T |
| F F | T | F | T T |

Since the conclusion of (11) is truth-functionally true, the is no tva on which the conclusion is false, so there is no tva on which all the premises are true and the conclusion is false. Hence, (11) is truth-functionally valid.

The situation with argument (12) is similar, but different. Here is its table:

| A B | A ∧ B | ¬A | B |
|-----|-------|-----|---|
| T T | T | F | T |
| T F | F | F | F |
| F T | F | T | T |
| F F | F | T | F |

Here the premises are inconsistent. There is no tva on which all the premises are true. So there is no tva on which all the premises are true and the conclusion is false. Hence, (12) is also truth-functionally valid.

## 3.3.1 **Exercises**

**A.** For each of the following, construct a truth table to determine whether the wffs in the pair are equivalent, contradictory, or neither. Write your determination under your table.

1. ¬(D ↔ E)                 (D ∧ ¬E) ∨ (¬D ∧ E)
2. I ∧ J                     ¬(J ∨ I)
3. D → (E → F)               D ∧ (E ∧ ¬F)

**B.** For each of the following, construct a truth table to determine whether the set of wffs is consistent or inconsistent. Write your determination under your table.

1. $\{(A \wedge B) \rightarrow C, \neg C \wedge A, A \rightarrow B\}$
2. $\{\neg(K \wedge L), \neg(K \vee L), K \rightarrow \neg L\}$

**C.** For each of the following, construct a truth table to test the entailment claim. Write your determination under your table.

1. $\{C \vee D, \neg E \rightarrow \neg D, C \rightarrow E\} \vDash E$
2. $\{A \rightarrow B, B\} \vDash A$

**D.** For each of the following, construct a truth table to determine whether the argument is valid or invalid. Write your determination under your table.

1. $\neg D \rightarrow C$
$B \rightarrow \neg C$
$\overline{\neg D \rightarrow \neg B}$

2. $G \leftrightarrow H$
$\neg H \wedge I$
$\overline{\neg G \wedge I}$

# 3.4  **Semantic Properties, Their Interrelations, and Simple Metalogic**

The generation and reading of truth tables is essentially a mechanical process. You should find that, with a little practice, you quickly gain facility in building tables and applying the definitions above to determine the presence or absence of the semantic properties discussed in the previous two sections. To have a firm grasp on those properties it is important to go beyond mere truth table construction and to be able to reason abstractly about the interrelations among the semantic properties.

For example: Suppose you have a wff, call it $\mathbb{P}$, and all you know about this wff is that it is truth-functionally contingent. We are not told how simple or complex it is, we do not know how many atomic components or connectives it has. All we are told is that it is truth-functionally contingent. Suppose you have another wff, $\mathbb{Q}$, but you don't know anything about $\mathbb{Q}$. Finally, consider the disjunction of the two, $\mathbb{P} \vee \mathbb{Q}$. What, if anything, can we determine about the semantic properties of $\mathbb{P} \vee \mathbb{Q}$? Justify your answer.

You might be tempted to try to generate a truth table at this point. But that would be of limited help, since we do not have specific information concerning the values of $\mathbb{P}$. Indeed, we don't know how many atomic components $\mathbb{P}$ might have, so we have no idea how many rows a truth table should have. Plus we know even less about $\mathbb{Q}$! How, then, can we answer the question?

We need to carefully unpack the information we are given. We were supposing that $\mathbb{P}$ is truth-functionally contingent. Look at the definition of truth-functional contingency. We don't know how many tvas are relevant to $\mathbb{P}$, nor do we have any sense of what the pattern of Ts and Fs would be if we could make a table. But we do know that there is at least one tva on which $\mathbb{P}$ is T, and at least one tva on which $\mathbb{P}$ is F. That is what it means for $\mathbb{P}$ to the truth-functionally contingent. We don't know anything about $\mathbb{Q}$, but we are being asked to consider the disjunction $P \vee Q$, and we know from the characteristic truth tables what it takes to make a disjunction true or false. In particular, for a disjunction to be true on a given tva, at least one of its disjuncts must be true. We know that $P$ is T on at least one tva, so on that same tva $P \vee Q$ must also be true, regardless of the value of $Q$. So $P \vee Q$ is T on at least one tva, and, according to our definitions, that means $P \vee Q$ is truth-functionally satisfiable. Could we deduce any more? Well, we also know that $P$ is F on at least one tva, but since we do not know anything about $Q$, this does not help us much, so we do not know any more.

What we have just done is use the information given, together with our knowledge of the semantic properties and the characteristic truth tables to construct an argument concluding that $P \vee Q$ is truth-functionally satisfiable. Indeed, we have shown that the disjunction of any contingent wff with any wff is itself satisfiable. In the metalanguage we made an argument about the semantic properties of certain forms of wffs. We call this a semantic meta-proof, and we have just done some simple metalogic. We are reasoning abstractly within our metalanguage about our object language system.

**Metalogic:**

>*Metalogic* is the study of the properties of logical systems. In particular, it is systematic reasoning in a metalanguage about the properties of object language systems of logic.

Let's put this semantic meta-proof together in a more compact form. In an exercise, quiz, or exam you might be asked the following:

1. Suppose $P$ is truth-functionally contingent, what, if anything, do we know about $P \vee Q$? Justify your answer.

Here is a sample answer (I use some standard abbreviations below):

>$P \vee Q$ is T-Fly satisfiable. Since $P$ is T-Fly contingent, it is T on at least one tva. That means that the left disjunct of $P \vee Q$ is T on at least one tva, making the whole disjunction T on that tva. That is, $P \vee Q$ is T-Fly satisfiable.

Note that I state the conclusion both at the beginning and end of my reasoning. This is not strictly necessary, but it is good form. In between, I use the information given and my knowledge of truth-functional semantics to offer premises that support my conclusion. I am using logic to reason about logic. This is carried out in the metalanguage. This is a simple form of metalogic!

Here is another, more challenging, example:

2. Suppose the set $\{P_1, \ldots, P_n\}$ is T-Fly consistent, and $Q$ is T-Fly false. What, if anything, do we know about the set $\{P_1, \ldots, P_n, \neg Q\}$? Justify your answer.

The relevant bits of information here are that we are dealing with an initial set (of some unspecified number of wffs) that is then added to. If we are being asked about a set without any question of entailment, then the issue is whether the set in question is consistent or inconsistent. Plus we need to take account of what we are told about $Q$, for its negation is added to the original set. See if the following makes sense to you.

The set $\{\mathbb{P}_1, \ldots, \mathbb{P}_n, \neg\mathbb{Q}\}$ is consistent. Since the original set $\{\mathbb{P}_1, \ldots, \mathbb{P}_n\}$ is consistent, there is at least one tva on which all of $\mathbb{P}_1, \ldots, \mathbb{P}_n$ are T. Since $\mathbb{Q}$ is T-Fly false, it is F on every tva, but that means that its negation $\neg\mathbb{Q}$ is T on every tva. In particular, $\neg\mathbb{Q}$ is T on the very same tva on which all the $\mathbb{P}$s are T. So adding it to the consistent set does not disrupt consistency. So, the set $\{\mathbb{P}_1, \ldots, \mathbb{P}_n, \neg\mathbb{Q}\}$ is consistent.

The basic strategy here is to unpack the definitions and the information given to see what, if any, conclusions can be drawn. Justifying your response requires making clear the connection between the information given the definitions, the characteristic truth tables, and the conclusion. The exercises below contain many such problems. In learning to answer them you will be gaining a firmer grasp on the semantic properties and their interrelations. You will also be learning how to engage in some simple metalogic.

## 3.4.1 **Exercises**

**A.** The following questions test your understanding of the definitions of semantic properties of wffs.

1. If a wff $\mathbb{P}$ is truth-functionally true, is there any other property we know it has? Justify your answer based on the definitions.
2. If a wff $\mathbb{P}$ is truth-functionally false, is there any other property we know it has? Justify your answer.
3. If a wff $\mathbb{P}$ is truth-functionally contingent, is there any other property we know it has? Justify your answer.
4. Contingency was defined in terms of truth-functional truth and truth-functional falsity; could we define contingency in terms of any other pairs of properties? If so, do so.
5. Suppose $\mathbb{P}$ is a truth-functionally true wff, what, if anything, do we know about $\mathbb{P} \vee \mathbb{Q}$? Justify your answer.
6. Suppose $\mathbb{P}$ is a truth-functionally false wff, what, if anything, do we know about $\mathbb{P} \rightarrow \mathbb{Q}$? Justify your answer.
7. Suppose $\mathbb{P}$ is a falsifiable wff, what, if anything, do we know about $\mathbb{P} \rightarrow \mathbb{Q}$? Justify your answer.
8. Suppose $\mathbb{P}$ is truth-functionally true and $\mathbb{Q}$ is truth-functionally false, what, if anything, do we know about $\mathbb{P} \rightarrow \mathbb{Q}$? Justify your answer.
9. Suppose $\mathbb{P}$ is truth-functionally true and $\mathbb{Q}$ is falsifiable, what, if anything, do we know about $\mathbb{P} \rightarrow \mathbb{Q}$? Justify your answer.
10. Suppose $\mathbb{P}$ and $\mathbb{Q}$ are truth-functionally contingent, what, if anything, do we know about $\mathbb{P} \rightarrow \mathbb{Q}$? Justify your answer.

**B.** The following questions test your understanding of the definitions of semantic properties of wffs and sets of wffs.

1. Suppose $\mathbb{P}$ and $\mathbb{Q}$ are equivalent, does this tell us anything about $\mathbb{P} \rightarrow \mathbb{Q}$? Justify your answer.

2. Suppose $\mathbb{P}$ and $\mathbb{Q}$ are equivalent, does this tell us anything about $\mathbb{P} \leftrightarrow \mathbb{Q}$? Justify your answer.

3. Suppose $\mathbb{P}$ and $\mathbb{Q}$ are equivalent, does this tell us anything about $\mathbb{P} \wedge \mathbb{Q}$? Justify your answer.

4. Suppose $\mathbb{P}$ and $\mathbb{Q}$ are contradictory, does this tell us anything about $\mathbb{P} \rightarrow \mathbb{Q}$? Justify your answer.

5. Suppose $\mathbb{P}$ and $\mathbb{Q}$ are contradictory, does this tell us anything about $\mathbb{P} \leftrightarrow \mathbb{Q}$? Justify your answer.

6. Suppose $\mathbb{P}$ and $\mathbb{Q}$ are contradictory, does this tell us anything about $\mathbb{P} \wedge \mathbb{Q}$? Justify your answer.

7. Suppose $\mathbb{P}$ is truth-functionally false, what do we know about the set $\{\mathbb{P}\}$? Justify your answer.

8. Suppose $\mathbb{P}$ is truth-functionally contingent, what do we know about the set $\{\mathbb{P}\}$? Justify your answer.

9. Suppose $\Gamma$ is inconsistent, does this tell us anything about whether $\Gamma \vDash \mathbb{P}$? Justify your answer.

10. $\vDash \mathbb{P}$ iff $\mathbb{P}$ is a truth-functionally true wff. True or false? Justify your answer.

11. $\vDash \neg\mathbb{P}$ iff $\mathbb{P}$ is a truth-functionally false wff. True or false? Justify your answer.

12. Suppose $\mathbb{P}$ is a truth-functionally true wff, what, if anything, does this tell us about an argument with $\mathbb{P}$ as conclusion? Justify your answer.

13. Suppose $\{\mathbb{P}_1, \ldots, \mathbb{P}_n, \neg\mathbb{Q}\}$ is inconsistent, does this tell us anything about the following argument? Justify your answer.

$\mathbb{P}_1$

$\vdots$

$\mathbb{P}_n$

$\overline{\mathbb{Q}}$

14. Suppose the above argument is valid, what, if anything, does this tell us about the conditional $(\mathbb{P}_1 \wedge \ldots \wedge \mathbb{P}_n) \rightarrow \mathbb{Q}$? Justify your answer.

15. Suppose $\mathbb{P}$ and $\mathbb{Q}$ are equivalent, and $\mathbb{P} \vDash \mathbb{R}$, do we know whether $\mathbb{Q} \vDash \mathbb{R}$? Justify your answer.

## 3.5 **Truth Trees**

Truth trees provide an additional way to test wffs and sets of wffs for various truth-functional properties. Rather than laying out all possible truth value assignments of a set of wffs, as in a truth table, we generate a list of truth value assignments that would satisfy (make true) the wffs in question. The list has the form of a downward branching tree, as we decompose more complex wffs into those components that must be true in order for the wffs above to be true. Sometimes it will be possible to satisfy all of the wffs in question—they are consistent. In these cases the tree will be called "open". Other times it is not possible to satisfy all of the wffs in question—they are inconsistent—and the tree will be called "closed". Whether a tree is open or closed will allow us to determine a variety of logical properties.

To begin to see how this works, consider the following set of wffs:

(1)  $\{K \wedge \neg J, H \rightarrow J, \neg\neg H\}$

Is the set consistent or inconsistent? Let's see if we can find a tva that would make all the member wffs true, thereby showing the set is consistent.

First, we list the members of the set, forming the trunk of our tree:

$$K \wedge \neg J$$
$$H \rightarrow J$$
$$\neg\neg H$$

The first wff is a conjunction. For it to be true, both conjuncts must be true. So we will list each conjunct below the trunk as our first branch, thus:

$$K \wedge \neg J \checkmark$$
$$H \rightarrow J$$
$$\neg\neg H$$
$$|$$
$$K$$
$$\neg J$$

Note that we check off the first wff to show that we have decomposed it. Our aim is to decompose all wffs until we have only atomic wffs (as in 'K') or negations of atomic wffs (as in '¬J'). What this indicates, so far, is that for 'K ∧ ¬J' to be true 'K' must be assigned T and 'J' must be assigned F (to make '¬J' T).

We still have the second and third wffs in the trunk to decompose, so we are not done. To make 'H → J' true, it must be that either the antecedent, 'H', is assigned F ('¬H' must be assigned T) or the consequent, 'J', is assigned T. Since we have two options here, the branch splits in two:

$$K \wedge \neg J \checkmark$$
$$H \rightarrow J \checkmark$$
$$\neg\neg H$$
$$|$$
$$K$$
$$\neg J$$
$$\diagup\diagdown$$
$$\neg H \quad J$$

Again, we check off the wff we have just decomposed. Our search for a tva that will make all the members of the set true has now resulted in two alternatives. Whenever we decompose a wff from here on out the results must be added to every branch that can be traced back to that wff. In particular, when we decompose '¬¬H' here the result must be appended to the end of both branches. Obviously for '¬¬H' to be true, 'H' must be assigned T. So we add 'H' to the end of each branch.

$$
\begin{array}{c}
\text{K} \land \lnot\text{J} \checkmark \\
\text{H} \rightarrow \text{J} \checkmark \\
\lnot\lnot\text{H} \checkmark \\
| \\
\text{K} \\
\lnot\text{J} \\
\diagup\diagdown \\
\lnot\text{H} \quad \text{J} \\
| \qquad | \\
\text{H} \quad \text{H}
\end{array}
$$

Each branch represents an *attempt* to assign truth values to make the set consistent, so here we have two attempts. So is the set consistent? Does either of these attempts succeed? If we look at the left branch first, it tells us to assign T to 'K', F to 'J', and *both* T *and* F to 'H'. So we have a problem. This branch contains a contradiction, so it does not represent a possible tva. 'H' cannot be both T and F. We call this a closed branch, and we will mark it with an '✗'. Look now at the right branch. It also contains a contradiction: it has both 'J' and '¬J', telling us to assign both T and F to 'J', which we cannot do. So this branch is closed as well, and we mark it also with an '✗'.

$$
\begin{array}{c}
\text{K} \land \lnot\text{J} \checkmark \\
\text{H} \rightarrow \text{J} \checkmark \\
\lnot\lnot\text{H} \checkmark \\
| \\
\text{K} \\
\lnot\text{J} \\
\diagup\diagdown \\
\lnot\text{H} \quad \text{J} \\
| \qquad | \\
\text{H} \quad \text{H} \\
✗ \quad\ ✗
\end{array}
$$

Since every branch is closed, the tree as a whole is closed (definitions below). What this means is that there is no tva on which every wff of the original set is true. Every attempt to produce one resulted in a contradiction. So the set is inconsistent. Had even just one of the branches been open, the whole tree would be called open. That branch would indicate a tva on which every wff in the original set is true, showing consistency for the set.

We can carry out similar processes for any finite set of wffs of **S**. To do so systematically, we need decomposition rules to handle every form of wff that is not either an atomic wff or a negated atomic wff. Nine rules will suffice. We will have one rule for each binary connective (so, 4); one rule for each negated binary connective $(4+4)$; and one rule to handle double negation $(4+4+1=9)$.

We saw how to deal with double negation in the example above. The rule is called Double Negation Decomposition $(\neg\neg D)$:

$$\neg\neg D$$
$$\neg\neg\mathbb{P} \checkmark$$
$$|$$
$$\mathbb{P}$$

This rule form tells us what to do whenever we have a wff of form $\neg\neg\mathbb{P}$ in the trunk or current branch of a tree. In order to make the double negation of $\mathbb{P}$ true, we must make $\mathbb{P}$, itself, true, so we write it below on every branch leading out of it and we check off the original.

We also saw how to deal with a conjunction. We need to make both conjuncts true, so we list them below on every branch leading out of the conjunction itself and check off the original. This is Conjunction Decomposition $(\wedge D)$:

$$\wedge D$$
$$\mathbb{P}\wedge\mathbb{Q} \checkmark$$
$$|$$
$$\mathbb{P}$$
$$\mathbb{Q}$$

Remember to check off the wff you have just decomposed, and if multiple branches lead out of the current wff at any distance below, then the result of decomposition must be written under all of those branches.

Here are the remaining rules, with brief explanations. If you are familiar with the characteristic truth tables, these rules should make sense.

In order to make a negated conjunction true, one or both conjuncts must be false, so the tree must branch to accommodate both options. Again, we check off the original. Negated Conjunction Decomposition $(\neg\wedge D)$:

$$\neg\wedge D$$
$$\neg(\mathbb{P}\wedge\mathbb{Q}) \checkmark$$
$$\diagup\diagdown$$
$$\neg\mathbb{P} \quad \neg\mathbb{Q}$$

Here is Disjunction Decomposition $(\vee D)$, which also branches since, at minimum, only one or the other disjunct need be true:

∨D
_____

$\mathbb{P} \vee \mathbb{Q}$ ✓

$\mathbb{P}$   $\mathbb{Q}$

To make a negated disjunction true, we must make each disjunct false, hence Negated Disjunction Decomposition (¬∨D):

¬∨D
_____

$\neg(\mathbb{P} \vee \mathbb{Q})$ ✓
|
$\neg\mathbb{P}$
$\neg\mathbb{Q}$

We saw above how to deal with a conditional. It must be the case that either the antecedent is false or the consequent is true. Conditional Decomposition (→D):

→D
_____

$\mathbb{P} \rightarrow \mathbb{Q}$ ✓

$\neg\mathbb{P}$   $\mathbb{Q}$

Here is Negated Conditional Decomposition (¬→D). We must have a true antecedent and a false consequent:

¬→D
_____

$\neg(\mathbb{P} \rightarrow \mathbb{Q})$ ✓
|
$\mathbb{P}$
$\neg\mathbb{Q}$

For a biconditional to be true we need either both sides true or both sides false. Biconditional Decomposition (↔D):

↔D
_____

$\mathbb{P} \leftrightarrow \mathbb{Q}$ ✓

$\mathbb{P}$   $\neg\mathbb{P}$
$\mathbb{Q}$   $\neg\mathbb{Q}$

For negated biconditional we need the sides to have opposite truth values. This gives us two options. Negated Biconditional Decomposition (¬↔D):

$$\neg \leftrightarrow D$$
$$\overline{\neg(\mathbb{P} \leftrightarrow \mathbb{Q})} \; \checkmark$$

$$\begin{array}{cc} \mathbb{P} & \neg\mathbb{P} \\ \neg\mathbb{Q} & \mathbb{Q} \end{array}$$

When faced with a set of wffs we need to evaluate, we iterate application of these rules to complete a tree by breaking down all wffs in the trunk and on the branches until we have only atomic wffs or negations thereof. Remember that if a wff has many branches leading out of it, we must put the results of decomposition at the end of each branch. Then we put an '✗' under every branch that contains a contradiction (both an atomic $\mathbb{P}$ and its negation $\neg\mathbb{P}$). These are called closed branches. If all branches are closed, then the tree as a whole is closed, and the set is inconsistent—there is no way for all members to be true. If at least one branch is open, then the tree as a whole is open, and the set is consistent—there is at least one way for all members to be true.

Here is another example:

(2) $\{E \wedge \neg B, E \rightarrow \neg\neg C, \neg(C \rightarrow B)\}$

We set up the trunk of the tree by listing the wffs of the set, like so:

$$E \wedge \neg B$$
$$E \rightarrow \neg\neg C$$
$$\neg(C \rightarrow B)$$

Now, we need to decompose each wff in the trunk. We can do so in any order we choose, but it is strategically wise to decompose wffs that do not branch before those that do. This will usually minimize how much we have to write and keep the tree simpler. Let's start with the first wff. Decomposing it leaves us with the following:

$$E \wedge \neg B \; \checkmark$$
$$E \rightarrow \neg\neg C$$
$$\neg(C \rightarrow B)$$
$$|$$
$$E$$
$$\neg B$$

We could now decompose the second wff, but since it branches it will keep things simpler to skip it and attack the third. So we get:

$$E \land \neg B \checkmark$$
$$E \to \neg\neg C$$
$$\neg(C \to B) \checkmark$$
$$|$$
$$E$$
$$\neg B$$
$$|$$
$$C$$
$$\neg B$$

Now we will address the second wff:

$$E \land \neg B \checkmark$$
$$E \to \neg\neg C \checkmark$$
$$\neg(C \to B) \checkmark$$
$$|$$
$$E$$
$$\neg B$$
$$|$$
$$C$$
$$\neg B$$

$$\neg E \qquad \neg\neg C$$

Finally, the right branch still needs some work, so we apply the ¬¬D rule:

$$E \land \neg B \checkmark$$
$$E \to \neg\neg C \checkmark$$
$$\neg(C \to B) \checkmark$$
$$|$$
$$E$$
$$\neg B$$
$$|$$
$$C$$
$$\neg B$$

$$\neg E \qquad \neg\neg C \checkmark$$
$$|$$
$$C$$

Reading carefully, you will see that the left branch closes, since it has both 'E' and '¬E'. So we mark it appropriately. The right branch, however, is open.

$$E \land \neg B \checkmark$$
$$E \rightarrow \neg\neg C \checkmark$$
$$\neg(C \rightarrow B) \checkmark$$
$$|$$
$$E$$
$$\neg B$$
$$|$$
$$C$$
$$\neg B$$

$$\neg E \qquad \neg\neg C \checkmark$$
$$\times \qquad |$$
$$\qquad C$$

Thus, since it has at least one open branch, the tree as a whole is open and we have shown that the set is consistent. In particular, the open branch indicates a tva on which every member of the original set is true. We simply need to read it off the branch: 'E' is T, 'B' is F, and 'C' is T. It does not matter that '¬B' and 'C' appear twice on the branch. This often happens.

---

**Note:** A complete table of tree rules appears in Appendix C

---

## 3.5.1 **Tests with Truth Trees**

Here are some definitions of terms we have been using:

**Truth Tree:**
> A *truth tree* for a set of wffs is a structure that lists all the wffs, and all the well-formed components that must be true in attempting to show the set truth-functionally consistent.

**Closed Branch:**
> A branch on a truth tree is a *closed branch* iff the branch contains both some atomic wff $\mathbb{P}$ and its negation $\neg\mathbb{P}$.

**Open Branch:**
> A branch on a truth tree is an *open branch* iff it is not closed.

**Closed Tree:**
> A truth tree is a *closed tree* iff every branch is closed.

**Open Tree:**
> A truth tree is an *open tree* iff at least one branch is open.

Below is a list of the various tests we can perform using truth trees. Recall that what we are always doing with a truth tree is testing some set of wffs for consistency. Thus to test for other properties, we must understand them in terms of the consistency or inconsistency of the appropriate set.

We have already seen use of trees to check for consistency or inconsistency of a set. Here are the definitions.

**Truth Tree Test for Truth-Functionally Consistent Set of Wffs:**
    A set $\Gamma$ is T-Fly consistent iff it has an open tree.

**Truth Tree Test for Truth-Functionally Inconsistent Set of Wffs:**
    A set $\Gamma$ is T-Fly inconsistent iff it has a closed tree.

Here are tree tests for properties of individual wffs.

**Truth Tree Test for Truth-Functionally True Wff:**
    A wff $\mathbb{P}$ is T-Fly true iff $\{\neg\mathbb{P}\}$ has a closed tree.

**Truth Tree Test for Truth-Functionally False Wff:**
    A wff $\mathbb{P}$ is T-Fly false iff $\{\mathbb{P}\}$ has a closed tree.

**Truth Tree Test for Truth-Functionally Contingent Wff:**
    A wff $\mathbb{P}$ is T-Fly contingent iff $\{\mathbb{P}\}$ has an open tree and $\{\neg\mathbb{P}\}$ has an open tree.

**Truth Tree Test for Truth-Functionally Satisfiable Wff:**
    A wff $\mathbb{P}$ is T-Fly satisfiable iff $\{\mathbb{P}\}$ has an open tree.

**Truth Tree Test for Truth-Functionally Falsifiable Wff:**
    A wff $\mathbb{P}$ is T-Fly falsifiable iff $\{\neg\mathbb{P}\}$ has an open tree.

Consider the following wffs:

(3) $(D \vee \neg E) \leftrightarrow (G \vee E)$
(4) $(A \wedge B) \rightarrow ((\neg A \vee \neg B) \rightarrow C)$

A truth tree for the set containing just wff (3) will look like this:

$$(D \lor \neg E) \leftrightarrow (G \lor E) \checkmark$$

$$
\begin{array}{cc}
D \lor \neg E \checkmark & \neg(D \lor \neg E) \checkmark \\
G \lor E \checkmark & \neg(G \lor E) \checkmark
\end{array}
$$

$$
\begin{array}{ccc}
D & \neg E & \neg D \\
& & \neg\neg E \checkmark
\end{array}
$$

$$
\begin{array}{cccc}
G & E & G & E \\
& & & \\
& & \text{✗} & \neg G
\end{array}
$$

$$\neg E$$

$$E$$

✗

This tree is open, so it tells us that (3) is satisfiable and not truth-functionally false. To get a more precise characterization, we need to construct a tree for the set containing the negation of (3). Thus:

$$\neg((D \lor \neg E) \leftrightarrow (G \lor E)) \checkmark$$

$$
\begin{array}{cc}
D \lor \neg E \checkmark & \neg(D \lor \neg E) \checkmark \\
\neg(G \lor E) \checkmark & G \lor E \checkmark
\end{array}
$$

$$
\begin{array}{cc}
\neg G & \neg D \\
\neg E & \neg\neg E \checkmark
\end{array}
$$

$$
\begin{array}{cc}
D \quad \neg E & E
\end{array}
$$

$$G \quad E$$

This tree is also open, so it tells us that (3) is falsifiable and not truth-functionally true. The two trees together tell us that wff (3) is contingent, satisfiable, and falsifiable.

To determine which of the five semantic properties a wff has, one will often have to build a tree for both it and its negation. If, however, the first tree you construct shows that the wff in question is either truth-functionally true or truth-functionally false, then you need not construct the other tree. We can illustrate this with wff (4). One might suspect it is truth-functionally true, so we should test for that first. To do so we build a tree for its negation:

$$\neg((A \wedge B) \rightarrow ((\neg A \vee \neg B) \rightarrow C)) \checkmark$$
$$|$$
$$A \wedge B \checkmark$$
$$\neg((\neg A \vee \neg B) \rightarrow C) \checkmark$$
$$|$$
$$A$$
$$B$$
$$|$$
$$\neg A \vee \neg B \checkmark$$
$$\neg C$$

$$\neg A \qquad \neg B$$
$$\textbf{✗} \qquad \textbf{✗}$$

Since the tree for the negation of (4) is closed, (4) is truth-functionally true. This implies that (4) is satisfiable. It also implies that it is not falsifiable, so it is not contingent and not truth-functionally false. So we do not need to construct another tree. Had we constructed a tree for (4) itself, we would have found it open, indicating only that it is satisfiable, but not determining whether or not it is falsifiable. We would have had to also construct the tree above to be sure that it is truth-functionally true. This would not be an error, but if you suspect the wff in question might be truth-functionally true or truth-functionally false, you might be able to save yourself a bit of work.

Tree tests for pairs of wffs are pretty straightforward.

**Truth Tree Test for Truth-Functionally Equivalent Pair of Wffs:**
Wffs $\mathbb{P}$ and $\mathbb{Q}$ are T-Fly equivalent iff $\{\neg(\mathbb{P} \leftrightarrow \mathbb{Q})\}$ has a closed tree.

**Truth Tree Test for Truth-Functionally Contradictory Pair of Wffs:**
Wffs $\mathbb{P}$ and $\mathbb{Q}$ are T-Fly contradictory iff $\{\mathbb{P} \leftrightarrow \mathbb{Q}\}$ has a closed tree.

The rationale for equivalence is that if $\mathbb{P}$ and $\mathbb{Q}$ are truth-functionally equivalent, then their biconditional, $\mathbb{P} \leftrightarrow \mathbb{Q}$, is truth-functionally true. Hence we will get a closed tree for the negation of their biconditional. If, on the other hand, $\mathbb{P}$ and $\mathbb{Q}$ are truth-functionally contradictory, then their biconditional is truth-functionally false. Hence we will get a closed tree for that biconditional. Consider the following pair:

(5) $I \rightarrow \neg J$ $\qquad\qquad$ $\neg J \vee \neg I$

Let's check to see if they are equivalent. This requires a tree for the negation of their biconditional:

$$\neg((I \to \neg J) \leftrightarrow (\neg J \vee \neg I)) \checkmark$$

$$
\begin{array}{ccc}
I \to \neg J \checkmark & & \neg(I \to \neg J) \checkmark \\
\neg(\neg J \vee \neg I) \checkmark & & \neg J \vee \neg I \checkmark \\
| & & | \\
\neg\neg J \checkmark & & I \\
\neg\neg I \checkmark & & \neg\neg J \checkmark \\
| & & | \\
J & & J \\
| & & \\
I & & \neg I \quad \neg J \\
& & \mathsf{X} \quad \mathsf{X} \\
\neg I \quad \neg J & & \\
\mathsf{X} \quad \mathsf{X} & &
\end{array}
$$

Since this tree is closed, the two wffs in (5) are equivalent. Note that the tree for their biconditional would have been open. This would show only that they are not contradictory. It would tell us nothing about whether or not they are equivalent. We would have had to do the same tree above. So, again, strategic guesswork can save you a bit of work.

Suppose we have a pair of wffs that yield open trees for both their biconditional and its negation. This implies that the pair is neither truth-functionally equivalent nor truth-functionally contradictory. Consider:

(6) $M \wedge S$          $\neg M \vee S$

Testing this pair for equivalence or contradiction requires two trees:

$$(M \wedge S) \leftrightarrow (\neg M \vee S) \checkmark$$

$$
\begin{array}{ccc}
M \wedge S \checkmark & & \neg(M \wedge S) \checkmark \\
\neg M \vee S \checkmark & & \neg(\neg M \vee S) \checkmark \\
| & & | \\
M & & \neg\neg M \checkmark \\
S & & \neg S \\
& & | \\
\neg M \quad S & & M \\
\mathsf{X} & & \\
& & \neg M \quad \neg S \\
& & \mathsf{X}
\end{array}
$$

$$\neg((M \wedge S) \leftrightarrow (\neg M \vee S)) \checkmark$$

$$
\begin{array}{ccc}
M \wedge S \checkmark & & \neg(M \wedge S) \checkmark \\
\neg(\neg M \vee S) \checkmark & & \neg M \vee S \\
|. & & \\
M & \neg M & \neg S \\
S & & \\
| & \neg M \quad S & \neg M \quad S \\
\neg\neg M \checkmark & & \mathsf{X} \\
\neg S & & \\
| & & \\
\neg M & & \\
\mathsf{X} & &
\end{array}
$$

The tree on the left shows that the wffs in pair (6) are not contradictory, while the tree on the right shows that they are not equivalent. Such cases will always require two trees.

Finally, we come to truth-functional entailment and truth-functional validity. Recall from the previous section that we could define these in terms of inconsistency. A set $\Gamma$ entails a target $\mathbb{P}$ iff the set containing all the members of $\Gamma$ and $\neg\mathbb{P}$ is inconsistent. Similarly, an argument is truth-functionally valid iff the set containing all the premises and the negation of the conclusion is inconsistent. Thus:

**Truth Tree Test for Truth-Functionally Entails:**
   A set of wffs, $\Gamma$, T-Fly entails a target wff, $\mathbb{P}$, iff $\Gamma$ together with $\neg\mathbb{P}$ has a closed tree.

**Truth Tree Test for Truth-Functionally Valid Argument:**
   An argument is T-Fly valid iff the set consisting of all and only the premises and the negation of the conclusion has a closed tree.

Here is a quick example.

(7)  $\{(A \vee B) \vee C, A \rightarrow D\} \vDash \neg D \rightarrow (B \vee C)$

(8)  $(A \vee B) \vee C$
      $A \rightarrow D$
      $\overline{\neg D \rightarrow (B \vee C)}$

Note that the members of the set and the target in the entailment claim (7) are the same as the premises and conclusion of argument (8). The same tree can be used to test each. We must build a tree for the set (premises) together with the negation of the target (conclusion).

This tree is closed, so it shows both that the entailment claim (7) is true, and it shows that the argument (8) is truth-functionally valid.

Note that, in the tree above, the work under the 'D' on the right branch could be left undone. Since we already have a contradiction between 'D' and '¬D' on that branch, we could place an '✗' under the 'D' and call that branch done. Consult with your instructor to see whether you are required to construct complete trees or whether you may shorten them as described.

We close with an example of an invalid argument.

(9)  B → C
    A → B
    ¬A
    —————
    ¬C

Here is the tree:

This tree is open, so there is at least one tva on which the premises and negation of the conclusion are all true—that is, at least one tva where all the premises are true while the conclusion is false—so the argument (9) is truth-functionally invalid.

## 3.5.2 Exercises

**A.** For each of the following, construct the appropriate truth tree(s) to determine which of the five semantic properties the wff has. Write your determination under your tree.

    **1.** A ∨ ¬A
    **2.** A ∧ ¬A
    **3.** F → ¬(G ∨ H)
    **4.** ¬(J ∧ K) ↔ (J → ¬K)
    **5.** (D → (D → D)) ↔ D

**B.** For each of the following, construct the appropriate truth tree(s) to determine whether the wffs in the pair are equivalent, contradictory, or neither. Write your determination under your tree.

   **1.** $\neg(D \leftrightarrow E)$                 $(D \wedge \neg E) \vee (\neg D \wedge E)$
   **2.** $I \wedge J$                       $\neg(J \vee I)$
   **3.** $D \rightarrow (E \rightarrow F)$           $D \wedge (E \wedge \neg F)$

**C.** For each of the following, construct a truth tree to determine whether the set of wffs is consistent or inconsistent. Write your determination under your tree.

   **1.** $\{(A \wedge B) \rightarrow C, \neg C \wedge A, A \rightarrow B\}$
   **2.** $\{\neg(K \wedge L), \neg(K \vee L), K \rightarrow \neg L\}$

**D.** For each of the following, construct a truth tree to test the entailment claim. Write your determination under your tree.

   **1.** $\{C \vee D, \neg E \rightarrow \neg D, C \rightarrow E\} \vDash E$
   **2.** $\{A \rightarrow B, B\} \vDash A$

**E.** For each of the following, construct a truth tree to determine whether the argument is valid or invalid. Write your determination under your tree.

   **1.** $\neg D \rightarrow C$                  **2.** $G \leftrightarrow H$
        $B \rightarrow \neg C$                         $\neg H \wedge I$
        $\overline{\neg D \rightarrow \neg B}$                      $\overline{\neg G \wedge I}$

# 3.6 **Chapter Glossary**

**Argument of $S$:**

   An *argument* of $S$ is a finite set of two or more wffs of $S$, one of which is the conclusion, while the others are premises. (81)

**Closed Branch:**

   A branch on a truth tree is a *closed branch* iff the branch contains both some atomic wff $\mathbb{P}$ and its negation $\neg \mathbb{P}$. (94)

**Closed Tree:**

   A truth tree is a *closed tree* iff every branch is closed. (94)

**Interpretation of $S$:**

   See Truth Value Assignment.

**Metalogic:**

   *Metalogic* is the study of the properties of logical systems. In particular, it is systematic reasoning in a metalanguage about the properties of object language systems of logic. (85)

**Model:**

   A tva is a *model* for (or *models*) a wff $\mathbb{P}$ (or set of wffs $\Gamma$) iff the wff $\mathbb{P}$ (or all wffs in $\Gamma$) are true on that tva. (72)

**Open Branch:**

A branch on a truth tree is an *open branch* iff it is not closed. (94)

**Open Tree:**

A truth tree is an *open tree* iff at least one branch is open. (94)

**Truth Table:**

A *truth table* for a wff or set of wffs is a structure that lists the wff or wffs, all relevant truth value assignments, and the truth value of each wff on each truth value assignment. (75)

**Truth Tree:**

A *truth tree* for a set of wffs is a structure that lists all the wffs, and all the well-formed components that must be true in attempting to show the set truth-functionally consistent. (94)

**Truth Value Assignment:**

A *truth value assignment* (or tva) is an assignment of the value true (abbreviated by a 'T') or the value false ('F'), but not both, to each of the atomic wffs of $S$. A truth value assignment is a function from the set of atomic wffs into the set $\{T, F\}$. Also called an interpretation. (71)

**Truth-Functionally Consistent Set of Wffs:**

A set $\Gamma$ of wffs of $S$ is *truth-functionally consistent* iff there is at least one truth value assignment on which all members of $\Gamma$ are true. We also say that $\Gamma$ has a *model* or is *modeled*. $\Gamma$ is *truth-functionally inconsistent* iff it is not truth-functionally consistent. (79)

**Truth Table Test:** A set $\Gamma$ of wffs is T-Fly consistent iff, in the set's joint truth table, there is at least one row in which all the members of $\Gamma$ have a T under their main connectives. $\Gamma$ is T-Fly inconsistent iff there is no such row. (79)

**Truth Tree Test:** A set $\Gamma$ is T-Fly consistent iff it has an open tree. $\Gamma$ is T-Fly inconsistent iff it has a closed tree. (95)

**Truth-Functionally Contingent Wff:**

A wff $\mathbb{P}$ of $S$ is *truth-functionally contingent* iff $\mathbb{P}$ is neither truth-functionally true nor truth-functionally false; i.e., iff it is false on at least one truth value assignment and true on at least one truth value assignment. (75)

**Truth Table Test:** $\mathbb{P}$ is T-Fly contingent iff, in its truth table, at least one F and at least one T appear in the column under the main connective. (76)

**Truth Tree Test:** A wff $\mathbb{P}$ is T-Fly contingent iff $\{\mathbb{P}\}$ has an open tree and $\{\neg\mathbb{P}\}$ has an open tree. (95)

**Truth-Functionally Contradictory Pair of Wffs:**

Wffs $\mathbb{P}$ and $\mathbb{Q}$ of $S$ are *truth-functionally contradictory* iff there is no truth value assignment on which they have the same truth value. (78)

**Truth Table Test:** Wffs $\mathbb{P}$ and $\mathbb{Q}$ are T-Fly contradictory iff, in their joint truth table, there is no row in which the values under their main connectives are the same. (78)

**Truth Tree Test:** Wffs $\mathbb{P}$ and $\mathbb{Q}$ are T-Fly contradictory iff $\{\mathbb{P} \leftrightarrow \mathbb{Q}\}$ has a closed tree. (97)

## Truth-Functionally Entails:

A set $\Gamma$ of wffs of $S$ *truth-functionally entails* a wff $\mathbb{P}$ iff there is no truth value assignment on which all the members of $\Gamma$ are true and $\mathbb{P}$ is false. (Every model of $\Gamma$ is a model of $\mathbb{P}$.) (80)

**Truth Table Test:** A set $\Gamma$ T-Fly entails a wff $\mathbb{P}$ iff there is no row of their joint truth table on which all the members of $\Gamma$ have a T under their main connectives and $\mathbb{P}$ has an F under its main connective. (80)

**Truth Tree Test:** A set of wffs, $\Gamma$, T-Fly entails a target wff, $\mathbb{P}$, iff $\Gamma$ together with $\neg\mathbb{P}$ has a closed tree. (99)

## Truth-Functionally Equivalent Pair of Wffs:

Wffs $\mathbb{P}$ and $\mathbb{Q}$ of $S$ are *truth-functionally equivalent* iff there is no truth value assignment on which they differ in truth value (every model of $\mathbb{P}$ is a model of $\mathbb{Q}$, and every model of $\mathbb{Q}$ is a model of $\mathbb{P}$). (77)

**Truth Table Test:** Wffs $\mathbb{P}$ and $\mathbb{Q}$ are T-Fly equivalent iff, in their joint truth table, there is no row in which the values under their main connectives differ. (78)

**Truth Tree Test:** Wffs $\mathbb{P}$ and $\mathbb{Q}$ are T-Fly equivalent iff $\{\neg(\mathbb{P} \leftrightarrow \mathbb{Q})\}$ has a closed tree. (97)

## Truth-Functionally False Wff:

A wff $\mathbb{P}$ of $S$ is *truth-functionally false* iff $\mathbb{P}$ is false on every truth value assignment (no tva is a model). (75)

**Truth Table Test:** $\mathbb{P}$ is T-Fly false iff, in its truth table, only Fs appear in the column under the main connective. (75)

**Truth Tree Test:** A wff $\mathbb{P}$ is T-Fly false iff $\{\mathbb{P}\}$ has a closed tree. (95)

## Truth-Functionally Falsifiable Wff:

A wff $\mathbb{P}$ of $S$ is *truth-functionally falsifiable* iff $\mathbb{P}$ is not truth-functionally true; i.e., it is false on at least one truth value assignment. (76)

**Truth Table Test:** $\mathbb{P}$ is T-Fly falsifiable iff, in its truth table, at least one F appears in the column under the main connective. (76)

**Truth Tree Test:** A wff $\mathbb{P}$ is T-Fly falsifiable iff $\{\neg\mathbb{P}\}$ has an open tree. (95)

## Truth-Functionally Satisfiable Wff:

A wff $\mathbb{P}$ of $S$ is *truth-functionally satisfiable* iff $\mathbb{P}$ is not truth-functionally false; i.e., it is true on at least one truth value assignment. We also say that the wff has a *model* or is *modeled*. (76)

**Truth Table Test:** $\mathbb{P}$ is T-Fly satisfiable (has a model) iff, in its truth table, at least one T appears in the column under the main connective. (76)

**Truth Tree Test:** A wff $\mathbb{P}$ is T-Fly satisfiable iff $\{\mathbb{P}\}$ has an open tree. (95)

### Truth-Functionally True Wff:

A wff $\mathbb{P}$ of *S* is *truth-functionally true* iff $\mathbb{P}$ is true on every truth value assignment (every tva is a model). (75)

**Truth Table Test:** $\mathbb{P}$ is T-Fly true iff, in its truth table, only Ts appear in the column under the main connective. (75)

**Truth Tree Test:** A wff $\mathbb{P}$ is T-Fly true iff $\{\neg\mathbb{P}\}$ has a closed tree. (95)

### Truth-Functionally Valid Argument:

An argument of *S* is *truth-functionally valid* iff there is no truth value assignment on which all the premises are true and the conclusion is false. An argument of *S* is *truth-functionally invalid* iff it is not truth-functionally valid. (81)

**Truth Table Test:** An argument is T-Fly valid iff there is no row of their joint truth table on which all the premises have a T under their main connectives and the conclusion has an F under its main connective. If there is such a row, the argument is T-Fly invalid. (81)

**Truth Tree Test:** An argument is T-Fly valid iff the set consisting of all and only the premises and the negation of the conclusion has a closed tree. If the tree is open, the argument is T-Fly invalid. (99)

# 4 *SD*: **Natural Deduction in** *S*

## 4.1 **The Basic Idea**

While we can use truth tables to test for various semantic properties, there are at least two reasons to develop a slightly different method of examining the properties of wffs and sets of wffs. First, truth table tests do not at all mimic the way we reason naturally in day to day contexts. It would be a rare person who, upon reading or hearing a bit of reasoning expressed in conversation, breaks out pencil and paper and generates a truth table to test for validity. Second, because the size of a table increases exponentially with the increase of atomic wffs, tables, though always finite, become unwieldy when sentences, sets, or arguments involve many atomic wffs. Thus, we might like to find some other way of evaluating wffs and sets thereof—one that mimics a more natural way of reasoning and can (sometimes, at least!) avoid the unwieldiness of truth tables. Of course, this method will also involve pencil and paper, but the patterns involved follow a more natural flow of reasoning.

An illustration is easiest. Consider the following argument:

> If Santacruz is injured, then she will not play. And if Santacruz doesn't play, then Khumalo must play. What's more, if Khumalo and Nakata both play, then Jones must be moved into the center. In fact, Santacruz is injured and Nakata will play. Thus, Khumalo will play and Jones will be moved into the center.

This argument is truth-functionally valid. With 5 atomic wffs needed to symbolize the claims, a truth table showing this would require $2^5 = 32$ lines. Not terribly long, but long enough. Moreover we can probably see it is valid by reasoning through it in something like the following fashion:

| | | |
|---|---|---|
| (1) | If Santacruz is injured, then she will not play | premise |
| (2) | If Santacruz does not play, then Khumalo will play | premise |
| (3) | If Khumalo and Nakata both play, then Jones will be moved into the center | premise |
| (4) | Santacruz is injured and Nakata will play | premise |
| (5) | Santacruz is injured | from (4) |
| (6) | Santacruz will not play | from (1), (5) |
| (7) | Khumalo will play | from (2), (6) |

|      |                                                        |             |
|------|--------------------------------------------------------|-------------|
| (8)  | Nakata will play                                       | from (4)    |
| (9)  | Khumalo and Nakata will play                           | from (7), (8) |
| (10) | Jones will be moved into the center                    | from (3), (9) |
| (11) | Khumalo will play and Jones will be moved into the center | from (7), (10) |

Note how, from line (5) on, each new line is arrived at via a sort of mini-argument with some previous lines as premises. The idea is that since each step along the way is an instance of a valid form, then the ultimate conclusion validly follows from the original premises. We will develop a natural deduction system, a set of derivation rules for *S*, called *SD*, in which we can proceed step by step from certain wffs toward others, and, in a manner roughly parallel to that above, demonstrate validity of arguments (among other things).

Continuing our example, the derivation in *SD* corresponding to the line of reasoning above looks like this (in lieu of a translation key, compare to the above):

| 1  | $I \rightarrow \neg S$        | P             |              |
|----|-------------------------------|---------------|--------------|
| 2  | $\neg S \rightarrow K$        | P             |              |
| 3  | $(K \wedge N) \rightarrow J$  | P             |              |
| 4  | $I \wedge N$                  | P             | $\vdash K \wedge J$ |
| 5  | $I$                           | 4 $\wedge$E   |              |
| 6  | $\neg S$                      | 1, 5 $\rightarrow$E |        |
| 7  | $K$                           | 2, 6 $\rightarrow$E |        |
| 8  | $N$                           | 4 $\wedge$E   |              |
| 9  | $K \wedge N$                  | 7, 8 $\wedge$I |             |
| 10 | $J$                           | 3, 9 $\rightarrow$E |        |
| 11 | $K \wedge J$                  | 7, 10 $\wedge$I |            |

The vertical line that runs the length of the derivation is called a *scope line*. In particular, this (the only scope line appearing) is the *main scope line*. The horizontal line below line 4 sets off the four *primary assumptions* (each justified by P). That this horizontal connects to the main scope line tells us that everything below it and to the right of the main scope line depends on those four primary assumptions. To the right of line 4, the single turnstile, '$\vdash$', serves to indicate the wff to be derived, the *target* or *goal* wff. On the right are the *justifications*, with relevant line numbers and rule employed.

The rules, as we shall soon see, are specified purely in terms of formal syntax, but, of course, are chosen with an eye to their mimicking semantically valuable inferences— i.e., valid ones. *SD* will consist of eleven rules of inference, an Introduction rule and an Elimination rule for each connective, and one additional rule.

**Note:** A complete table of derivation rules appears in Appendix C

## 4.1.1 **Reiteration—R**

The simplest rule of all, Reiteration (R for short) allows you to rewrite any line at any later point in the derivation (subject to subderivation restrictions). Here is the schematic representation of the rule:

R
════════════

$i$ | $\mathbb{P}$

▷ | $\mathbb{P}$                    $i$ R

As with all the rule schemas, a scope line is shown, and the input line(s) for the rule appear at the top of the scope line, while the output line appears at the bottom indicated by a triangle and the justification. What this schematic tells us is that if you have a wff $\mathbb{P}$ on some line number $i$ of your derivation, you may rewrite $\mathbb{P}$ on some later line, citing as justification line $i$ and rule R. The semantic parallel of this rule would be an argument that has its premise as its conclusion. This, of course, would be useless (because obviously circular) in most argumentative contexts. In the derivation system we are developing, however, the Reiteration rule will play a limited but important role (mainly in subderivations, to be explored later). Here is a basic example:

| | | | |
|---|---|---|---|
| 1 | A | P | |
| 2 | D∧B | P | ⊢ |
| 3 | A | 1 R | |
| 4 | D∧B | 2 R | |
| 5 | D∧B | 2 R | |
| 6 | D∧B | 5 R | |
| 7 | A | 3 R | |
| 8 | A | 1 R | |

This is not a very interesting derivation (indeed, I did not bother specifying a target), but we will see some situations where we want this rule. Note that the wff in question need not be atomic—this is the case for all the rules, which is the point of using metavariables. Moreover, the rule can be repeated, and the same rule can be applied to the same line more than once.

## 4.1.2 **Wedge Rules—∧I, ∧E**

One of the most basic natural language inferences, know as Conjunction, is the joining of two sentences into a conjunction:

Santacruz plays
Santacruz scores a goal
_____
So, Santacruz plays and Santacruz scores a goal

Another, known as Simplification, is when one of the component sentences is inferred from a conjunction:

> Santacruz plays and Nakata plays
> _____
> So, Nakata plays

These are easily mimicked in our system by Wedge Introduction (∧I) and Wedge Elimination (∧E). Here are the schemas:

∧I                                            ∧E

| $i$ | ℙ        |         | | $i$ | ℙ∧ℚ |        |
|-----|----------|---------| |-----|-----|--------|
| $j$ | ℚ        |         | |     |     |        |
|     |          |         | | ▷   | ℙ   | $i$ ∧E |
| ▷   | ℙ∧ℚ    | $i, j$ ∧I | |     | *-or-* |     |
|     |          |         | | ▷   | ℚ   | $i$ ∧E |

The Wedge Introduction rule simply tells you that you can insert a wedge between any two wffs and write the result on a later line. Thus, each of the following applications of ∧I is correct.

| 1 | S          | P        |                |
|---|------------|----------|----------------|
| 2 | G          | P        | ⊢ (S∧G)∧S    |
| 3 | S∧G      | 1, 2 ∧I  |                |
| 4 | G∧S      | 1, 2 ∧I  |                |
| 5 | S∧S      | 1, 1 ∧I  |                |
| 6 | (S∧G)∧S | 1, 3 ∧I  |                |

Note that, as with all rule schemas with multiple input lines, the order of the input is irrelevant. I.e., the input wff that appears first in your derivation need not appear as the left conjunct of the result—compare lines 3 and 4. Note as well that we cite the relevant lines in numerical order followed by the rule abbreviation. Line 5 reminds us that ℙ and ℚ need not be distinct wffs. Also, we must reinsert the proper parentheses in cases like line 6. Moreover, there is no bar against reusing a line previously used (e.g., line 1 is used repeatedly).

Wedge Elimination allows us to break off either one of the conjuncts of some conjunction (a wff with the wedge as its main connective) and write it on a line by itself.

| 1 | (S∧N)∧J | P       | ⊢ S∧J |
|---|------------|---------|---------|
| 2 | S∧N      | 1 ∧E    |         |
| 3 | J          | 1 ∧E    |         |
| 4 | S          | 2 ∧E    |         |
| 5 | S∧J      | 3, 4 ∧I |         |

Note that in line 2 we drop the parentheses per our convention. An important point to stress is that while we know that lines 4 and 5 validly follow from the primary assumption in line 1, the rules of **SD** do not allow us to jump straight from line 1 to 4 or 5. ∧E allows us to derive one or the other conjuncts of a wff. The wff in line 4 is not a conjunct of line 1. The intermediate step taken in line 2 is necessary for deriving line 4, and this is indicated in the justification of line 4. ∧I allows us to conjoin two separate wffs already appearing in the derivation. But neither 'J' nor 'S' appears as a separate wff in line 1 (they are each atomic components of the wff in line 1, of course), so 5 cannot be derived immediately from 1. Rather, the intermediate steps taken in 3 and 4 are necessary to 5, and, again, this is indicated in the justification of line 5. Line 2 is also necessary for 5, but not immediately, so it is not listed in the justification of 5. But we can easily read the derivation backwards, seeing that 4 is necessary for 5, and 2 is necessary for 4, we know that 2 is necessary for 5. Only the immediately relevant lines are cited in the justification of a step in the derivation, and the form of the rule used dictates which lines must be cited.

The preceding discussion demonstrates the technical point that, while our derivation rules are semantically motivated, they are specified purely in terms of the syntax of the wffs involved. So correct execution of a rule or whole derivation is a question, not of semantics and validity, but one of syntax and symbolic form.

## 4.1.3  Arrow Rules— →I, →E

One of the most frequent forms of inference we perform is Modus Ponens—from a conditional and its antecedent we infer the consequent (see page 11). E.g.,

> If Santacruz is injured, then she will not play
> Santacruz is injured
> _____
> So, Santacruz will not play

The Arrow Elimination rule, →E, syntactically mimics Modus Ponens. Here is the form:

$$\Rightarrow\text{E}$$

$$
\begin{array}{c|l}
i & \mathbb{P} \to \mathbb{Q} \\
j & \mathbb{P} \\
\\
\triangleright & \mathbb{Q} \qquad\qquad i, j \to\text{E}
\end{array}
$$

So, given a conditional (a wff with the arrow as its main connective) and a wff that exactly matches the antecedent of the conditional, we can derive the consequent of the wff. Thus, suppose we are trying to derive 'G' from the primary assumptions below.

```
1 | A → B                    P
2 | S → ¬N                   P
3 | K ∧ S                    P
4 | ¬N → (K → G)             P          ⊢ G
5 |   K                      3 ∧E
6 |   S                      3 ∧E
7 |   ¬N                     2, 6 →E
8 |   K → G                  4, 7 →E
9 |   G                      5, 8 →E
```

Note that even though we have 'G' as the consequent of 'K → G' in line 4, and we have 'K' in line 5, we cannot at line 6 derive 'G'. This is because the main connective of line 4 is the first arrow and the antecedent of line 4 is '¬N', which, of course, does not match 'K'. That is, we cannot use the 'K' from line 5 in an application of →E to get 'G' until we get 'K → G' on a line by itself. And this takes some work. First we derive 'S' from line 3 by ∧E. Doing so gives us a line 6 that exactly matches the antecedent of line 2. We can then use →E on 2 and 6 to get '¬N' (the consequent of 2) in line 7. Since that matches the antecedent of 4, we can apply →E to 4 and 7 to get line 8. In line 8 we now have a wff with 'K' as the antecedent and 'G' as the consequent. Thus we reach our goal with one final application of →E to lines 5 and 8. Note that in this derivation we did not even use line 1—this happens sometimes, it is not a problem. Finally, we will never be able to derive 'B' from line 1, because we need 'A' to get 'B' and we aren't given 'A' by itself, nor can we get it from anywhere else.

There is the temptation to treat conditionals like conjunctions, and just pull the antecedent or consequent down to a new line (as ∧E allows with a conjunction), but this is a grave error, and if permitted would allow us to derive whatever we want, making the system useless. Thinking that you could get 'A' or 'B' from line 1 are examples of such an error. Remember: to perform →E you need two lines; one must have → as its main connective, and the other must exactly match the antecedent. This then allows you to derive the consequent on a line by itself.

Arrow Introduction is the first of a number of rules that involve subderivations. The basic idea is that we provisionally make a new assumption (an *auxiliary assumption* that is marked off by a new scope line and horizontal), derive some result (dependent on the auxiliary assumption) within the subderivation, end the subderivation, and then derive some further result (not dependent on the auxiliary assumption) on the basis of the subderivation. In the case of Arrow Introduction, we ultimately derive a conditional with the auxiliary assumption, $\mathbb{P}$, as antecedent, and some further (not necessarily distinct) wff, $\mathbb{Q}$, as consequent. Here is the rule schema:

$$→I$$

```
i |   | P                      A
  |   |
j |   | Q
▷ | P → Q                    i–j →I
```

A new scope line is started with a single assumption, labeled 'A' for 'auxiliary'. That scope line is extended until we *discharge* the assumption by applying →I, at which point we have a conditional that does not depend on the auxiliary assumption. An example:

> If Santacruz is injured, then she will not play
> If Santacruz does not play, then she will not score any goals
> If Santacruz does not play, then Jones will play
> If Jones plays, she will score a goal
> ———————————————————————————
> So, if Santacruz is injured, then she will not score a goal but Jones will

One might reason this out as follows:

| | | |
|---|---|---|
| (1) | If Santacruz is injured, then she will not play | premise |
| (2) | If Santacruz does not play, then she will not score any goals | premise |
| (3) | If Santacruz does not play, then Jones will play | premise |
| (4) | If Jones plays, she will score a goal | premise |
| (5) |     Suppose Santacruz is injured | assumption |
| (6) |     Santacruz will not play | (1), (5) |
| (7) |     Santacruz will not score | (2), (6) |
| (8) |     Jones will play | (3), (6) |
| (9) |     Jones will score | (4), (8) |
| (10) |     Santacruz won't score and Jones will | (7), (9) |
| (11) | If Santacruz is injured, then she will not score a goal but Jones will | (5)–(10) |

Note the provisional supposition (5) that Santacruz is injured. On the basis of this (as shown by indentation) we conclude (10) that Santacruz will not score but Jones will. We cannot stop at line (10) because it depends not just on the original premises, but also on the provisional supposition in (5). (10) simply does not follow validly from (1)–(4). But in line (11) we dispense with (or *discharge*) the provisional supposition in (5) by making it the antecedent and (10) the consequent of line (11). Thus, the conditional in line (11) does, indeed, follow from (1)–(4). In natural language contexts (as well as some symbolic logic books) this procedure is called Conditional Proof. Provisionally assume the antecedent, draw some initial subconclusion on that basis, then discharge the assumption by taking it as the antecedent of a conditional with the subconclusion as consequent. Here it is in **SD**:

| | | | |
|---|---|---|---|
| 1 | $I \to \neg S$ | P | |
| 2 | $\neg S \to \neg G$ | P | |
| 3 | $\neg S \to J$ | P | |
| 4 | $J \to R$ | P | $\vdash I \to (\neg G \land R)$ |
| 5 |   $I$ | A | |
| 6 |   $\neg S$ | 1, 5 →E | |
| 7 |   $\neg G$ | 2, 6 →E | |
| 8 |   $J$ | 3, 6 →E | |
| 9 |   $R$ | 4, 8 →E | |
| 10 |   $\neg G \land R$ | 7, 9 ∧I | |
| 11 | $I \to (\neg G \land R)$ | 5–10 →I | |

Note that we cite the subderivation *as a whole*. The scope line of the subderivation shows that, in addition to depending on the four primary assumptions, lines 5–10 also depend on the auxiliary assumption in line 5. The schema for →I allows us to end the subderivation and discharge the auxiliary assumption by deriving a conditional wff on the main scope line. Once we have done this and we are back out on the main scope line, anything within the scope of the subderivation is off limits, for it would depend on the relevant auxiliary assumption, and therefore should not be on the main scope line. For instance, we could not add a line 12 in which we use R to reiterate 'J' from line 8, nor could we write a line 12 that derives '¬G ∧ R' from 5 and 11 by →E, for 5 is off limits.

Strictly speaking you may start a subderivation by assuming any wff you please (however simple or complex and even if that wff doesn't appear in the derivation so far), but only certain auxiliary assumptions will lead where you are trying to go. Indeed, your goal and the form of the rule dictate what you should assume. If you think you are going to execute →I to derive some conditional, then start a subderivation by assuming the antecedent of the conditional you want. Furthermore, one must exit all subderivations before the final line of the main derivation. Box 4.1 on page 124 addresses subderivations.

The arrow rules are not as complex as they might at first seem, and Section 4.2 discusses useful strategies for dealing with all the rules introduced in this section.

## 4.1.4  Hook Rules— ¬I, ¬E

Both of the hook rules employ subderivations, indeed the introduction and elimination rules for the hook are nearly identical, essentially being syntactic versions of Reductio ad Absurdum (see page 11). Thus I will display both, but only illustrate one:

| ¬I | | | | ¬E | | | |
|---|---|---|---|---|---|---|---|
| $i$ | $\mathbb{P}$ | | A | $i$ | ¬$\mathbb{P}$ | | A |
| | $\mathbb{Q}$ | | | | $\mathbb{Q}$ | | |
| $j$ | ¬$\mathbb{Q}$ | | | $j$ | ¬$\mathbb{Q}$ | | |
| ▷ | ¬$\mathbb{P}$ | | $i$–$j$ ¬I | ▷ | $\mathbb{P}$ | | $i$–$j$ ¬E |

The basic idea is that a wff taken as an auxiliary assumption leads to a syntactic contradiction, the subderivation is closed, the assumption discharged, and a wff is derived that matches the assumption except for having a hook attached or removed.[1] Again, a natural language illustration is useful.

> Santacruz doesn't play only if Jones and Khumalo do
> But if Nakata and Khumalo play, then Jones won't
> In fact, Nakata will play
> _____
> So, obviously Santacruz will play

---

1. Can you think of a way we could syntactically state a single hook rule to take care of both elimination and introduction?

We might reason through this as follows:

| | | |
|---|---|---|
| (1) | If Santacruz does not play then Jones and Khumalo do | premise |
| (2) | If Nakata and Khumalo play, then Jones won't | premise |
| (3) | Nakata will play | premise |
| (4) |     Suppose Santacruz does not play | assumption |
| (5) |     Jones and Khumalo play | (1), (4) |
| (6) |     Khumalo plays | (5) |
| (7) |     Nakata and Khumalo play | (3), (6) |
| (8) |     Jones will not play | (2), (7) |
| (9) |     Jones will play | (5) |
| (10) | Santacruz plays | (4)–(9) |

We show that the conclusion follows by showing that, given the premises we have, the assumption of the conclusion's negation leads to a contradiction, so the conclusion itself must be true. The hook rules give us a syntactic version of this process.

| | | | |
|---|---|---|---|
| 1 | $\neg S \rightarrow (J \wedge K)$ | P | |
| 2 | $(N \wedge K) \rightarrow \neg J$ | P | |
| 3 | N | P | $\vdash S$ |
| 4 |   $\neg S$ | A | |
| 5 |   $J \wedge K$ | 1, 4 $\rightarrow$E | |
| 6 |   K | 5 $\wedge$E | |
| 7 |   $N \wedge K$ | 3, 6 $\wedge$I | |
| 8 |   $\neg J$ | 2, 7 $\rightarrow$E | |
| 9 |   J | 5 $\wedge$E | |
| 10 | S | 4–9 $\neg$E | |

The order of $Q$ and $\neg Q$ is irrelevant. We simply need, in the scope of the subderivation, some wff and the exact same wff except with a hook at left. Then we are able to end the subderivation and add a hook to or remove a hook from the assumed wff. Thus, in a sense, you need to assume the opposite of what you want. As with Arrow Introduction, the forms of these rules dictate what you should assume.

## 4.1.5  Vee Rules — $\vee$I, $\vee$E

On the one hand, $\vee$I is second only to R in simplicity. On the other hand, $\vee$E is the most complex rule of all. Both of them are tricky, one because of its simplicity, the other because of its complexity. Here they are:

VI                                          ∨E
─────────────────────                       ─────────────────────

$i$ | ℙ                                      $i$ | ℙ ∨ ℚ
                                            $j$ | | ℙ                 A
▷ | ℙ ∨ ℚ         $i$ ∨I                     ─────
    -or-                                     $k$ | | ℝ
▷ | ℚ ∨ ℙ         $i$ ∨I                     $l$ | | ℚ                 A
                                            ─────
                                            $m$ | | ℝ
                                            ▷ | ℝ      $i, j{-}k, l{-}m$ ∨E

Vee Introduction bothers some because it seems like a strange step to make. It allows us to derive, from the input wff ℙ, the disjunction of that wff with any other wff ℚ (whether ℚ appears in the derivation or not). In its natural language form it is rather a silly inference. If I know that, say, Santacruz will play, I usually do not need to infer the logically weaker and less informative claim that either Santacruz will play or Nakata will play. But suppose I have the following argument:

> Santacruz will play
> If either Santacruz or Nakata plays, then Khumalo will not play
> ─────────────────────────────────────────────────────────────
> So Khumalo will not play

We could reason through this as follows:

| (1) | Santacruz will play | premise |
| (2) | If either Santacruz or Nakata plays, then Khumalo will not play | premise |
| (3) | Either Santacruz will play or Nakata will play | (1) |
| (4) | Khumalo will not play | (2), (3) |

Semantically, and in our day to day thinking, we would jump right to Khumalo not playing, without the apparent detour though the disjunction in (3). But, unless we want to unnecessarily complicate our derivation rules, when proceeding syntactically we must form a wff matching the antecedent of the relevant conditional. (This is just an illustration, not the only situation in which ∨I is important.) Thus, the corresponding derivation:

| 1 | S | P | |
| 2 | (S ∨ N) → ¬K | P | ⊢ ¬K |
| 3 | S ∨ N | 1 ∨I | |
| 4 | ¬K | 2, 3 →E | |

The point is that we cannot derive line 4 without the intermediate step in line 3. Even though you might just see that 4 follows, we need a line which exactly matched the antecedent of line 2 in order to apply →E.

People often forget about this rule, because it is not a natural inference to make, and it doesn't come up all that often. Just remember: if you need a disjunction, and you already have one of the disjuncts, do ∨I.

∨E, despite being syntactically complex, is simply a syntactic version of what is often called Constructive Dilemma. Suppose I know that either Santacruz or Nakata will play (but perhaps not which of them, or whether both, will play). Suppose I can also show that if we assume Santacruz plays then the Falcons will score, and that if we assume Nakata plays then the Falcons will score. All this together leads to the conclusion that the Falcons will score.

> Either Santacruz or Nakata will play
> If Santacruz plays then the Falcons will score
> If Nakata plays then the Falcons will score
> _____
> The Falcons will score

Reasoning naturally:

| (1) | Either Santacruz or Nakata will play | premise |
|---|---|---|
| (2) | If Santacruz plays then the Falcons will score | premise |
| (3) | If Nakata plays then the Falcons will score | premise |
| (4) | Suppose Santacruz plays | assumption |
| (5) | The Falcons will score | (2), (4) |
| (6) | Suppose Nakata plays | assumption |
| (7) | The Falcons will score | (3), (6) |
| (8) | So the Falcons will score | (1), (4)–(5), (6)–(7) |

Here is a symbolic version of the example:

| 1 | $S \vee N$ | P | |
|---|---|---|---|
| 2 | $S \rightarrow F$ | P | |
| 3 | $N \rightarrow F$ | P | $\vdash F$ |
| 4 | $\quad$ S | A | |
| 5 | $\quad$ F | 2, 4 $\rightarrow$E | |
| 6 | $\quad$ N | A | |
| 7 | $\quad$ F | 3, 6 $\rightarrow$E | |
| 8 | F | 1, 4–5, 6–7 $\vee$E | |

Think about it this way: each subderivation shows a path to get to the same conclusion from two different starting points. The disjunction tells us that at least one (and possibly both) of those starting points will be used—at least one of those paths will be taken. So, even though we don't know which path will ultimately be taken, we know we will arrive safely at the conclusion. In the justification of the result of ∨E it is important to cite the line containing the disjunction (line 1, above), because it is a necessary part of the rule.

Without the disjunction—without knowing that we will indeed take one or both of the paths—we cannot arrive at the conclusion.

Though ∨E is the most complex of our rules, the "two paths arriving at the same conclusion" metaphor will help you remember how it works.[2]

## 4.1.6  Double Arrow Rules — ↔I, ↔E

Once you understand the arrow rules, the double arrow rules are no problem. Double Arrow Introduction requires two subderivations, one in which you assume the left side of the biconditional and derive the right, and one in which you assume the right and derive the left. Essentially you are doing two Arrow Introductions. Double Arrow Elimination is just like Arrow Elimination, except that it is bidirectional. Given a biconditional and one side of it, you can derive the other side.

| ↔I | | | | ↔E | | | |
|---|---|---|---|---|---|---|---|
| $i$ | $\mathbb{P}$ | | A | $i$ | $\mathbb{P} \leftrightarrow \mathbb{Q}$ | | |
| | | | | $j$ | $\mathbb{P}$ | | |
| $j$ | $\mathbb{Q}$ | | | | | | |
| $k$ | $\mathbb{Q}$ | | A | ▷ | $\mathbb{Q}$ | | $i, j \leftrightarrow$E |
| | | | | | *-or-* | | |
| $l$ | $\mathbb{P}$ | | | $i$ | $\mathbb{P} \leftrightarrow \mathbb{Q}$ | | |
| ▷ | $\mathbb{P} \leftrightarrow \mathbb{Q}$ | | $i{-}j,\ k{-}l \leftrightarrow$I | $j$ | $\mathbb{Q}$ | | |
| | | | | ▷ | $\mathbb{P}$ | | $i, j \leftrightarrow$E |

I'll give one example employing both rules.

| | | | |
|---|---|---|---|
| 1 | $E \leftrightarrow N$ | P | |
| 2 | $E \leftrightarrow G$ | P | |
| 3 | $(S \wedge D) \leftrightarrow G$ | P | $\vdash N \leftrightarrow (S \wedge D)$ |
| 4 | $\quad N$ | A | |
| 5 | $\quad E$ | 1, 4 ↔E | |
| 6 | $\quad G$ | 2, 5 ↔E | |
| 7 | $\quad (S \wedge D)$ | 3, 6 ↔E | |
| 8 | $\quad (S \wedge D)$ | A | |
| 9 | $\quad G$ | 3, 8 ↔E | |
| 10 | $\quad E$ | 2, 9 ↔E | |
| 11 | $\quad N$ | 1, 10 ↔E | |
| 12 | $N \leftrightarrow (S \wedge D)$ | 4–7, 8–11 ↔I | |
| 13 | $(S \wedge D) \leftrightarrow N$ | 4–7, 8–11 ↔I | |

2. All due respect to Robert Frost, of course.

Note that both 12 (the actual target) and 13 (not the target) are admissible steps based on the same two subderivations (order of input is irrelevant). Also note that the two subderivations need not be immediately consecutive—that is, there might have been some steps to be executed between lines 7 and 8. Further, while these two subderivations are basically inverses of each other, this will not always be the case. Sometimes they will be very similar, and sometimes different—even different in length. It all depends on whether different work will need to be done in getting from one side of the target biconditional to the other.

## 4.1.7 Exercises

A. Complete each of the following derivations. You will need only ∧I, ∧E, and →E (though you might not need all of them at once).

1.　1 | F 　　　　　　　　 P
　　 2 | E∧G 　　　　　　 P 　　　 ⊢ F∧G

2.　1 | A 　　　　　　　　 P
　　 2 | A→B 　　　　　　 P 　　　 ⊢ B∧A

3.　1 | A∧E 　　　　　　　 P
　　 2 | B∧C 　　　　　　　 P
　　 3 | (A∧B)→D 　　　 P 　　　 ⊢ E∧D

4.　1 | A∧B 　　　　　　　 P
　　 2 | (B∧A)→C 　　　 P
　　 3 | C→D 　　　　　　 P 　　　 ⊢ D

5.　1 | (D∧G)∧H 　　　 P
　　 2 | (H∧D)→I 　　　 P
　　 3 | (I∧G)→J 　　　 P 　　　 ⊢ J

**B.** Complete each of the following derivations. You will need R, ∧I, ∧E, →E, →I, ¬E, ¬I.

**1.**  1 │ A → B                    P
       2 │ B → C                    P        ⊢ A → C

**2.**  1 │ A → B                    P
       2 │ ¬B                       P        ⊢ ¬A

**3.**  1 │ ¬F ∧ G                   P
       2 │ ¬C → D                   P
       3 │ D → F                    P        ⊢ C

**4.**  1 │ G → H                    P
       2 │ (H ∧ G) → K              P
       3 │ K → L                    P        ⊢ G → (K ∧ L)

**5.**  1 │ A → B                    P
       2 │ ¬B                       P        ⊢ A → ¬C

**C.** Complete each of the following derivations. All eleven rules are now in play.

**1.**  1 │ A                        P
       2 │ (A ∨ B) → C              P        ⊢ C

**2.**  1 │ T ↔ K                    P
       2 │ J → T                    P
       3 │ J ∨ K                    P        ⊢ T

**3.**   1 | $A \leftrightarrow B$           P

   2 | $C \leftrightarrow B$ ·         P        $\vdash A \leftrightarrow C$

**4.**   1 | $(G \lor H) \to I$         P

   2 | $K \to H$              P        $\vdash K \to (I \lor (M \land N))$

**5.**   1 | $A \land K$               P

   2 | $A \to (I \lor J)$        P

   3 | $(I \land K) \to L$        P

   4 | $L \leftrightarrow J$             P        $\vdash L \land A$

## 4.2 **Derivations: Strategies and Notes**

Unlike for truth tables, there is no algorithm for derivations. Doing derivations requires a bit of ingenuity, sometimes more than a bit. But, there are two main strategies that help a lot. You may already pursue one or the other strategy without realizing it. The first strategy, working forward, is actually less effective than the second strategy, working backward, but most effective of all is a combination of these two methods.

   Working forward basically consists of looking at the primary assumptions you have, and starting to apply derivation rules with the hope that you'll suddenly "see" how to arrive at your goal wff. For simple derivations, you may see the route to be taken immediately, before you apply any derivation rules. This is good. But for more complex derivations, the working forward strategy may leave you floundering around making no progress, or, worse, it may lead you off in a long and unproductive direction. These are situations the working backward strategy can help avoid.

### Goal Analysis — Working Backward

Working backward, or the method of goal analysis, is a much more structured way to approach a complex derivation. Basically, the idea is to look at the form of your goal wff, and if possible, let that tell you what your final step will be as well as what input lines you need to make that step. Those input lines then become your new (sub)goals, and you repeat the process, working your way back to the primary assumptions. This is done, of course, while paying attention to what you have in your primary assumptions, but you are not simply doing whatever comes to mind (as in the working forward method). Here is a simple example.

```
1 │ C → A          P
2 │ B ∧ D          P        ⊢ C → (A ∧ B)
  │
```

Since the target is a conditional, and does not appear as a component of one of the primary assumptions, then it is likely we can get it via →I. To do this, we need a subderivation starting with the antecedent of the target, and ending with the consequent. Thus, we write the target at the bottom of the main scope line as our main goal wff (leaving room to fill in the rest of the derivation), and then we fill in the top and bottom of the subderivation for the anticipated →I:

```
1 │ C → A          P
2 │ B ∧ D          P        ⊢ C → (A ∧ B)
3 │ │ C            A
  │ │
  │ │
i │ │ (A ∧ B)      ??
j │ C → (A ∧ B)    3−i →I
```

We have a significant part of the derivation completed, and though it was not dictated by the form of the target wff, it was very strongly suggested.

Now, our new (sub)goal is the wff in line *i*. Of course, what we need to achieve that is suggested by its form. Since it is a ∧-form we likely will apply ∧I. So we need to have 'A' on a line by itself and 'B' on a line by itself:

```
1 │ C → A          P
2 │ B ∧ D          P        ⊢ C → (A ∧ B)
3 │ │ C            A
g │ │ A            ??
h │ │ B            ??
i │ │ (A ∧ B)      g, h ∧I
j │ C → (A ∧ B)    3−i →I
```

All that remains is to see if we can derive our new subgoals in lines *g*, and *h*. Clearly these come from →E and ∧E, respectively. Since this is such a simple example, all we need to do is insert proper justifications, and fill out the line numbers. Thus:

```
1 │ C → A          P
2 │ B ∧ D          P        ⊢ C → (A ∧ B)
3 │ │ C            A
4 │ │ A            1, 3 →E
5 │ │ B            2 ∧E
6 │ │ (A ∧ B)      4, 5 ∧I
7 │ C → (A ∧ B)    3−6 →I
```

Often you will pursue a combination of backwards and forwards work. In many cases you will carry this process out in your head. As another example, here is exercise **C.5.** from the previous section:

| 1 | A ∧ K | P | |
|---|-------|---|---|
| 2 | A → (I ∨ J) | P | |
| 3 | (I ∧ K) → L | P | |
| 4 | L ↔ J | P | ⊢ L ∧ A |

Since the goal is a conjunction, our last step will probably be ∧I. So we can fill in the goal and subgoals as follows:

| 1 | A ∧ K | P | |
|---|-------|---|---|
| 2 | A → (I ∨ J) | P | |
| 3 | (I ∧ K) → L | P | |
| 4 | L ↔ J | P | ⊢ L ∧ A |
| | | | |
| h | A | ?? | |
| i | L | ?? | |
| j | L ∧ A | h, i ∧I | |

Now 'A' clearly can be gotten from 1 by ∧E, so we may as well enter that as line 5 (looking at line 2, we see that 'A' may be useful even before the last step, so it will help to derive it right up front). Since 'K' is just sitting there, we may as well grab that down too (a little bit of working forward).

| 1 | A ∧ K | P | |
|---|-------|---|---|
| 2 | A → (I ∨ J) | P | |
| 3 | (I ∧ K) → L | P | |
| 4 | L ↔ J | P | ⊢ L ∧ A |
| 5 | A | 1 ∧E | |
| 6 | K | 1 ∧E | |
| | | | |
| i | L | ?? | |
| j | L ∧ A | 5, i ∧I | |

Now we need to get 'L'. It is not terribly clear how we will get it. It appears in the consequent of 3, as well as on the left side of 4. Thus, if we could get either '(I ∧ K)' or 'J', then we could use either →E or ↔E to get 'L'. Since I already have 'K' on line 6, it becomes a matter of getting either 'I' or 'J'. If you pick one and try it, you'll soon see that there is no way to get either of them individually. But if you look at the consequent of 2, you'll see that we can get their disjunction. Now, knowing that 'I' will lead to 'L' and that 'J' will lead to 'L', and that we can get '(I ∨ J)', we are in a perfect position to try ∨E. Here's how it will look:

```
1 │ A ∧ K                    P
2 │ A → (I ∨ J)              P
3 │ (I ∧ K) → L              P
4 │ L ↔ J                    P          ⊢ L ∧ A
5 │ A                        1 ∧E
6 │ K                        1 ∧E
7 │ I ∨ J                    2, 5 →E
b │ │ I                      A
  │ │
  │ │
c │ │ L                      3, ? →E
d │ │ J                      A
e │ │ L                      4, d ↔E
i │ L                        7, b−c, d−e ∨E
j │ L ∧ A                    5, i ∧I
```

Line 7 is easily derived as indicated, and it is obvious that the second subderivation is completed via ↔E. The only question is how we get 'L' from 'I' in the first subderivation. But we've already said that will come from '(I ∧ K)'; and since we have each of those conjuncts individually (lines 6 and *b*), we need only apply ∧I, and →E. Here it is, with numbers filled in:

```
1  │ A ∧ K                   P
2  │ A → (I ∨ J)             P
3  │ (I ∧ K) → L·            P
4  │ L ↔ J                   P          ⊢ L ∧ A
5  │ A                       1 ∧E
6  │ K                       1 ∧E
7  │ I ∨ J                   2, 5 →E
8  │ │ I                     A
9  │ │ I ∧ K                 6, 8 ∧I
10 │ │ L                     3, 9 →E
11 │ │ J                     A
12 │ │ L                     4, 11 ↔E
13 │ L                       7, 8−10, 11−12 ∨E
14 │ L ∧ A                   5, 13 ∧I
```

## Subderivations

As you may have intuited, subderivations can be nested. There is no limitation to this, so long as the subderivations are closed in the opposite order of their opening (a sort of "first in–last out" principle). Here is an example:

| | | | |
|---|---|---|---|
| 1 | $A \rightarrow (C \wedge D)$ | P | |
| 2 | $D \rightarrow E$ | P | |
| 3 | $B \rightarrow \neg(E \vee F)$ | P | $\vdash A \rightarrow \neg B$ |
| 4 | $\quad$ A | A | |
| 5 | $\quad$ C $\wedge$ D | 1, 4 $\rightarrow$E | |
| 6 | $\quad\quad$ B | A | |
| 7 | $\quad\quad$ $\neg$(E $\vee$ F) | 3, 6 $\rightarrow$E | |
| 8 | $\quad\quad$ D | 5 $\wedge$E | |
| 9 | $\quad\quad$ E | 2, 8 $\rightarrow$E | |
| 10 | $\quad\quad$ E $\vee$ F | 9 $\vee$I | |
| 11 | $\quad\quad$ $\neg$B | 6–10 $\neg$I | |
| 12 | $A \rightarrow \neg B$ | 4–11 $\rightarrow$I | |

Note that, during the course of the inner subderivation, anything above and on a continuous scope line to the left is fair game. Once we close that inner subderivation, however, the lines within it may not be used. This is because, when we close a subderivation by applying one of the relevant rules, we discharge or deactivate its assumption. Since the assumption of a closed subderivation is no longer active, it and anything within its scope is off limits. Thus, if we had wanted to use 'D' as an input to a rule after line 10, then we could not do so by appealing to line 8, nor could we reiterate line 8 (that would require appealing to it). Rather, we would have to derive it anew from line 5. But, of course, line 5 is dependent on line 4, so once we close the first subderivation, line 5 is off limits as well.

Box 4.1 presents some guidelines for the construction of subderivations.

A further twist is that certain wffs can be derived that do not depend on any primary assumptions. This means you start a derivation with only a goal and a main scope line—no primary assumptions and no horizontal. Obviously, subderivations will have to be involved, for some assumptions (even if only auxiliary) must be made for us to have materials to work with.

Here is an example:

$$1 \quad \bigg| \qquad\qquad\qquad \vdash A \rightarrow ((B \leftrightarrow A) \rightarrow B)$$

Since we have no primary assumptions, we must make some auxiliary assumptions and use subderivations. Fortunately the form of the goal will dictate what we must do—since we are after a conditional we'll need a subderivation that assumes the antecedent and arrives at the consequent:

**Box 4.1: Guidelines for Subderivations**

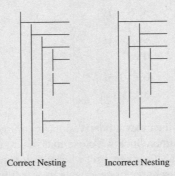

Correct Nesting        Incorrect Nesting

- Subderivations are closed in reverse order of opening
- New subderivations must be to the right of the current scope line
- Two subderivations are permitted at the same level as long as the first is closed before the second is opened
- No two subderivations may be opened or closed on the same line
- Only one auxiliary assumption per subderivation is allowed, but the assumption may be a complex wff
- Subderivations should be used with an appropriate rule in mind
- Except for the first subderivation in ↔I and ∨E, subderivations should only be closed by applying an appropriate rule
- All subderivations must be closed before the last line of the main derivation

$$\begin{array}{ll}
1 & \quad\text{A} \qquad\qquad\qquad\qquad\;\text{A} \qquad \vdash \text{A} \to ((\text{B} \leftrightarrow \text{A}) \to \text{B})
\end{array}$$

$$\begin{array}{ll}
i & \quad (\text{B} \leftrightarrow \text{A}) \to \text{B} \qquad\quad ?? \\
j & \text{A} \to ((\text{B} \leftrightarrow \text{A}) \to \text{B}) \quad 1{-}i \;{\to}\text{I}
\end{array}$$

Line *i* is now the subgoal, and, again, its conditional form virtually dictates our action. Assume the antecedent and derive the consequent.

$$\begin{array}{ll}
1 & \quad\text{A} \qquad\qquad\qquad\qquad\;\text{A} \qquad \vdash \text{A} \to ((\text{B} \leftrightarrow \text{A}) \to \text{B}) \\
2 & \qquad \text{B} \leftrightarrow \text{A} \qquad\qquad\;\;\, \text{A}
\end{array}$$

$$\begin{array}{ll}
h & \qquad \text{B} \qquad\qquad\qquad\quad\; ?? \\
i & \quad (\text{B} \leftrightarrow \text{A}) \to \text{B} \qquad\quad 2{-}h \;{\to}\text{I} \\
j & \text{A} \to ((\text{B} \leftrightarrow \text{A}) \to \text{B}) \quad 1{-}i \;{\to}\text{I}
\end{array}$$

At this point we are essentially done. Line *h* clearly comes from 1 and 2 by ↔E. Hence, all we need do is fill in the line numbers and the justification of *h*:

$$\begin{array}{ll}
1 & \quad\text{A} \qquad\qquad\qquad\qquad\;\text{A} \qquad \vdash \text{A} \to ((\text{B} \leftrightarrow \text{A}) \to \text{B}) \\
2 & \qquad \text{B} \leftrightarrow \text{A} \qquad\qquad\;\;\, \text{A} \\
3 & \qquad \text{B} \qquad\qquad\qquad\quad\; 1, 2 \leftrightarrow\text{E} \\
4 & \quad (\text{B} \leftrightarrow \text{A}) \to \text{B} \qquad\quad 2{-}3 \;{\to}\text{I} \\
5 & \text{A} \to ((\text{B} \leftrightarrow \text{A}) \to \text{B}) \quad 1{-}4 \;{\to}\text{I}
\end{array}$$

The wff in line 5 is on the main scope line, yet no primary assumptions are attached to that line, hence the wff in line 5 does not depend on any assumptions whatsoever. The assumptions we employed in deriving line 5 are all discharged.

## 4.3 **Proof Theory in** *SD*

Similar to our use of truth tables to test for semantic properties of (sets of) wffs, we can use derivations to test for certain syntactic or *proof-theoretic* properties of wffs. Here are some preliminary definitions.

**Derivation in *SD*:**

A *derivation in SD* is a finite sequence of wffs of *S* such that each wff is either an assumption with scope indicated or justified by one of the rules of *SD*.

**Derivable in *SD*, $\Gamma \vdash \mathbb{P}$:**

A wff $\mathbb{P}$ of *S* is *derivable in SD* from a set $\Gamma$ of wffs of *S* iff there is a derivation in *SD* the primary assumptions of which are members of $\Gamma$ and $\mathbb{P}$ depends on only those assumptions.

We use the single turnstile to assert or deny derivability:[3]

$$\Gamma \vdash \mathbb{P}$$
$$\Gamma \nvdash \mathbb{P}$$
$$\vdash \mathbb{P}$$

The first line asserts that $\mathbb{P}$ is derivable from $\Gamma$, while the second line denies this. The third line above asserts that $\mathbb{P}$ is derivable from the empty set.

To show that $\Gamma \vdash \mathbb{P}$ we produce a derivation, all the primary assumptions of which are members of $\Gamma$, and then derive $\mathbb{P}$. For example, in the previous section we showed (among other things) that

$$\{C \to A,\ B \land D\} \vdash C \to (A \land B) \qquad \text{(page 120)}$$
$$\vdash A \to ((B \leftrightarrow A) \to B) \qquad \text{(page 125)}$$

As another example, we'll produce a derivation to show that

$$\{G \land L,\ (H \lor L) \leftrightarrow K\} \vdash G \land K$$

| | | | |
|---|---|---|---|
| 1 | $G \land L$ | P | |
| 2 | $(H \lor L) \leftrightarrow K$ | P | $\vdash G \land K$ |
| 3 | G | 1 $\land$E | |
| 4 | L | 1 $\land$E | |
| 5 | $H \lor L$ | 4 $\lor$I | |
| 6 | K | 2, 5 $\leftrightarrow$E | |
| 7 | $G \land K$ | 3, 6 $\land$I | |

Note that, given the definition of $\Gamma \vdash \mathbb{P}$, not all of the members of $\Gamma$ need to be primary assumptions. There may sometimes be wffs in $\Gamma$ that do not figure in the derivation of $\mathbb{P}$. This is not a problem—being derivable from $\Gamma$ does not require that all the wffs of $\Gamma$ be used, only that all the wffs used as primary assumptions be members of $\Gamma$. For instance, the following derivation shows that

$$\{A,\ A \to B,\ C,\ D,\ E\} \vdash B$$

| | | | |
|---|---|---|---|
| 1 | A | P | |
| 2 | $A \to B$ | P | |
| 3 | D | P | $\vdash B$ |
| 4 | B | 1, 2 $\to$E | |

Wffs 'C', 'D', and 'E' are all irrelevant to deriving 'B'. Whether or not they appear as primary assumptions is indifferent (and so we could have eliminated line 3). What is important is that those wffs that do appear as primary assumptions are members of the set in question.

---

3. Strictly speaking the single turnstile should be subscripted with an indication of the relevant derivation system—the above should be $\Gamma \vdash_{SD} \mathbb{P}$, etc. I shall omit such subscripts except where clarity is threatened.

**Valid in *SD*:**

An argument of *S* is *valid in SD* iff the conclusion is derivable from the set consisting of only the premises, otherwise it is invalid in *SD*.

On page 120 in the previous section we showed that the argument

$$C \rightarrow A$$
$$B \wedge D$$
$$\overline{C \rightarrow (A \wedge B)}$$

is valid in *SD*. That is, we produced a derivation the primary assumptions of which included only the premises above, and the final line of which was the conclusion. As another example,

$$A \vee B$$
$$A \rightarrow E$$
$$B \rightarrow D$$
$$E \leftrightarrow D$$
$$\overline{D}$$

is shown to be valid in *SD* by the following derivation:

| | | |
|---|---|---|
| 1 | A ∨ B | P |
| 2 | A → E | P |
| 3 | B → D | P |
| 4 | E ↔ D | P        ⊢ D |
| 5 |   A | A |
| 6 |   E | 2, 5 →E |
| 7 |   D | 4, 6 ↔E |
| 8 |   B | A |
| 9 |   D | 3, 8 →E |
| 10 | D | 1, 5–7, 8–9 ∨E |

**Theorem of *SD*:**

A wff $\mathbb{P}$ is a *theorem of SD* iff $\mathbb{P}$ is derivable from the empty set; i.e., iff $\vdash \mathbb{P}$.

On page 125 we showed that

$$\vdash A \rightarrow ((B \leftrightarrow A) \rightarrow B)$$

It is also the case that

$$\vdash (A \wedge B) \rightarrow (A \vee C)$$

as is shown by

$$\vdash (A \wedge B) \to (A \vee C)$$

| 1 | $A \wedge B$ | A |
| 2 | A | 1 $\wedge$E |
| 3 | $A \vee C$ | 2 $\vee$I |
| 4 | $(A \wedge B) \to (A \vee C)$ | 1–3 $\to$I |

Note that the goal sentence rests on a main scope line containing no assumptions. This shows that it has been derived from the empty set, and, hence, is a theorem of *SD*.

**Equivalent in *SD*:**

Two wffs $\mathbb{P}$ and $\mathbb{Q}$ are *equivalent in SD* iff they are interderivable in *SD*; i.e., iff both $\mathbb{P} \vdash \mathbb{Q}$ and $\mathbb{Q} \vdash \mathbb{P}$.

For example, the following derivations show that 'A $\to$ B' and '$\neg$B $\to$ $\neg$A' are equivalent in *SD*.

| 1 | $A \to B$ | P    $\vdash \neg B \to \neg A$ |
| 2 | $\neg B$ | A |
| 3 | A | A |
| 4 | B | 1, 3 $\to$E |
| 5 | $\neg B$ | 2 R |
| 6 | $\neg A$ | 3–5 $\neg$I |
| 7 | $\neg B \to \neg A$ | 2–6 $\to$I |

| 1 | $\neg B \to \neg A$ | P    $\vdash A \to B$ |
| 2 | A | A |
| 3 | $\neg B$ | A |
| 4 | $\neg A$ | 1, 3 $\to$E |
| 5 | A | 2 R |
| 6 | B | 3–5 $\neg$E |
| 7 | $A \to B$ | 2–6 $\to$I |

Note that two derivations are required. Here the two derivations follow roughly the same pattern, but this will not always be the case.

**Inconsistent in *SD*:**

A set $\Gamma$ of wffs is *inconsistent in SD* iff, for some wff $\mathbb{P}$, both $\Gamma \vdash \mathbb{P}$ and $\Gamma \vdash \neg \mathbb{P}$.

For example, we show that the following set is inconsistent in *SD*.

$$\{F \to (G \wedge J), \ L \leftrightarrow (J \wedge H), \ F \wedge H, \ G \to \neg L\}$$

```
1 │ F → (G ∧ J)        P
2 │ L ↔ (J ∧ H)        P
3 │ F ∧ H              P
4 │ G → ¬L             P        ⊢ ℙ, ¬ℙ
  ├─────────────
5 │ F                  3 ∧E
6 │ H                  3 ∧E
7 │ G ∧ J              1, 5 →E
8 │ G                  7 ∧E
9 │ J                  7 ∧E
10│ J ∧ H              6, 9 ∧I
11│ L                  2, 10 ↔E
12│ ¬L                 4, 8 →E
```

Note that no particular wff and its negation are required, just some wff and its negation. (We could have derived 'G' and '¬G', though it would take more steps.) While the wff, ℙ, and its negation, ¬ℙ, must be on the main scope line, they do not need to be on adjacent lines. (We could have derived '¬L' immediately after line 8.) Because no specific goal is given when constructing a derivation to show inconsistency, we have to combine working forward with working backward. Scanning the primary assumptions often reveals a possible contradiction, and then we can try working backwards from there. If scanning the primaries does not help, we can start applying rules to the primaries and see if that suggests anything. Again, derivations are more interesting than truth tables, because they require some strategy and ingenuity.

## 4.3.1 Exercises

**A.** Complete each of the following derivations.

**1.**
```
1 │ F ∧ G              P
2 │ H ↔ F              P        ⊢ G ∧ H
  ├─────────
```

**2.**
```
1 │ A ∧ (B ∧ C)        P
2 │ (A ∧ B) → D        P        ⊢ C ∧ D
  ├─────────────
```

**3.**
```
1 │ F → K              P
2 │ F → (G ∧ J)        P
3 │ (K ∧ J) → H        P        ⊢ F → (G ∧ H)
  ├─────────────
```

**4.**  

| | | | |
|---|---|---|---|
| 1 | $\neg L \rightarrow (G \wedge C)$ | P | |
| 2 | $D \rightarrow (\neg L \wedge J)$ | P | |
| 3 | $G \rightarrow \neg J$ | P | $\vdash \neg D$ |

**B.** For each of the following, construct a derivation in *SD* to show that the derivability claim holds.

  **1.** $\{I \wedge J, (J \vee H) \rightarrow G\} \vdash F \vee G$

  **2.** $\{\neg H \leftrightarrow (F \vee \neg D), J \wedge H\} \vdash D$

  **3.** $\{O \wedge (K \rightarrow M), (N \rightarrow M) \wedge (K \vee N)\} \vdash M \wedge O$

  **4.** $\{D \vee (K \wedge L), (E \wedge J) \leftrightarrow (F \wedge D), F \wedge (K \rightarrow E)\} \vdash E \wedge F$

**C.** For each of the following, construct a derivation to show that the argument is valid in *SD*.

  **1.** $C \rightarrow E$
  $(G \rightarrow I) \leftrightarrow (A \wedge E)$
  $\overline{(A \wedge C) \rightarrow (G \rightarrow I)}$

  **2.** $A \rightarrow B$
  $\neg B$
  $\overline{\neg A}$

  **3.** $A \rightarrow (D \vee F)$
  $G \leftrightarrow \neg D$
  $F \rightarrow \neg G$
  $\overline{A \rightarrow \neg G}$

  **4.** $(A \wedge D) \rightarrow E$
  $E \rightarrow (A \rightarrow D)$
  $\overline{A \rightarrow (D \leftrightarrow E)}$

**D.** For each of the following, construct a derivation to show that the wff is a theorem of *SD*.

  **1.** $\vdash \neg (A \wedge \neg A)$

  **2.** $\vdash A \rightarrow (B \rightarrow A)$

  **3.** $\vdash (B \rightarrow C) \rightarrow ((A \rightarrow B) \rightarrow (A \rightarrow C))$

  **4.** $\vdash D \vee \neg D$

**E.** For each of the following pairs, construct derivations to show that the wffs are equivalent in *SD*.

  **1.** $A \rightarrow B$ $\qquad\qquad$ $\neg B \rightarrow \neg A$

  **2.** $A$ $\qquad\qquad$ $\neg\neg A$

3. $A \rightarrow (B \rightarrow C)$          $(A \wedge B) \rightarrow C$
4. $\neg A \vee \neg B$               $\neg(A \wedge B)$

**F.** For each of the following sets of wffs, construct a derivation to show that the set is inconsistent in *SD*.

1. $\{C \wedge A,\ B \leftrightarrow A,\ B \rightarrow \neg C\}$
2. $\{(\neg C \vee \neg D) \rightarrow \neg E,\ \neg D \wedge E\}$
3. $\{S \vee T,\ \neg T,\ \neg S\}$
4. $\{A \leftrightarrow \neg A\}$

## 4.4 *SDE*, an Extension to *SD*

It turns out that the derivation system *SD* allows *only* truth-functionally valid inferences, and, indeed, it allows *all* truth-functionally valid inferences. Thus, it allows all and only those inferences that a truth table would show valid. As a result, the semantic concepts of truth-functional truth, equivalence, entailment, validity, inconsistency, etc. pick out exactly the same wffs and sets of wffs as the syntactic concepts of theorem of *SD*, equivalent in *SD*, derivable in *SD*, valid in *SD*, inconsistent in *SD*, etc. What more could we want from our derivation system?

### 4.4.1 The Inference Rules of *SDE*

Well, for certain inferences it can be a bit unwieldy. For instance, three common, intuitive, and useful patterns of inference are not represented as immediate inferences in the eleven rules of *SD*. These are Modus Tollens, which allows one to infer $\neg \mathbb{P}$ from $\mathbb{P} \rightarrow \mathbb{Q}$ and $\neg \mathbb{Q}$; Hypothetical Syllogism, which allows the inference of $\mathbb{P} \rightarrow \mathbb{R}$ from $\mathbb{P} \rightarrow \mathbb{Q}$ and $\mathbb{Q} \rightarrow \mathbb{R}$; and Disjunctive Syllogism, which allows $\mathbb{P}$ from $\mathbb{P} \vee \mathbb{Q}$ and $\neg \mathbb{Q}$. It is not that we cannot carry out a series of steps to make such inferences in *SD*—we can—but they are long and involved, we cannot carry them out in a single step.

We may, therefore, like to extend *SD* by introducing some new rules. Since *SD* allows all and only the inferences that are truth-functionally valid, we do not want our extension to allow any new inferences. We only want to make some of the things we can already do easier. In other words, we want our extension to be conservative. We do not want to be able to derive anything we couldn't already derive, we merely want to abbreviate certain lengthy processes.

Thus, let us stipulate that the system *SDE* consists of the eleven rules of *SD* plus the rules to be introduced in this section. Here are the three new inference rules:

| MT | | | | HS | | |
|---|---|---|---|---|---|---|
| $i$ | $\mathbb{P} \rightarrow \mathbb{Q}$ | | | $i$ | $\mathbb{P} \rightarrow \mathbb{Q}$ | |
| $j$ | $\neg \mathbb{Q}$ | | | $j$ | $\mathbb{Q} \rightarrow \mathbb{R}$ | |
| | | | | | | |
| $\triangleright$ | $\neg \mathbb{P}$ | $i, j$ MT | | $\triangleright$ | $\mathbb{P} \rightarrow \mathbb{R}$ | $i, j$ HS |

DS

$$\begin{array}{c|c}
i & \mathbb{P} \vee \mathbb{Q} \\
j & \neg\mathbb{P} \\
\\
\triangleright & \mathbb{Q} \qquad i, j\,\text{DS}
\end{array}$$

-or-

$$\begin{array}{c|c}
i & \mathbb{P} \vee \mathbb{Q} \\
j & \neg\mathbb{Q} \\
\\
\triangleright & \mathbb{P} \qquad i, j\,\text{DS}
\end{array}$$

The idea is that we cannot do anything new, we can just do some things more easily. How can we be sure that we haven't introduced a new element that allows an invalid inference? We merely need to show that anything the three new rules allow, can also be done via the resources of *SD* alone. I will demonstrate this for MT.

Any time I perform the following step in *SDE*:

$$\begin{array}{c|c}
i & \mathbb{P} \to \mathbb{Q} \\
j & \neg\mathbb{Q} \\
\\
 & \neg\mathbb{P} \qquad i, j\,\text{MT}
\end{array}$$

I could replace it, achieving the same outcome, with steps requiring only the rules of *SD*:

$$\begin{array}{c|c|l}
i & \mathbb{P} \to \mathbb{Q} & \\
j & \neg\mathbb{Q} & \\
\\
n & \quad \mathbb{P} & A \\
n+1 & \quad \mathbb{Q} & i, n \to\!\text{E} \\
n+2 & \quad \neg\mathbb{Q} & j\,\text{R} \\
 & \neg\mathbb{P} & n-n+2\ \neg\text{I}
\end{array}$$

Strictly speaking, I have not actually given a derivation, for no wffs of *S* appear; rather I have given a *derivation form*. The use of metavariables instead of specific wffs shows that this applies to any wffs $\mathbb{P}$ and $\mathbb{Q}$ of *S*. So, obviously, anything I can do with MT can also be done without MT via the *SD* rules alone. I do, however, save myself three steps, and a subderivation. Moreover, a very common and intuitive inference is syntactically mimicked by a single step. So MT may be looked on as a convenient abbreviation of a longer process. Try to show that HS and DS are conservative extensions as well. You'll find that DS really does save some steps.

Here are some example derivations involving the new rules.

$$\begin{array}{c|l|ll}
1 & \neg A \to C & P & \\
2 & A \to B & P & \\
3 & \neg B & P & \vdash C \\
\hline
4 & \neg A & 2, 3\,\text{MT} & \\
5 & C & 1, 4 \to\!\text{E} &
\end{array}$$

```
1 | P → (Q ∨ R)        P
2 | (N ∧ O) → P        P
3 | M → (N ∧ O)        P        ⊢ M → (Q ∨ R)
4 | M → P              2, 3 HS
5 | M → (Q ∨ R)        1, 4 HS
```

```
1 | ¬D ∧ ¬C            P
2 | (A ∧ B) ∨ E        P
3 | (A ∧ B) → D        P        ⊢ ¬C ∧ E
4 | ¬D                 1 ∧E
5 | ¬(A ∧ B)           3, 4 MT
6 | E                  2, 5 DS
7 | ¬C                 1 ∧E
8 | ¬C ∧ E             6, 7 ∧I
```

## 4.4.2  Exercises

Complete each of the following derivations using the rules of *SDE* so far introduced (the *SD* rules plus MT, HS, DS).

**1.**
```
1 | C ∨ ¬B             P
2 | (¬A ∨ B) ∧ ¬C      P        ⊢ ¬A
```

**2.**
```
1 | J → L              P
2 | K → J              P
3 | L → M              P        ⊢ ¬M → ¬K
```

**3.**
```
1 | (G ∨ H) ∧ ¬J                P
2 | (¬I → J) ∧ (H → ¬I)         P        ⊢ G
```

**4.**
```
1 | K → C              P
2 | H → (S ∨ T)        P
3 | S → K              P        ⊢ H → (C ∨ T)
```

### 4.4.3 **The Replacement Rules of** *SDE*

There are other types of inferences that can usefully be abbreviated by allowing ourselves to substitute equivalent wffs or well-formed components of wffs. For instance, suppose we have the following situation:

| | | | |
|---|---|---|---|
| 1 | $A \wedge B$ | P | |
| 2 | $(B \wedge A) \rightarrow C$ | P | $\vdash C$ |
| 3 | A | 1 $\wedge$E | |
| 4 | B | 1 $\wedge$E | |
| 5 | $B \wedge A$ | 3, 4 $\wedge$I | |
| 6 | C | 2, 5 $\rightarrow$E | |

Note that we cannot simply execute $\rightarrow$E because, despite our semantic sense that '$A \wedge B$' and '$B \wedge A$' are equivalent, they are not of the same syntactic form. Hence the rearrangements of lines 3−5. But the following two derivation forms show that any wff of the form $\mathbb{P} \wedge \mathbb{Q}$ is equivalent in *SD* to its commutation, $\mathbb{Q} \wedge \mathbb{P}$:

| | | |
|---|---|---|
| $i$ | $\mathbb{P} \wedge \mathbb{Q}$ | |
| $n$ | $\mathbb{P}$ | $i$ $\wedge$E |
| $n+1$ | $\mathbb{Q}$ | $i$ $\wedge$E |
| | $\mathbb{Q} \wedge \mathbb{P}$ | $n, n+1$ $\wedge$I |

| | | |
|---|---|---|
| $i$ | $\mathbb{Q} \wedge \mathbb{P}$ | 2, 3 $\wedge$I |
| $n$ | $\mathbb{Q}$ | $i$ $\wedge$E |
| $n+1$ | $\mathbb{P}$ | $i$ $\wedge$E |
| | $\mathbb{P} \wedge \mathbb{Q}$ | $n, n+1$ $\wedge$I |

This licenses the intersubstitutability of the two forms of wffs so that we can introduce a rule such as this:

| Com | Commutation |
|---|---|

$$\mathbb{P} \wedge \mathbb{Q} \quad \triangleleft\triangleright \quad \mathbb{Q} \wedge \mathbb{P}$$
$$\mathbb{P} \vee \mathbb{Q} \quad \triangleleft\triangleright \quad \mathbb{Q} \vee \mathbb{P}$$

Such a rule tells us that whenever a wff or well-formed component of a wff has the form on the left of the double triangle, we can substitute for it the form on the right side, and vice versa. This allows us to shorten the derivation above to 4 lines:

| | | | |
|---|---|---|---|
| 1 | $A \wedge B$ | P | |
| 2 | $(B \wedge A) \rightarrow C$ | P | $\vdash C$ |
| 3 | $B \wedge A$ | 1 Com | |
| 4 | C | 2, 3 $\rightarrow$E | |

Here, we applied Com to line 1 to generate a wff that matches the antecedent of line 2. This shortens the derivation by two lines. It is important to keep in mind that replacement rules apply not just to complete wffs, but also to well-formed components of wffs. This means we can apply Com to the antecedent of line 2 to generate a wff the antecedent of which matches line 1. This also shortens the original derivation by two lines:

$$
\begin{array}{lll}
1 & A \land B & P \\
2 & (B \land A) \to C & P \qquad \vdash C \\
\hline
3 & (A \land B) \to C & 2 \text{ Com} \\
4 & C & 1, 3 \to E
\end{array}
$$

In some cases a rule as simple as Com can save us much more than two lines. Compare the following two derivations, the first of which does not use Com:

$$
\begin{array}{lll}
1 & A \to (B \land C) & P \\
2 & (C \land B) \to D & P \qquad \vdash A \to D \\
3 & \quad A & A \\
4 & \quad B \land C & 1, 3 \to E \\
5 & \quad B & 4 \land E \\
6 & \quad C & 4 \land E \\
7 & \quad C \land B & 5, 6 \land I \\
8 & \quad D & 2, 7 \to E \\
9 & A \to D & 3\text{--}8 \to I
\end{array}
$$

$$
\begin{array}{lll}
1 & A \to (B \land C) & P \\
2 & (C \land B) \to D & P \qquad \vdash A \to D \\
\hline
3 & (B \land C) \to D & 2 \text{ Com} \\
4 & A \to D & 1, 3 \text{ HS}
\end{array}
$$

Without commuting either the consequent of line 1 or the antecedent of line 2, we are unable to use HS and must go through a lengthy subderivation. Use of Com saves us all this work.

Two important differences between replacement rules and inference rules need to be reiterated. First, replacement rules are bidirectional, but inference rules are unidirectional. Second, while inference rules require that the wff on a given line (i.e., the *whole* wff) match the form in the rule, replacement rules require only that some well-formed component of a wff on a given line match the form in the rule. That is, though inference rules work only on whole lines, replacement rules work on whole lines or well-formed subcomponents.

This is but one example of many replacement rules. The full set of *SDE* replacement rules appears in Box 4.2 as well as in Appendix C.

```
┌─────────────────────────────────────────────────────────────────────┐
│                  Box 4.2: Replacement Rules of SDE                    │
```

| Com            Commutation | Assoc          Association |
|---|---|
| $\mathbb{P} \wedge \mathbb{Q}$  ◁▷  $\mathbb{Q} \wedge \mathbb{P}$ <br> $\mathbb{P} \vee \mathbb{Q}$  ◁▷  $\mathbb{Q} \vee \mathbb{P}$ | $\mathbb{P} \wedge (\mathbb{Q} \wedge \mathbb{R})$  ◁▷  $(\mathbb{P} \wedge \mathbb{Q}) \wedge \mathbb{R}$ <br> $\mathbb{P} \vee (\mathbb{Q} \vee \mathbb{R})$  ◁▷  $(\mathbb{P} \vee \mathbb{Q}) \vee \mathbb{R}$ |
| **Impl**          Implication | **DN**          Double Negation |
| $\mathbb{P} \rightarrow \mathbb{Q}$  ◁▷  $\neg \mathbb{P} \vee \mathbb{Q}$ | $\mathbb{P}$  ◁▷  $\neg\neg \mathbb{P}$ |
| **DeM**          De Morgan | **Idem**          Idempotence |
| $\neg(\mathbb{P} \wedge \mathbb{Q})$  ◁▷  $\neg \mathbb{P} \vee \neg \mathbb{Q}$ <br> $\neg(\mathbb{P} \vee \mathbb{Q})$  ◁▷  $\neg \mathbb{P} \wedge \neg \mathbb{Q}$ | $\mathbb{P}$  ◁▷  $\mathbb{P} \wedge \mathbb{P}$ <br> $\mathbb{P}$  ◁▷  $\mathbb{P} \vee \mathbb{P}$ |
| **Trans**          Transposition | **Exp**          Exportation |
| $\mathbb{P} \rightarrow \mathbb{Q}$  ◁▷  $\neg \mathbb{Q} \rightarrow \neg \mathbb{P}$ | $\mathbb{P} \rightarrow (\mathbb{Q} \rightarrow \mathbb{R})$  ◁▷  $(\mathbb{P} \wedge \mathbb{Q}) \rightarrow \mathbb{R}$ |
| **Dist**          Distribution | **Equiv**          Equivalence |
| $\mathbb{P} \wedge (\mathbb{Q} \vee \mathbb{R})$  ◁▷  $(\mathbb{P} \wedge \mathbb{Q}) \vee (\mathbb{P} \wedge \mathbb{R})$ <br> $\mathbb{P} \vee (\mathbb{Q} \wedge \mathbb{R})$  ◁▷  $(\mathbb{P} \vee \mathbb{Q}) \wedge (\mathbb{P} \vee \mathbb{R})$ | $\mathbb{P} \leftrightarrow \mathbb{Q}$  ◁▷  $(\mathbb{P} \rightarrow \mathbb{Q}) \wedge (\mathbb{Q} \rightarrow \mathbb{P})$ <br> $\mathbb{P} \leftrightarrow \mathbb{Q}$  ◁▷  $(\mathbb{P} \wedge \mathbb{Q}) \vee (\neg \mathbb{P} \wedge \neg \mathbb{Q})$ |

### Proof Theory in *SDE* and Notes on Derivations

Proof theory in *SDE* parallels proof theory in *SD*. The definitions of 'derivation of *SDE*', 'derivable in *SDE*', 'valid in *SDE*', 'theorem of *SDE*', 'equivalent in *SDE*', and 'inconsistent in *SDE*', are identical to those above, but with '*SDE*' substituted for '*SD*'.

Remember the two differences between inference and replacement rules:

- Replacement rules are bidirectional, inference rules are not

- Replacement rules apply to both wffs and well-formed components of wffs, inference rules apply only to wffs (whole lines)

Finally, the *SDE* rules, in order to give us more and shorter ways to complete derivations, introduce a huge amount of redundancy. Thus, as you will likely notice, there are many more ways to finish a given derivation than there were in *SD*.[4] There is something to be said for brevity and elegance—the shorter and simpler the derivation, the better—and

---

4. Of course, even in *SD* there is always more than one way to finish a derivation; the point is that *SDE* gives us yet more options.

there may not always be a unique shortest solution. We will not, however, insist on finding the shortest derivation, though you might like to try.

*SDE* is supposed to make derivations shorter and easier. In particular, we can often avoid ¬I, ¬E, and the dreaded ∨E subderivations. But one possible effect of giving us so many options is an increased difficulty in determining which options apply, which rule to use. In addition to pursuing goal analysis as usual, a useful strategy is to look for patterns in the wffs and well-formed subcomponents that match the patterns in the replacement rules, doing this with an eye to achieving the (sub)goal, or generating patterns that fit the form of inference rules in order to reach the (sub)goal.

As always, the very best strategy is to practice.

## 4.4.4 Exercises

**A.** Complete each of the following derivations using *SDE*. Many can be completed in a very small number of steps with the aid of the replacement rules.

1.  1 │ H ∨ (G ∧ F)            P        ⊢ (F ∧ G) ∨ H

2.  1 │ A → B                  P        ⊢ B ∨ ¬A

3.  1 │ ¬A → B                 P
    2 │ ¬B                     P        ⊢ A

4.  1 │ ¬(K ∧ J)              P
    2 │ J                      P        ⊢ ¬K

5.  1 │ (A ∨ C) ∨ B           P
    2 │ ¬A                     P
    3 │ ¬C                     P        ⊢ B

6.  1 │ (Z ∧ Z) ∨ Z           P        ⊢ Z

**7.**  1 | $I \rightarrow (J \wedge K)$          P
      2 | $J \rightarrow (K \rightarrow L)$       P       $\vdash I \rightarrow L$

**8.**  1 | $F \leftrightarrow G$                P
      2 | $\neg (F \wedge G)$                 P       $\vdash \neg F \wedge \neg G$

**9.**  1 | $F \rightarrow G$                 P
      2 | $\neg F \rightarrow \neg H$           P       $\vdash \neg G \rightarrow \neg H$

**10.**  1 | $(A \vee B) \wedge (A \vee C)$      P
       2 | $\neg (B \wedge C)$                P       $\vdash A$

**B.** For each of the following, construct a derivation in *SDE* to show that the derivability claim holds.

   **1.** $\{(A \wedge B) \rightarrow C, \ C \rightarrow D, \ \neg (D \vee E)\} \vdash \neg A \vee \neg B$

   **2.** $\{\neg C\} \vdash \neg (B \wedge C)$

   **3.** $\{(A \vee B) \rightarrow (C \wedge D), \ C \rightarrow \neg D\} \vdash \neg B$

   **4.** $\{F \rightarrow G, \ H \leftrightarrow G, \ I \vee \neg H, \} \vdash \neg F \vee I$

**C.** For each of the following, construct a derivation to show that the argument is valid in *SDE*.

   **1.** $D \rightarrow (E \rightarrow F)$
       $\neg F$
       $\overline{\neg E \vee \neg D}$

   **2.** $\neg (D \vee \neg A)$
       $(\neg A \vee \neg B) \vee (\neg A \vee \neg C)$
       $\overline{\neg (B \wedge C)}$

   **3.** $\neg J \wedge \neg K$
       $(L \vee K) \vee (J \wedge M)$
       $\overline{L}$

   **4.** $C \vee D$
       $(\neg E \wedge \neg H) \vee F$
       $\neg D \vee E$
       $\neg F \vee G$
       $\overline{C \vee G}$

**D.** For each of the following, construct a derivation to show that the wff is a theorem of **SDE**.

   **1.** $A \lor \neg A$

   **2.** $(B \rightarrow C) \leftrightarrow \neg (B \land \neg C)$

**E.** For each of the following pairs, construct derivations to show that the wffs are equivalent in **SDE**.

   **1.** $H \land \neg G$                 $\neg(\neg H \lor G)$

   **2.** $\neg(D \leftrightarrow E)$           $(\neg D \land E) \lor (D \land \neg E)$

**F.** For each of the following sets of wffs, construct a derivation to show that the set is inconsistent in **SDE**.

   **1.** $\{(L \lor \neg M) \land (L \lor N), J \leftrightarrow (N \land \neg M), \neg L \land \neg J\}$

   **2.** $\{(\neg A \lor \neg B) \leftrightarrow (\neg C \lor \neg D), (C \land D) \land (\neg B \lor \neg C)\}$

## 4.5 **Chapter Glossary**

**Derivable in *SD*, $\Gamma \vdash \mathbb{P}$:**

   A wff $\mathbb{P}$ of *S* is *derivable in SD* from a set $\Gamma$ of wffs of *S* iff there is a derivation in *SD* the primary assumptions of which are members of $\Gamma$ and $\mathbb{P}$ depends on only those assumptions. (125)

**Derivation in *SD*:**

   A *derivation in SD* is a finite sequence of wffs of *S* such that each wff is either an assumption with scope indicated or justified by one of the rules of *SD*. (125)

**Equivalent in *SD*:**

   Two wffs $\mathbb{P}$ and $\mathbb{Q}$ are *equivalent in SD* iff they are interderivable in *SD*; i.e., iff both $\mathbb{P} \vdash \mathbb{Q}$ and $\mathbb{Q} \vdash \mathbb{P}$. (128)

**Inconsistent in *SD*:**

   A set $\Gamma$ of wffs is *inconsistent in SD* iff, for some wff $\mathbb{P}$, both $\Gamma \vdash \mathbb{P}$ and $\Gamma \vdash \neg \mathbb{P}$. (128)

**Theorem of *SD*:**

   A wff $\mathbb{P}$ is a *theorem of SD* iff $\mathbb{P}$ is derivable from the empty set; i.e., iff $\vdash \mathbb{P}$. (127)

**Valid in *SD*:**

   An argument of *S* is *valid in SD* iff the conclusion is derivable from the set consisting of only the premises, otherwise it is invalid in *SD*. (127)

# Part III

# Quantificational Logic

# 5 **The Language** *P*

## 5.1 **Introducing** *P*

### 5.1.1 **Quantificational Logic**

The language *S* allowed us to formally express a wide variety of natural language statements, and enabled our investigation of various truth-functional properties of wffs, sets
of wffs, and arguments. But *S* has its shortcomings. First of all, there is more structure
to natural language than truth-functional structure. Sentences we treated as simple have
internal structure which can be logically important. Second, although we could deal with
quantity terms in a restricted way in *S*, the method was unwieldy and there was no way
to deal with unnamed objects or infinite sets of objects. Third, there are plenty of valid
arguments which, because they are not valid in virtue of truth-functional structure, cannot
be properly represented in *S*. Consider the following example from Chapter 1:

> All humans are mortal
> Socrates is human
> _____
> Socrates is mortal

In order to represent this argument in *S*, we would have to choose three different statement
letters as atomic wffs, because none of the sentences contains truth-functional connectives,
and they are all different from each other. The result might look as follows:

> A
> H
> ___
> M

Clearly, this argument is not valid in *S*. The tva that assigns T to 'A' and 'H', and F to
'M', shows this. But, equally as clearly, the original argument is valid. Hence, its validity
is not a result of its truth-functional structure. Indeed, it has no truth-functional structure.
But the sentences of the original argument do have plenty of structure, and it is in virtue of
at least some of that structure that it is valid. Indeed, you could probably do a decent job
of explaining the role of the quantifier 'all', the predicates 'is human' and 'is mortal', and
the name 'Socrates' in the validity of the argument. (Give it a try.) It is, however, a more
difficult task to explore the logical properties of quantifier-predicate structures generally.
That is what this part of the book is about—quantificational logic.

**Quantificational Logic:**

> *Quantificational Logic* is the logic of sentences involving quantifiers, predicates, and names. It investigates the properties that arguments, sentences, and sets of sentences have in virtue of their quantificational structure.[1]

What we need, then, is to develop a formal language that represents the role of quantifier terms ('all', 'some', etc.), predicates, and names. The basic non-logical units of this language will be predicates, and so we name the language '$P$'. Before developing the formal language $P$ it will be useful to discuss some of the structure of natural language sentences (English, in our case).

## 5.1.2  Predicates and Singular Terms

If one looks even casually at the three sentences of the example argument, certain points of structure are obvious. The second premise and the conclusion are of simple subject-predicate structure, with subject position filled by a proper noun, designating an individual. The first sentence may seem to be of simple subject predicate structure as well; with 'All humans' in the subject position, and 'are mortal' as the predicate. This would be as if we are asserting mortality of some individual object: all humans. And this would be to treat 'all humans' much like a proper name. But this is not precisely how we will analyze it. It will be easier and more accurate for our purposes to treat instances of such words as 'all', 'every', 'any', 'some', etc., as a special category called *quantifiers*. And we will treat sentences involving these words as more complex than those like the second premise and conclusion.

You are probably quite familiar with much of the structure mentioned above, but it is important to have a common way of talking about it. Consider the following sentences:

> Socrates is human
> Plato is human
> Socrates is bald
> The inventor of Velcro is bald
> The inventor of Velcro kills Plato
> Socrates kills Socrates -*or*- Socrates kills himself
> Seven is less than twenty
> Socrates introduces Plato to the inventor of Velcro
> The least prime number is between seven and twenty

We can break these sentences up into the singular terms and predicates appearing in them.

**Singular Term:**

> A *singular term* is a word or phrase that designates or is supposed to designate some individual object. Natural language singular terms are either proper nouns or

---

1. Quantificational logic is also frequently called predicate logic. I prefer the former since the quantifiers are what generate the interesting logical properties. Note that the name of our language '$P$' reminds us of the basic non-logical unit, predicates (predicate letters to be precise).

definite descriptions (a phrase that is supposed to designate an object via a unique description of it).[2]

The singular terms above are:

Socrates
Plato
the inventor of Velcro
seven
twenty
the least prime number

The third and sixth are definite descriptions, the others are proper nouns.
   We are also interested in predicates.

**Predicate:**

A *predicate* is a series of words with one or more blanks that yields a sentence when all its blanks are filled with singular terms. Conversely, we could think of a predicate as what remains after removing one or more singular terms from a sentence.

From above:

| | |
|---|---|
| _____ is human | _____ is less than ... |
| _____ is bald | _____ introduces ... to - - - |
| _____ kills ... | _____ is between ... and - - - |

Obviously the same predicate can produce different sentences when different singular terms are inserted, and the same singular term(s) can produce different sentences when inserted into different predicates. Note also, that predicates have different numbers of blanks. For instance the first two predicates above are *one-place* or *unary* predicates. The second two are *two-place* or *binary*. The third two are *three-place* or *ternary*. More generally we can speak of *n-place* or *n-ary* predicates.

## 5.1.3   Predicate Letters and Individual Constants in *P*

Our language *P* will have basic components analogous to predicates and proper nouns (definite descriptions are treated as complex in *P*, and will be constructed later from basic components). *Predicate letters* will be uppercase letters 'A' through 'Z', and, like

---

2. In natural language, of course, the exact object designated depends on the particular context of use. 'Socrates' could as much designate a soccer player as a philosopher, depending on the context of the discussion. Moreover, some singular terms (e.g., 'Pegasus', 'the King of the USA') may fail to designate any object. The phrase 'the inventor of Velcro' may well fail to designate an object, because (for all I know) there was more than one person involved in the invention. In English contexts, unless otherwise noted, we will simply assume that all singular terms designate some unique object. As we develop the language *P* we will require that our analogue of proper nouns always designate a unique object. Definite descriptions have a special treatment in *P*, and may fail to designate.

natural language predicates, will have a certain number of "blanks" or places that when properly filled produce a wff or sentence of *P*. Just as proper nouns can be inserted into natural language predicates, one way of filling the places in a predicate letter is to insert *individual constants* into the spaces. The lowercase letters 'a' through 'u', will be our individual constants (to save breath and ink, I will also call these 'constants'). Finally, we will see lowercase 'v' through 'z' in the role of variables a little later on.

As with *S*, we need to provide a translation key or interpretation for the components of *P*. Since *P* is more complex than *S*, the interpretations will be correspondingly more complex. Rather than simple truth value assignments, or assignments of natural language sentences to statement letters, an interpretation of *P* consists of three parts. First is a *universe of discourse*. This is simply a specification of the range of things we'll be talking about, be it people, numbers, rocks, or what have you. Second, we need to assign natural language predicates to predicate letters. (In Chapter 6 we will see a more formal way of interpreting predicate letters via the specification of extensions.) Third, we need to specify which objects from the universe of discourse our constants will be naming.

**Interpretation of *P*:**

An *interpretation of P* consists of 3 components:

(1) a non-empty universe of discourse, **UD**, specifying the things about which we'll be talking

(2) an assignment of truth values or natural language statements to statement letters and of natural language predicates to predicate letters via a translation key

(3) an assignment of objects from the **UD** to constants via a translation key such that

(a) every individual constant is assigned to an object, and

(b) no individual constant is assigned to more than one object

For the time being it is sufficient to say that each constant gets matched up with a proper noun, and no constant gets matched up with more than one proper noun. Note that it is permissible for two different constants to be assigned the same interpretation (this is the case with 'l' and 'p' below). If we think of constants as proper nouns or "names", then our requirements imply that every "name" must designate[3] one and only one object, but that two "names" may designate the same object.

Here is an example:

**UD:** The set containing all people and all positive integers

B①: ① is bald                   B①②③: ① is between ② and ③
H①: ① is human                  I①②③: ① introduces ② to ③
K①②: ① kills ②
L①②: ① is less than ②                       a: Aristotle

___
3. Remember, we are assuming that natural language proper nouns designate uniquely within a specified context. See note 2.

l: Plato  
p: Plato  
s: Socrates  

f: five  
e: eleven  
t: twenty  

In the left column, the predicate letter appears to the left of its "blanks", and those blanks (and their order) are indicated by small circled numbers. The situation is simple for one-place predicates. To translate a sentence of English, we simply fill in the blank with the constant corresponding to an English proper noun and we have a sentence of *P*. Given our interpretation, the following translated into *P*

(1) Socrates is human  
(2) Plato is human  
(3) Socrates is bald  

become

(1′) Hs  
(2′) Hp  
(2″) Hl  
(3′) Bs  

Note that, since 'l' and 'p' are both interpreted as designating Plato, both (2′) and (2″) count as correct translations of (2).

With predicate letters of two or more places we must be careful to keep track of the order of the blanks in the English predicate and be sure to insert the various constants into the proper positions of the predicate letter. Thus,

(4) Socrates kills Plato  
(5) Plato is killed by Socrates  
(6) Plato kills Socrates  
(7) Socrates kills Socrates  
(8) Five is less than twenty  
(9) Socrates introduces Plato to Aristotle  
(10) Eleven is between five and twenty  

become,

(4′) Ksp  
(5′) Ksp  
(6′) Kps  
(7′) Kss  
(8′) Lft  
(9′) Ispa  
(10′) Beft

Note that, despite the difference in wording of (4) and (5), they receive identical translations into *P* in (4′) and (5′).

While the blanks are always labeled in numerical order in the predicate letter of *P*, they need not be so ordered in the English predicate. That is, we could have defined an alternate two-place predicate: 'K₁①②' as '② kills ①'. The difference between 'K' and 'K₁' is just that in the original the killer is designated in blank ①, and the victim in ②, while in 'K₁' the victim is designated in ①, with the killer designated in ②.

Of course, the usual truth-functional compounding, familiar from *S*, can be applied to sentences of *P*. Treating each filled-in predicate letter of *P* as we did the statement letters of *S*,

(11)  Plato killed Socrates and Socrates killed Plato
(12)  Plato did not kill Socrates
(13)  If Socrates is bald then so is Plato
(14)  Plato is bald and human
(15)  If five is less than eleven and eleven is less than twenty, then eleven is between five and twenty

become,

(11′)  $Kps \land Ksp$
(12′)  $\neg Kps$
(13′)  $Bs \rightarrow Bp$
(14′)  $Bp \land Hp$
(15′)  $(Lfe \land Let) \rightarrow Beft$

Accounting for predicate-noun structure in this way is only one of the advantages of *P* over *S*. The real power is in the way *P* will handle sentences with quantifier terms and related pronouns.

## 5.1.4  **Pronouns and Quantifiers**

You are familiar with pronouns of laziness—i.e., pronouns used to stand in for repeated singular terms. As Albert Andreas Armadillo sings in "Rufus Xavier Sarsaparilla":

> You see, a pronoun was made to take the place of a noun,
> 'Cause saying all those nouns over and over can really wear you down!

When pronouns are used in this way—standing in for nouns present elsewhere—we can always reinsert the omitted nouns, and do without the pronouns. For instance, the members of each of the following pairs are more or less interchangeable:

• Socrates was human, and he killed himself

• Socrates was human, and Socrates killed Socrates

- Da Vinci painted the *Mona Lisa* and now he is considered a great artist and it is considered a masterpiece

- Da Vinci painted the *Mona Lisa* and now Da Vinci is considered a great artist and the *Mona Lisa* is considered a masterpiece

- Rufus Xavier Sarsaparilla found a kangaroo that followed him home and now it belongs to him

- Rufus Xavier Sarsaparilla found a kangaroo that followed Rufus home, and now that kangaroo [the kangaroo Rufus found] belongs to Rufus Xavier Sarsaparilla[4]

Because we can always do without them in favor of the original nouns, pronouns used in this way are called *eliminable* pronouns.

But not every use of a pronoun can be eliminated in favor of reinserting some noun. Consider the following:

(1) Any student caught cheating will have her name published and she will be expelled

If we were to try to replace 'her' and 'she' with some noun, we would either have to pick a name at random,

(2) Any student caught cheating will have Susie's name published and Susie will be expelled

or treat 'any student' as the referent of the pronouns, resulting in,

(3) Any student caught cheating will have any student's name published and any student will be expelled

It is debatable whether (3) even makes sense. Supposing it does, the practices described by (2) and (3) are rather unjust. Why should Susie suffer? Why some random student? Obviously, neither (2) nor (3) is equivalent to (1). (1) has a generality that (2) lacks, and a specificity that (3) lacks. (1) is more general than (2) because it dictates consequences that apply more broadly than to just Susie. At the same time (1) is more specific than (3), because it dictates precisely who will suffer the consequences—the particular student(s) who cheated. This is achieved by the connection of the pronouns 'her' and 'she' referring back to the quantity term 'any student'.

The pronouns in (1) are, thus, doing something different than just replacing a noun (*pace* Albert Andreas Armadillo). They are *ineliminable* pronouns, not mere pronouns

---

4. Strictly speaking, 'a kangaroo' is not a singular term, being an *in*definite description; but the information we are given in the first part of the sentences allows us to fix on a definite description, 'the kangaroo Rufus found', that we can either use or replace with a pronoun. Indeed, the Sarsaparilla example contains both eliminable and ineliminable pronouns. Here it is with only the ineliminable ones:

There is something such that Rufus Xavier Sarsaparilla found *it*, and *it* is a kangaroo, and *it* followed Rufus home, and now *it* belongs to Rufus Xavier Sarsaparilla.

of laziness. They are doing important work only they can do, and they do their work in conjunction with quantifier terms such as 'every', 'all', 'any', 'some', 'none', and so on. Thus, in order to formalize quantity claims in *P* we will need a way to represent this interplay between quantifier terms and ineliminable pronouns.

From a certain logical point of view, a statement like (1) is not just about students. It is, in a sense, about everything (or, everything in the universe of discourse). An illustration of how the sentence can be so general, and yet say something informative about students, will also help to illustrate the role of ineliminable pronouns. A fully general paraphrase of (1) would run thus:

(4) Every thing is such that, if *it* is a student and *it* is caught cheating, then *it* will have *its* name published and *it* will be expelled

The part of this sentence preceding the first comma lets us know that it is talking about everything. The middle part of the sentence, between the two commas, narrows the range (so to speak) by specifying a certain set of objects (i.e., cheating students). The final part asserts some further things about objects in that set (i.e., their names will be published and they will be expelled). This is all achieved via the quantity term 'every' and the way in which the pronouns link the sentence together.

Consider another example:

(5) Some student was caught cheating and she had her name published and she was expelled

Clearly, the following would be incorrect understandings of (5).

(6) Some student was caught cheating and Susie had Susie's name published and Susie was expelled
(7) Some student was caught cheating and some student had some student's name published and some student was expelled

Rather, the following is an accurate interpretation:

(8) Some thing is such that, *it* is a student and *it* was caught cheating and *it* had *its* name published and *it* was expelled

Here, the first part of both (5) and (8) show that we are talking about some thing (i.e., that there is *at least one* thing meeting the ensuing description). But since this thing is not specified by some proper noun, the ineliminable pronouns are needed to link the predicates back to the quantifier. Again, the combination of quantifier term and ineliminable pronouns gives us a unique combination of generality (at least one unnamed object is being talked about) and specificity (that object satisfies a certain description). *P* will contain formal apparatus for quantification that basically parallels the structure of natural language quantification with ineliminable pronouns. To this we now turn.

### 5.1.5  Variables and Quantifiers in *P*

In *P* the role of ineliminable pronouns is played by *individual variables*, the lowercase letters v, w, x, y, z. Individual variables (or, more briefly, variables), may be inserted into the blanks of predicate letters in just the same way that constants may be so inserted. Thus, the following are permissible:

(1)  Bx
(2)  Kxx
(3)  Kxp
(4)  Kxy
(5)  Ixyz
(6)  Ixxy

Unlike with constants, however, the result of filling out a predicate letter with variables is not considered a sentence of *P*. For consider how we would have to read (1)–(6):

(1′)  it is bald
(2′)  it kills it (the same it (itself))
(3′)  it kills Plato
(4′)  it kills it (a potentially different it)
(5′)  it introduces (potentially different) it to (potentially different) it
(6′)  it introduces itself to (potentially different) it

It is debatable whether (1′)–(6′), standing alone, are even meaningful. Were they filled out with some proper nouns or some quantifiers, we would have meaningful sentences. But since the referents of the pronouns are left entirely unspecified, there is a sense in which (1′)–(6′) are incomplete skeletons of, or schemas for, sentences. Their correlates in *P*, (1)–(6), also may be considered incomplete sentences. Officially, expressions such as (1)–(6) will be a subset of the wffs of *P*; namely, the *open wffs* (see Sec. 5.2.2).

In order to complete (or *close*) open wffs, we could either replace all the variables with constants, or we can apply the appropriate quantifiers to the open wffs. *P* will have two symbols for quantifiers:

∀

∃

The first, an upside-down 'A', is the universal quantifier symbol. The second, an upside-down 'E', is the existential quantifier symbol.[5] It may help to remember that the '∀' symbol is for '∀ll' and the '∃' symbol is for '∃xists'. When a variable is placed to the right and the result properly inserted between parentheses, the symbols form either universal or existential quantifiers. Here are the x-quantifiers, with the various ways of reading them:

---

5. Some people think of this as a backward 'E'. See if you can explain why it might also be considered upside down.

(∀x)  **Universal** x**-Quantifier**

all (objects, things) x (are such that)
for all x
every (thing) x (is such that)
each (thing) x (is such that)
any x (is such that)

(∃x)  **Existential** x**-Quantifier**

there exists at least one (object, thing) x (such that)
there is an x (such that)
(for) some (thing) x
at least one x (is such that)

Had we placed a y (or z, etc.) in the parentheses, we would have had universal y- and existential y-quantifiers. Which reading is best (and whether to include the parenthetical phrases) depends on the rest of the wff and the fluidity of expression desired. The top two readings for each quantifier are the most official, and should always work well enough.

Quantifiers are used by appending them to the left of an open wff (in some ways much like the hook symbol). To properly close a wff and create a sentence of *P* there must be exactly one quantifier for each distinct variable in the open wff. We will be very explicit about the syntax in the next section. Some simple illustrations will suffice for now.

(7)  (∀x)Bx

All things x are such that x is bald
For all x, x is bald
Everything is bald

(8)  (∀z)Bz

All things z are such that z is bald
For all z, z is bald
Everything is bald

(9)  (∃x)Bx

There exists at least one thing x such that x is bald
There is an x such that x is bald
Some x is bald
Something is bald

(10)  (∃w)Kwp

There is a w such that w kills Plato
Some w kills Plato
Something kills Plato

(11)  (∃w)Kpw

> There is a w such that Plato kills w
> Plato kills something

(12)  (∃y)Kyy

> There is a y such that y kills y
> Something kills itself

For simplicity, these examples involve only one variable and quantifier, though down the road we will construct very complex sentences of *P* using multiple variables and quantifiers. The first one or two readings offered for each example are more or less the official readings. The final reading for each example above is the most colloquial English. Of course, various readings (not all of which are shown) are acceptable. Notice that (7) and (8) are essentially the same, except for the variable used, and this similarity is reflected in the sameness of the colloquial reading.

You will likely have guessed that a universally quantified wff (∀x)Fx is true iff the condition expressed by the sub-wff Fx is true of (satisfied by) *every* object in the universe of discourse. An existentially quantified wff (∃x)Fx is true iff the condition expressed by the sub-wff Fx is true of (satisfied by) *at least one* object in the **UD**. Remember that we take 'some' to mean at least one, and that includes the possibility of two, three, a few, many, most, and so on, up to and including all. On this understanding of the quantifiers, then, (7) and (8) are false, but (9) is true. (10) and (11) are very likely false (though one never knows for sure). While (12) is true. Interpreting quantified formulas can get pretty complex, and we will elaborate on it later in this chapter and, especially, in Chapter 6.

Truth-functional connectives can, of course, be combined with quantifiers. Any predicate that is filled out with either (or both) constants and variables, and any quantified wff, can become part of a properly constructed sentence involving truth-functional connectives. Here are a few examples with interpretations, the full details of syntax being saved for the next section:

(13)  Bx ∧ Hx

> it is both bald and human

(14)  ¬(∀y)By

> It is not the case that for all y, y is bald
> Not everything is bald

(15)  (∃z)¬Bz

> There is a z that is not bald
> Something is not bald

(16)  (∃x)Bx ∧ (∃x)Hx

> There is an x that is bald, and there is an x (not necessarily the same x) that is human
>
> Something is bald, and something (not necessarily the same thing) is human

(17) $(\exists x)(Bx \wedge Hx)$

> There is an x that is both bald and human
>
> Something is both bald and human

These are just a few simple examples. Notice the difference between the readings of (16) and (17), and see if you can explain it. Before we discuss these or more complex examples in full detail, it will help to have a formal specification of the syntax of $P$.

## 5.2  **The Syntax of** $P$

### 5.2.1  **Defining the Language**

The language $P$ shall consist of the following symbols:

**Symbols of $P$:**

> **Predicate Letters:**
> $$A_1^0, B_1^0, \ldots, Z_1^0, A_2^0, B_2^0, \ldots, Z_2^0, A_3^0, \ldots$$
> $$A_1^1, B_1^1, \ldots, Z_1^1, A_2^1, B_2^1, \ldots, Z_2^1, A_3^1, \ldots$$
> $$A_1^2, B_1^2, \ldots, Z_1^2, A_2^2, B_2^2, \ldots, Z_2^2, A_3^2, \ldots$$
> $$A_1^3, B_1^3, \ldots, Z_1^3, A_2^3, B_2^3, \ldots, Z_2^3, A_3^3, \ldots$$
> $$\vdots \qquad \ddots$$
>
> That is, any uppercase letter with zero or positive integer superscript, $n$, indicating the number of places, and positive integer subscript, $k$, to give us an infinite (denumerable) supply.

> **Individual Terms:**
>
> > **Individual Constants:**
> > $$a, b, \ldots, u, a_1, b_1, \ldots, u_1, a_2, \ldots$$
> >
> > **Individual Variables:**
> > $$v, w, x, y, z, v_1, w_1, x_1, y_1, z_1, v_2, \ldots$$

> **Truth-Functional Connectives:**
> $$\neg \ \wedge \ \vee \rightarrow \leftrightarrow$$

> **Quantifier Symbols:**
> $$\forall \ \exists$$

> **Punctuation Marks:**
> $$(\ )$$

This completes the lexicon of $P$.

Let us now increase the stock of metavariables as follows:

**Metavariables:**

$\mathbb{A}, \mathbb{B}, \mathbb{C}, \ldots, \mathbb{Z}, \mathbb{A}_1, \ldots$

$\mathbb{A}_k^n, \mathbb{B}_k^n, \ldots, \mathbb{Z}_k^n$

$\mathbb{a}, \mathbb{b}, \mathbb{c}, \ldots, \mathbb{z}, \mathbb{a}_1, \ldots$

The first row above is our original stock which will be used to range over expressions and wffs. The second row will be used to range over predicate letters. The third row, in lowercase blackboard bold, will be used to range over individual terms (constants and/or variables).

**Expression of $P$:**

An *expression of $P$* is any finite sequence of the symbols of $P$.

**Quantifier of $P$:**

Where x ranges over individual variables, expressions of the form $(\forall x)$ are called *universal quantifiers*, while expressions of the form $(\exists x)$ are called *existential quantifiers*. We may also refer to a quantifier by the particular variable it contains—e.g., '$(\forall y_3)$' is a universal $y_3$-quantifier, while '$(\exists x)$' is an existential x-quantifier.

Now we can define 'wff of $P$':

**Well-Formed Formula, Wff, of $P$:**

Where $\mathbb{P}$ and $\mathbb{Q}$ range over expressions of $P$, $\mathbb{A}_k^n$ ranges over predicate letters of $P$, $\mathbb{t}$ ranges over individual terms of $P$, and x ranges over individual variables of $P$,

(1) Any expression with form $\mathbb{A}_k^0$ or $\mathbb{A}_k^n \mathbb{t}_1 \ldots \mathbb{t}_n$ is a wff of $P$

(2) If $\mathbb{P}$ and $\mathbb{Q}$ are wffs of $P$, then

    (a) $\neg \mathbb{P}$ is a wff of $P$

    (b) $(\mathbb{P} \wedge \mathbb{Q})$ is a wff of $P$

    (c) $(\mathbb{P} \vee \mathbb{Q})$ is a wff of $P$

    (d) $(\mathbb{P} \rightarrow \mathbb{Q})$ is a wff of $P$

    (e) $(\mathbb{P} \leftrightarrow \mathbb{Q})$ is a wff of $P$

(3) If $\mathbb{P}$ is a wff of $P$ that (i) contains at least one occurrence of x and (ii) contains no x-quantifiers, then

    (a) $(\forall x)\mathbb{P}$ is a wff of $P$

    (b) $(\exists x)\mathbb{P}$ is a wff of $P$

(4) Nothing is a wff of $P$ unless it can be shown so by a finite number of applications of clauses (1) through (3)

Here are some examples of wffs of $P$:

---

**Box 5.1: A Labeled Syntax Tree**

$$D_1^1x \qquad G_7^2by_1$$

3a $\quad$ 2a

$$(\forall x)D_1^1x \quad \neg G_7^2by_1$$

2d

$$((\forall x)D_1^1x \rightarrow \neg G_7^2by_1)$$

3b

$$(\exists y_1)((\forall x)D_1^1x \rightarrow \neg G_7^2by_1)$$

---

(1) $A_1^0$

(2) $A_1^2x_3b$

(3) $\neg A_1^2x_3b$

(4) $(\neg A_1^2x_3b \rightarrow \neg A_1^1w)$

(5) $(\forall x)B_1^2xy$

(6) $(\exists y)(\forall x)B_1^2xy$

(7) $(\forall x_1)D_3^2x_1c_2$

(8) $(\exists w)((\forall y_3)B_1^2y_3b_1 \rightarrow \neg E_5^1w)$

Expression (1) counts as a wff in virtue of clause (1) of the definition of a wff. '$A_1^0$' is a zero-place predicate letter followed zero individual terms—essentially this is a statement letter, familiar from the language $S$. We keep these around in case we need them. Expression (2) counts as a wff in virtue of clause (1) as well—'$A_1^2$' is a two-place predicate letter and is followed by two individual terms, '$x_3$' and '$b$'. (3) is the result of applying clause (2a) to (2), so it qualifies as a wff. (4) is the result of applying clause (2d) to (3) and '$\neg A_1^1w$'. (5) is the result of applying clause (3a) to the wff '$B_1^2xy$', and (6) is the result of applying clause (3b) to wff (5). You should be able to explain why (7) and (8) qualify as wffs, either by breaking them down or building them up according to the definition.

Box 5.1 illustrates the process of constructing a wff (or the process of justifying that an expression is a wff).

Here are some examples that are not wffs of $P$:

(9) $A_1^0cd$

(10) $Ax_3b$

(11) $\neg Ax_3b$

(12) $(\neg A_1^2x_3b \rightarrow \neg A_1^6w)$

(13) $(\forall z)B_1^2xy$

(14) $(\exists x)(\forall x)B_1^2xy$

(15) $(\forall c_2)(\forall x)D_3^2x_1c_2$

(16) $(\exists w)(\forall y_3)B_1^2y_3b_1 \rightarrow \neg E_5^1w$

Expression (9), though similar to (1), is a zero-place predicate with two terms attached. Thus (9) violates clause (1). (10) and (11), though much like (2) and (3), fail to be wffs because '$A$' is not a predicate letter—it lacks super- and subscript (though shortly we will adopt the convention of eliminating super- and subscripts). (12) fails to be a wff because the predicate letter in the consequent, '$A_1^6$' is a six-place predicate letter, but it is only followed by one term. Hence, the consequent of (12) violates clause (1) of the definition. (13) is similar to (5), but it violates clause (3) by appending a z-quantifier to a wff that does not contain 'z'. (14) is like (6), but involves the mistake of applying an existential x-quantifier to a wff already containing an x-quantifier, thus violating clause (3). (15) fails to be a wff because '$c_2$' is not a variable, and so '$(\forall c_2)$' is not a quantifier. (16) lacks the parentheses of (9), so the scope of the w-quantifier includes only the antecedent, but the antecedent does not contain 'w'.

## 5.2.2 Syntactic Concepts and Conventions

### Parentheses, Brackets, and Super-/Subscripts

It will be convenient to adopt some conventions concerning the syntax of *P*. First, as with *S*, in most contexts we will drop the outermost parentheses of wffs, though this does not include the parentheses of the quantifiers. Second, as in *S*, we reserve the option of using square brackets, '[' and ']', in the place of some pairs of parentheses (though not in quantifiers) in order to facilitate visual chunking of formulas. Third, where no confusion threatens, we allow ourselves to drop the super- and subscripts on predicate letters—this will avoid visual clutter, and the interpretation will always specify the number of places of the predicate letter. We will occasionally have to use a subscript to distinguish predicate letters involving the same capital letter—e.g., 'B' is different from '$B_1$' and '$B_{23}$'. And in some cases it will be important to be explicit about the number of places in the predicate via a superscript. Most of the time, however, we can avoid the super- and subscripts on predicate letters. Under this convention, (9) and (10) above would qualify as wffs. Moreover, we could rewrite (1)–(8) above as:

(1′) A

(2′) $Ax_3b$

(3′) $\neg Ax_3b$

(4′) $(\neg Ax_3b \rightarrow \neg Aw)$

(5′) $(\forall x)Bxy$

(6′) $(\exists y)(\forall x)Bxy$

(7′) $(\forall x)Dx_1c_2$

(8′) $(\exists w)[(\forall y_3)By_3b_1 \rightarrow \neg Ew]$

Note that subscripts on terms may be avoided simply by choosing terms without subscripts. No special convention is needed. Also note that this imposes a limit of 21 constants and 5 variables.

### Main Operator, Well-Formed Components, and Scope

It is worthwhile defining certain syntactic notions for later use.

**Atomic Formula, Molecular Formula:**
> Any wff that qualifies simply in virtue of clause (1) of the definition of a wff (that is, any wff that just is some predicate letter with the appropriate number of terms), is called an *atomic* formula. By analogy, all other wffs are *molecular*.

For example, (1′) and (2′) are atomic wffs, while (3′)–(8′) are molecular.

**Operator:**
> The truth-functional connectives and quantifiers are *operators*.

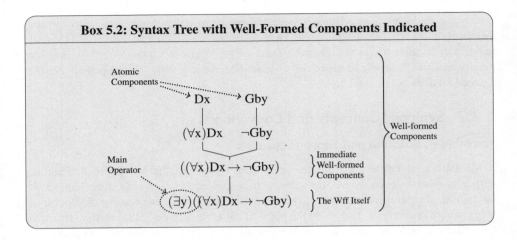

**Box 5.2: Syntax Tree with Well-Formed Components Indicated**

## Main Operator, Well-Formed Components:

Atomic wffs have no main operator. The *main operator* of a molecular wff ℝ is the operator appearing in the clause of the definition of a wff cited last in showing ℝ to be a wff. The *immediate well-formed components* of a molecular wff are the values of ℙ and ℚ (in the case of clause (2a) simply ℙ) in the last-cited clause of the definition of a wff. The *well-formed components* of a wff are the wff itself, its immediate well-formed components, and the well-formed components of its immediate well-formed components. The *atomic components* of a wff are the well-formed components that are atomic wffs.

Thus, (1′) and (2′), above, have no main operator. The main operator of (3′) is the hook. The main operator of (4′) is the arrow. In (5′) it is the universal x-quantifier. In (6′) it is the existential y-quantifier. See Box 5.2 for a detailed example.

## Scope:

The *scope* of an operator is that portion of the wff containing its immediate well-formed component(s).

In (6′), above, for instance, the scope of the existential y-quantifier is wff (5′). In (5′) the scope of the universal x-quantifier is the 'Bxy'. The scope of quantifiers parallels that of the hook.

See Box 5.3 for a graphical representation of scope.

## Free/Bound Variable, Substitution Instance

### Bound Variable, Free Variable:

An occurrence of a variable x in a wff ℙ is *bound* iff it is within the scope of an x-quantifier. An occurrence of a variable is *free* iff it is not bound.

---

**Box 5.3: Scope of Operators**

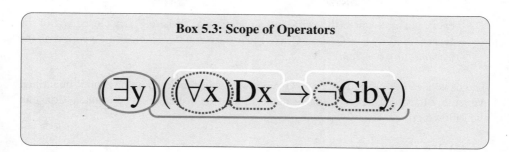

---

**Open Wff:**
>    A wff of *P* is *open* iff it contains at least one free occurrence of a variable. Otherwise it is a closed wff.

**Closed Wff, Sentence of *P*:**
>    A wff of *P* is *closed* iff it contains no free occurrences of variables. We also call such wffs *sentences* of *P*.

Consider the following wffs:

|       |                                        |        |
|-------|----------------------------------------|--------|
| (17)  | Daba                                   | closed |
| (18)  | Dxyx                                   | open   |
| (19)  | (∀x)Dxyx                               | open   |
| (20)  | (∃y)(∀x)Dxyx                           | closed |
| (21)  | Fx → Gx                                | open   |
| (22)  | (∀x)Fx → Gx                            | open   |
| (23)  | (∀x)Fx → (∀x)Gx                        | closed |
| (24)  | (∀x)(Fx → Gx)                          | closed |

Wff (17) is closed (is a sentence) since there are no variables, and so there are no free variables. In (18), both occurrences of 'x' and the occurrence of 'y' are free, so the wff is open. In contrast, the x-quantifier in (19) binds both occurrences of 'x', but since the 'y' is free, the sentence is still open. (20) is a closed wff, or a sentence, for all the variables are bound by quantifiers. In (21) neither occurrence of 'x' is bound, so the wff is open. (22) is also open, since the rightmost occurrence of 'x' is free (the scope of the universal x-quantifier includes only the antecedent). (23) is a closed wff (sentence) since the second universal quantifier binds the occurrence of 'x' in the consequent. (24) is also closed—the placement of the parentheses extends the scope of the quantifier over the whole conditional. Thus, the closed wffs or sentences are (17), (20), (23), and (24).

Note that, unlike (21)–(23) where the main operator is the arrow, the main operator of (24) is the universal x-quantifier. Remember the scope of quantifiers is much like that of the hook.

**Substitution Instance:**
>    Let Q(a/x) indicate the wff that is just like Q except for having the constant a in

every position where the variable x appears in $\mathbb{Q}$. Where $\mathbb{P}$ is a closed wff of the form $(\forall x)\mathbb{Q}$ or $(\exists x)\mathbb{Q}$, then $\mathbb{Q}(a/x)$ is a *substitution instance* of $\mathbb{P}$, with $a$ as the *instantiating constant*.

The basic idea here is quite simple. Given a wff whose main operator is an x-quantifier, we remove the quantifier and replace all the xs with the instantiating constant, $a$. Consider the following:

(25) $\mathbb{P} = (\exists y)\mathbb{Q} = (\exists y)(Bya \wedge Hcy)$

Here we see that $\mathbb{P}$ is a wff with an existential y-quantifier as the main operator. The wff itself is shown in full detail on the right. Given this wff, we can form various substitution instances of $\mathbb{P}$ by removing the y-quantifier and replacing each instance of 'y' with some constant:

(26) $\mathbb{Q}(a/y) = (Baa \wedge Hca)$
(27) $\mathbb{Q}(c/y) = (Bca \wedge Hcc)$
(28) $\mathbb{Q}(g/y) = (Bga \wedge Hcg)$

Note that forming a substitution instance is a purely syntactic operation. It does not express any inference or semantic transformation. Later, in Chapter 7, the notion of substitution instance will be very important.

### 5.2.3  Exercises

For each of the following, determine whether the expression is a wff of *P*; if it is not, write 'Not a wff'; if it is a wff, then draw its syntax tree—you do not need to label the branches. If it is a closed wff, or sentence, write 'closed' next to the base of the tree. If it is open, write 'open' and underline the free variables in the base of the tree. Assume the conventions concerning outer parentheses and optional brackets are in effect.

**1.** $(\forall z)Fxy$

**2.** $(\forall z)Fxz$

**3.** $(\forall x)(\forall z)Fxz$

**4.** $(\forall w)Aw \wedge Kjw$

**5.** $(\forall w)(Aw \wedge Kjw)$

**6.** $(\forall w)(Aw \wedge Kxw)$

**7.** $(\forall w)(Aw \wedge (\exists x)Kxw)$

**8.** $(\forall w)(\exists x)(Aw \wedge Kxw)$

## 5.3 **Simple Symbolizations**

It is important to have at least a rough understanding of quantificational semantics before proceeding to translate English into *P*. Consider the following rough definitions:

**Satisfaction and Truth in *P* (Informal):**
> Given an interpretation, where $\mathbb{F}x$ is a wff with only instances of the variable x free, and a is a constant:

(1) The denotation of a *satisfies* $\mathbb{F}x$ (or $\mathbb{F}x$ is *true of* the object named by a) iff $\mathbb{F}a$ is $\mathsf{T}$

(2) A universally quantified wff $(\forall x)\mathbb{F}x$ is $\mathsf{T}$ iff the condition expressed by the immediate subcomponent $\mathbb{F}x$ is satisfied by *every* object in the **UD**

(3) An existentially quantified wff $(\exists x)\mathbb{F}x$ is $\mathsf{T}$ iff the condition expressed by the immediate subcomponent $\mathbb{F}x$ is satisfied by *at least one* object in the **UD**

Chapter 6 will deal with the semantics of *P* in greater detail. A few examples will help us in beginning translations. We use the interpretation that begins on page 146:

(1) Hs
(2) $(\forall x)$Hx
(3) $(\exists x)$Hx

Formula (1) is $\mathsf{T}$ on the interpretation because Socrates satisfies the predicate 'Hx'. That is, is it true of Socrates that he is human. Formula (2), however, is $\mathsf{F}$ on the interpretation since it is not the case that every object in the **UD** satisfies 'Hx'. In particular, none of the positive integers is human. Finally, (3) is $\mathsf{T}$ on this interpretation, because there is some object in the **UD** that satisfies 'Hx'. Socrates is an example, so am I, and you (even though we are not named by a constant in the interpretation, we are still in the **UD**).[6]

## 5.3.1 **Non-categorical Claims**

The simplest quantificational claims assert that everything in the **UD** (universe of discourse) is such and such, or that something (at least one thing) in the **UD** is such and such. Only slightly more complicated are when negations are involved. We will use the following interpretation.

**UD**: The marbles in Fred's collection

B①: ① is blue  
G①: ① is green  
R①: ① is red  
Y①: ① is yellow  

C①: ① is cracked  
S①: ① is scratched  
B①②: ① is bigger than ②  
S①②: ① is smaller than ②  

---

6. I am, of course, assuming you are a human!

g:  The Green Giant                  r:  Big Red
o:  Old Yeller                        s:  Sky[7]

We can easily symbolize the following.

(1)  All the marbles are blue
(2)  No marbles are blue
(3)  Some marbles are blue
(4)  Some marbles are not blue

(1) is a straightforward universal claim, and since we have specified the **UD** to include all and only the marbles in Fred's collection, we need only say that everything is blue. I.e., given the **UD**, it is understood that we are referring to Fred's marbles. Thus, we may translate (1) as:

(1′)  $(\forall x)Bx$

Note that the choice of 'x' as variable is arbitrary. We could have chosen any other variable instead, so long as the variable in the quantifier matches the variable in the predicate. (1′) may be read 'Everything (in the **UD**) is blue'. But, again, given that the **UD** is specified as all and only Fred's marbles, we can equally read (1′) as 'All the marbles are blue', or 'Every marble is blue'. Obviously it is an appropriate translation of (1).

(3) is similarly straightforward:

(3′)  $(\exists x)Bx$

We may read it 'Something (in the **UD**) is blue', or 'Some marbles are blue', or 'There is at least one blue marble', or 'There are blue marbles'.

There are two natural ways of translating (2):

(2′)  $(\forall x)\neg Bx$
(2″)  $\neg(\exists x)Bx$

(2′) can be read 'All the marbles are not blue', or 'Every marble is non-blue', or a number of other variants. (2″) may be read 'It is not the case that there is at least one marble that is blue', or 'It is not the case that there exists a blue marble', or 'There are no blue marbles', or a number of other variants. These English claims are all equivalent to (2). Moreover (2′) and (2″) are quantificationally equivalent, so each is an acceptable translation of (3).

The situation is similar for (4):

(4′)  $(\exists x)\neg Bx$
(4″)  $\neg(\forall x)Bx$

---

7. These are names of particular marbles in the **UD**, but these are not all the marbles in the **UD**. We assume the collection contains some unnamed marbles.

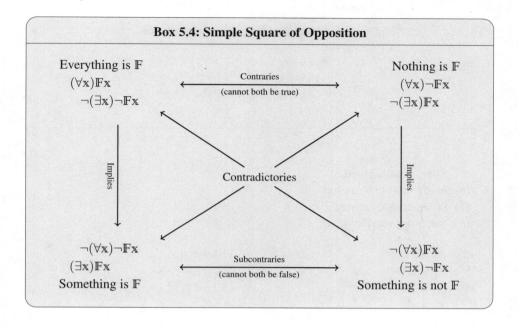

Box 5.4: Simple Square of Opposition

Everything is $\mathbb{F}$
$(\forall x)\mathbb{F}x$
$\neg(\exists x)\neg\mathbb{F}x$

Contraries
(cannot both be true)

Nothing is $\mathbb{F}$
$(\forall x)\neg\mathbb{F}x$
$\neg(\exists x)\mathbb{F}x$

Implies

Contradictories

Implies

$\neg(\forall x)\neg\mathbb{F}x$
$(\exists x)\mathbb{F}x$
Something is $\mathbb{F}$

Subcontraries
(cannot both be false)

$\neg(\forall x)\mathbb{F}x$
$(\exists x)\neg\mathbb{F}x$
Something is not $\mathbb{F}$

(4′) may be read as 'There is at least one marble that is not blue'. Whereas (4″) may be read 'It is not the case that all the marbles are blue'. Again, these English sentences are equivalent to (4), while (4′) and (4″) are quantificationally equivalent. So both are acceptable translations.

Note that (1) and (4) are contradictory, and this is reflected very clearly in the translations (1′) and (4″), the latter being the former with a hook attached. (3) and (2) are also contradictory, as illustrated by (3′) and (2″).

Indeed, we can lay out the general relationship between the basic forms in a Simple Square of Opposition, as in Box 5.4. In the square, a standard English reading appears in each corner, with appropriate translation forms in the same corner. The square is set up so that diagonally opposite corners are contradictory (i.e., must have opposite truth value). Note that the inner form in each corner is the negation of the outer form in the opposite corner. Moreover, as indicated by the arrows running vertically down the sides, wffs of the forms in the top corners imply wffs of the form directly below. Finally, the contrary (cannot both be true—horizontally across the top) and subcontrary (cannot both be false—horizontally across the bottom) relations are indicated.[8]

The pairings in each corner are equivalences, and may also be usefully listed as in Box 5.5.

These forms are the building blocks for symbolizing very simple quantificational claims, and there are relatively simple ways of compounding them. Here is another set of examples, very similar to the previous:

8. It is important to note that in the Modern Categorical Square of Opposition (Box 5.7) none of the horizontal or vertical relations obtain, only the diagonal relations of contradiction. This will be discussed below.

---

**Box 5.5: Quantifier-Negation Equivalences**

$$(\forall x)\neg \mathbb{F}x \quad \Leftrightarrow \quad \neg(\exists x)\mathbb{F}x$$
$$\neg(\forall x)\mathbb{F}x \quad \Leftrightarrow \quad (\exists x)\neg \mathbb{F}x$$
$$\neg(\forall x)\neg \mathbb{F}x \quad \Leftrightarrow \quad (\exists x)\mathbb{F}x$$
$$(\forall x)\mathbb{F}x \quad \Leftrightarrow \quad \neg(\exists x)\neg \mathbb{F}x$$

---

(5)  Every marble is red

(6)  Every marble is not red

(7)  Some marbles are red

(8)  Not every marble is red

| | |
|---|---|
| (5′) $(\forall z)Rz$ | (7′) $(\exists z)Rz$ |
| (5″) $\neg(\exists z)\neg Rz$ | (7″) $\neg(\forall z)\neg Rz$ |
| (6′) $(\forall z)\neg Rz$ | (8′) $\neg(\forall z)Rz$ |
| (6″) $\neg(\exists z)Rz$ | (8″) $(\exists z)\neg Rz$ |

First, consider (5) and (6). As was discussed in Chapter 2, the placement of the 'not' sometimes introduces ambiguity in English. Many people would understand (6) to be the denial of (5), and in certain contexts, with the proper emphasis, this could be plausible:

> Fred:  Every marble is red! (5)
> Kate:  No. Every marble is *not* red! (6)?

Here, given the context, Kate plausibly means that not every marble is red, as in (8). So, she is leaving open the possibility that one or more marble *is* red. This is suggested by her opening "No" and her emphasis on "*not*". But I think Kate is not being clear. If she wanted to deny (5), she should have said, "No. Not every marble is red." This is (8). In the absence of any context—or if we are being strictly logical and eliminating context where possible—we should interpret (6) as asserting of every marble that it is not red and translate it as (6′) or the equivalent (6″). These say that no marble is red. We take the English word order in (6)—in particular the placement of the 'every' (or 'all') prior to the 'not'—to indicate that the scope of the quantity term is greater than the scope of the 'not'. This yields (6′), which is also equivalent to (6″).

If you think carefully you will see that while (5) and (6) and their respective translations cannot both be true, they can both be false. Hence, they are not contradictory—the one is not the denial of the other—they are contraries (see Box 5.4). So, if Kate meant to deny (5), she would have done better to say, "No. Some of them are not red." This is (8″). In general, especially in writing, but even in speaking, we should try to avoid potential ambiguity, and use (6) only when we mean (6′). Better yet, say some are not.

Think of it this way, if Fred were to claim "All lawyers are crooks", and Kate responded "All lawyers are *not* crooks!", she is, perhaps, being too generous to lawyers, and is not clearly contradicting Fred. It would eliminate all potential ambiguity for Kate to say "*Not all lawyers are crooks!*" or "Some lawyers are not crooks!" This would be to deny that all are, but allow that some may be crooks.

So far we have looked only at single quantifiers combined with negation. The next step is to see how quantifiers combine with conjunction and disjunction.

The following two sentences, with translations, combine the universal quantification with conjunction.

(9)  Every marble is both cracked and scratched
(9′)  $(\forall x)(Cx \wedge Sx)$

(10)  Every marble is cracked and every marble is scratched
(10′)  $(\forall x)Cx \wedge (\forall x)Sx$

Note that in (9′) the main operator is the universal quantifier, while in (10′) it is the wedge. This reflects a difference in the wording of the English. It so happens that (9′) and (10′) are quantificationally equivalent—think this through and you'll see that every marble is cracked and scratched if and only if every marble is cracked and every marble is scratched.

This parallel equivalence does not hold for universal quantification combined with disjunction, as the following illustrate.

(11)  Every marble is either cracked or scratched
(11′)  $(\forall x)(Cx \vee Sx)$

(12)  Every marble is cracked or every marble is scratched
(12′)  $(\forall x)Cx \vee (\forall x)Sx$

As above, a difference in the main operators of the wffs reflects a difference in the wording of the English, but here there is no equivalence. (12) and (12′) imply (11) and (11′), but not vice versa—i.e. (11) and (11′) could be true while (12) and (12′) are false. Imagine that half the marbles are cracked and the other half are scratched. In that case (11) and (11′) are true, but each disjunct of (12) and (12′) is false (not all the marbles are cracked and not all the marbles are scratched), hence the whole claim is false.

Thus, we can say that the universal quantifier distributes across conjunction (equivalence holds for (9′) and (10′)), but not across disjunction (equivalence fails for (11′) and (12′)).

When we combine existential quantification with conjunction and disjunction we get the inverse result—the existential quantifier distributes across disjunction, but not across conjunction. Consider:

(13)  Some marble is both cracked and scratched
(13′)  $(\exists x)(Cx \wedge Sx)$

(14)  Some marble is cracked and some marble is scratched
(14′)  $(\exists x)Cx \wedge (\exists x)Sx$

---

**Box 5.6: Quantifiers, Conjunction, Disjunction**

$$(\forall x)\mathbb{F}x \wedge (\forall x)\mathbb{G}x \quad \Leftrightarrow \quad (\forall x)(\mathbb{F}x \wedge \mathbb{G}x)$$
$$(\forall x)\mathbb{F}x \vee (\forall x)\mathbb{G}x \quad \Rightarrow \quad (\forall x)(\mathbb{F}x \vee \mathbb{G}x)$$

$$(\exists x)\mathbb{F}x \wedge (\exists x)\mathbb{G}x \quad \Leftarrow \quad (\exists x)(\mathbb{F}x \wedge \mathbb{G}x)$$
$$(\exists x)\mathbb{F}x \vee (\exists x)\mathbb{G}x \quad \Leftrightarrow \quad (\exists x)(\mathbb{F}x \vee \mathbb{G}x)$$

---

Again, the difference in English word order is reflected in the translations, but while (13′) implies (14′), the converse does not hold. (14′) can be true while (13′) is false. Imagine there is just one marble that is cracked, and just one marble, a *different* one, that is scratched. In that case (14) and (14′) are true, but (13) and (13′) are false. Thus, while each is an appropriate translation of its English sentence, (13′) and (14′) are not equivalent.

We do get equivalence when existential quantification is combined in the parallel fashion with disjunction.

(15)  Some marble is either cracked or scratched
(15′)  $(\exists x)(Cx \vee Sx)$

(16)  Some marble is cracked or some marble is scratched
(16′)  $(\exists x)Cx \vee (\exists x)Sx$

(15) is equivalent to (16), and (15′) is equivalent to (16′). If at least one marble is either cracked or scratched (possibly both), then clearly it must be the case that either at least one marble is cracked, or at least one marble is scratched (possibly both). So (15) and (15′) imply (16) and (16′). Conversely, if it is the case that either at least one is cracked or at least one is scratched (possibly both), then it must be the case that at least one marble is cracked or scratched (possibly both). So (16) and (16′) imply (15) and (15′). The existential quantifier distributes across disjunction. Box 5.6 schematizes these relationships.

The following examples elaborate on the basic theme.

(17)  Either some marble is cracked or all marbles are not cracked
(17′)  $(\exists w)Cw \vee (\forall x)\neg Cx$

(18)  If Old Yeller is cracked, then there is a cracked marble
(18′)  $Co \rightarrow (\exists y)Cy$

(19)  If Big Red is red, then not all the marbles are blue
(19′)  $Rr \rightarrow \neg(\forall z)Bz$

(20)  If some marble is cracked, then all marbles are scratched
(20′)  $(\exists x)Cx \rightarrow (\forall x)Sx$

(20″) $(\exists x)Cx \rightarrow (\forall y)Sy$

(21) There is no marble bigger than The Green Giant
(21′) $\neg(\exists z)Bzg$
(21″) $(\forall z)\neg Bzg$

(22) If Big Red is bigger than Sky, then there is a marble smaller than Big Red
(22′) $Brs \rightarrow (\exists x)Sxr$

(23) Big red is bigger than some marble if and only if there is some marble smaller than Big Red
(23′) $(\exists x)Brx \leftrightarrow (\exists x)Sxr$
(23″) $(\exists y)Bry \leftrightarrow (\exists x)Sxr$

(24) No marble is smaller than Sky
(24′) $\neg(\exists w)Sws$
(24″) $(\forall w)\neg Sws$

(25) No marble is smaller than itself
(25′) $\neg(\exists x)Sxx$
(25″) $(\forall x)\neg Sxx$

(26) No marble is both scratched and cracked
(26′) $\neg(\exists x)(Sx \wedge Cx)$
(26″) $(\forall x)\neg(Sx \wedge Cx)$

(27) Some marble is scratched and some marble cracked
(27′) $(\exists x)Sx \wedge (\exists x)Cx$
(27″) $(\exists x)Sx \wedge (\exists y)Cy$

Most of these are pretty clear. Note how 'there is no marble...' and 'no marbles...' (in (21), (24)–(26)) may either be translated as negated existential claims or as universal denials. This is in accord with the equivalence in the upper right corner of the Simple Square of Opposition (Box 5.4). In each case the choice of variable is arbitrary, except that in the translations of (25) and (26), we must use the same variable in each position to ensure that we have a closed wff and that the predicates are applied to the same object (reflexively in the case of (25), and jointly in the case of (26)). Note that when a sentence contains two quantifiers whose scopes do not overlap we may choose the same variable, as in (20′), (23′), and (27′). In such cases we are not, however, *required* to use the same variable, as in (20″), (23″), and (27″). In each of these examples, the first and second translations are equivalent. Take, for example, (27′) and (27″). They say exactly the same thing, and, despite the reuse of 'x' in the one and the change to 'y' in the other, there is no commitment in either formula to the scratched marble being the same as or different from the cracked marble. (27) and its translations would be true in either case.

## 5.3.2 Exercises

Using the interpretation given, translate each of the following into *P*.

**UD**: Cats

G①: ① is grey                L①②: ① likes ②
O①: ① is orange             M①②: ① a better mouser than ②
T①: ① has a tail                b: Boots
M①: ① meows                f: Felix
W①: ① has whiskers          s: Sylvester
B①②: ① is bigger than ②      t: Tom

**1.** Felix is grey and has whiskers

**2.** If Felix is bigger than Sylvester, then Felix is a better mouser than Sylvester

**3.** All cats have whiskers

**4.** Some cats have whiskers

**5.** Some cats do not have whiskers

**6.** No cats have whiskers

**7.** Not all cats are orange

**8.** All cats are not orange

**9.** All cats either meow or have whiskers

**10.** Either all cats meow or all cats have whiskers

**11.** If some cats have whiskers, then some have tails

**12.** Some cats are orange but don't have tails

**13.** If Felix is a better mouser than Tom, then there is no better mouser than Felix

**14.** Boots doesn't like Sylvester if and only if no cat likes Sylvester

**15.** If Sylvester likes Boots, then all cats like Boots

**16.** If Tom does not like himself, then there is a cat that does not like itself

**17.** Some cats like Felix, but not all do

**18.** Every cat likes itself

**19.** If every cat likes itself, then there is a cat that likes Felix

**20.** There are no cats that don't have tails

## 5.3.3 **Categorical Claims**

In the previous subsection we were dealing with simple quantificational claims that spoke either of every member of the **UD** or of some one or more non-specified members of the **UD**. Often we will want to speak specifically about subsets of the **UD**—categories smaller than the whole **UD**—and assert things about only that category of objects. Such claims are called categorical claims. Symbolizing categorical claims will require a slightly more complex quantificational construction than we saw in the previous section. We continue using the interpretation begun on page 161.

Consider the following claims:

(1)  All the blue marbles are cracked
(2)  No blue marbles are cracked
(3)  Some blue marble is cracked
(4)  Some blue marbles are not cracked

We cannot symbolize (1) in any of the following ways.

$(1^X)$  $(\forall x)Bx \land (\forall x)Cx$          **INCORRECT**
$(1^{XX})$  $(\forall x)(Bx \land Cx)$          **INCORRECT**

These are **INCORRECT** translations of (1) because each asserts that every marble is both blue and cracked. This is consistent with (1), but not equivalent. What we want to do is assert of those marbles that are blue that they (the blue ones) are cracked. We can do this by employing a quantified conditional. Here, indicated by asterisks, are two useful paraphrases, intermediate between colloquial English and **P**:

(1\*)  For every thing in the **UD** [every marble in this case], if *it* is blue, then *it* (the same it) is cracked
(1\*\*)  All x are such that, if x is blue, then x is cracked

Thus paraphrased, we can easily see how to translate (1):

(1′)  $(\forall x)(Bx \rightarrow Cx)$

The situation is similar for (2). Since saying that no blue marbles are cracked is equivalent to the following,

(2\*)  All blue marbles are not cracked
(2\*\*)  All x are such that, if x is blue, then x is not cracked

we can translate (2) as:

(2′)  $(\forall x)(Bx \rightarrow \neg Cx)$

The other two cases are even simpler. They each assert that there is at least one object in the **UD** that meets a certain description. Hence, they could be paraphrased as follows:

(3\*)  There is at least one x such that, x is blue and x is cracked

(4\*)  There is at least one x such that, x is blue and x is not cracked

These make obvious the following translations:

(3′)  $(\exists x)(Bx \wedge Cx)$

(4′)  $(\exists x)(Bx \wedge \neg Cx)$

Indeed, (1)–(4) are in what is called standard categorical form. Each form has a standard letter used to label the form, a standard expression in English, and a standard translation into $P$.

**A:**  All $\mathbb{F}$ are $\mathbb{G}$
    $(\forall x)(\mathbb{F}x \to \mathbb{G}x)$

**E:**  No $\mathbb{F}$ are $\mathbb{G}$
    $(\forall x)(\mathbb{F}x \to \neg\mathbb{G}x)$

**I:**  Some $\mathbb{F}$ are $\mathbb{G}$
    $(\exists x)(\mathbb{F}x \wedge \mathbb{G}x)$

**O:**  Some $\mathbb{F}$ are not $\mathbb{G}$
    $(\exists x)(\mathbb{F}x \wedge \neg\mathbb{G}x)$

These are the four basic categorical forms, and many complex quantificational claims can be constructed from them or variations on them. Like the simple forms in the previous subsection, the relationships between **A**, **E**, **I**, and **O** forms can be efficiently represented on a square of opposition. Box 5.7 shows the Modern Categorical Square of Opposition. Again, in each corner is shown an English schema grouped with appropriate translations. As with the simple square, the diagonally opposite corners are contradictory. So a negated **A** is equivalent to an **O**, and vice versa; while a negated **I** is equivalent to an **E**, and vice versa. But in the Modern Categorical Square, none of the horizontal or vertical relations hold.[9] This will be discussed further on.

## The Existential and The Arrow

You might have noticed that the Modern Categorical Square does not contain any formula combining the existential quantifier with the arrow. For example, one might think we can translate (3), the **I** form from above, as follows:

(3)  Some blue marble is cracked

$(3^X)$  $(\exists x)(Bx \to Cx)$             **INCORRECT**

This is a syntactical parallel of the **A** form, '$(\forall x)(\mathbb{F}x \to \mathbb{G}x)$', but with the existential quantifier. Formula $(3^X)$ does not, however, correctly translate (3). It gets the truth conditions wrong. Any adequate translation of (3), such as (3′), should be true iff there exists at least one marble that is both blue and cracked. But $(3^X)$ will be true even if there are *no* marbles that are both blue and cracked. Thus, $(3^X)$ gets it all wrong! To see why, recall that since $(3^X)$ has an existential quantifier, it is true iff at least one object satisfies its immediate subcomponent. Next, remember that the material conditional, '$\to$', is true whenever

---

9. This, of course, is unlike the Simple Square of Opposition (Box 5.4), or the Aristotelian Square of Opposition (not shown).

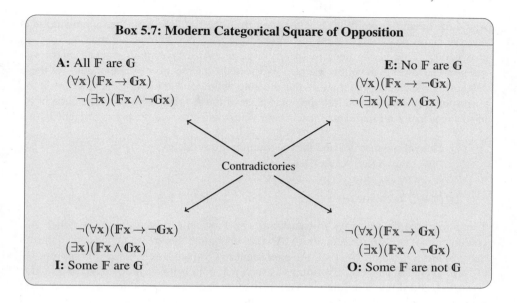

**Box 5.7: Modern Categorical Square of Opposition**

**A:** All $\mathbb{F}$ are $\mathbb{G}$
$(\forall x)(\mathbb{F}x \to \mathbb{G}x)$
$\neg(\exists x)(\mathbb{F}x \wedge \neg\mathbb{G}x)$

**E:** No $\mathbb{F}$ are $\mathbb{G}$
$(\forall x)(\mathbb{F}x \to \neg\mathbb{G}x)$
$\neg(\exists x)(\mathbb{F}x \wedge \mathbb{G}x)$

Contradictories

$\neg(\forall x)(\mathbb{F}x \to \neg\mathbb{G}x)$
$(\exists x)(\mathbb{F}x \wedge \mathbb{G}x)$
**I:** Some $\mathbb{F}$ are $\mathbb{G}$

$\neg(\forall x)(\mathbb{F}x \to \mathbb{G}x)$
$(\exists x)(\mathbb{F}x \wedge \neg\mathbb{G}x)$
**O:** Some $\mathbb{F}$ are not $\mathbb{G}$

the antecedent is false. Thus, the subcomponent of $(3^X)$, 'Bx $\to$ Cx' will be true of any marble that is not blue. Such a marble would make the antecedent 'Bx' false, thereby making the whole conditional true. Thus, the existence of a non-blue marble makes $(3^X)$ true, whether or not there are any marbles that are blue and cracked. If we read $(3^X)$, it says 'There is at least one marble which is such that *if* it is blue, *then* it is cracked'. Take a marble that is red, not blue. Given our understanding of the material conditional, that red marble is such that *if* it is blue, *then* it is cracked. So that red marble makes $(3^X)$ true. One way to think of this is that the combination of the existential and the arrow makes it *too easy* for the formula to be true under the wrong circumstances. As a result, we will rarely see a formula with this combination of the existential and arrow.

## Categorical Variations

There are various equivalent ways of stating categorical claims in English, but if you keep in mind the four basic forms almost everything else is a simple variation on these.

Consider the following, all of which are **A** claims.

(5) Cracked marbles are also scratched
(5′) $(\forall z)(Cz \to Sz)$

Here, the only complication is that the quantifier word 'all' has been omitted, but clearly the claim refers to all cracked marbles.

(6) The yellow cracked marbles are also scratched
(6′) $(\forall y)((Yy \wedge Cy) \to Sy)$

Again the quantifier word has been omitted from the English. A further complication is that the claim picks out, not all the cracked marbles, but the cracked marbles that are also yellow. That is, we need to specify a category narrower than cracked marbles, a category narrower than yellow marbles. We need the intersection, or conjunction, of these categories; the category of marbles that are both yellow and cracked. Hence, we insert the corresponding conjunction into the antecedent of the **A** form, as in (6′). The paraphrase of (6′) reads: 'for all marbles y, if y is both yellow and cracked, then y is scratched'.

(7)  The yellow marbles and the red marbles are scratched
(7′)  $(\forall w)(Yw \rightarrow Sw) \wedge (\forall w)(Rw \rightarrow Sw)$
(7″)  $(\forall w)((Yw \rightarrow Sw) \wedge (Rw \rightarrow Sw))$
(7‴)  $(\forall w)((Yw \vee Rw) \rightarrow Sw)$

There are three natural ways of translating (7). First, in (7′) we treat the sentence as a conjunction of two simple **A** claims—'All the yellow marbles are scratched and all the red marbles are scratched'. Next, (7″) is equivalent to (7′) (see Box 5.6), so it is appropriate as well. (7″) may be read: 'Everything is such that both if it is a yellow marble then it is scratched, and if it is a red marble then it is scratched'. There is, third, a more compact way of symbolizing it. Note that (7) is referring not just to yellow marbles, nor just to red marbles. But neither is it referring to the narrower category of those that are both yellow and red. Hence the strategy of (6′) does not apply.[10] Rather, (7) refers to the *broader* category that consists of the union, or disjunction, of the two categories. Hence, we insert the corresponding disjunction into the antecedent of the **A** form, as in (7‴). This may be read, 'for all marbles w, if w is either yellow or red, then w is scratched'. Being quantificationally equivalent, (7′), (7″), and (7‴) are all acceptable.

(8)  All the yellow marbles but the cracked ones are scratched
(8′)  $(\forall x)((Yx \wedge \neg Cx) \rightarrow Sx)$

Here, we have a narrowing of the antecedent category, as in (6). The 'but' functions as an 'and not' or 'except', taking the category of yellow marbles and narrowing it to those that are yellow and not cracked. Hence the conjunction and negation that appear in the antecedent of (8′), yielding 'for all marbles x, if x is both yellow and not cracked, then x is scratched'.

Similar considerations regarding the narrowing or broadening of the antecedent category apply to **E** claims. Consider the following, which parallel (6)–(8) above.

(9)  None of the yellow cracked marbles are scratched
(9′)  $(\forall y)((Yy \wedge Cy) \rightarrow \neg Sy)$

(10)  No yellow marbles and no red marbles are scratched
(10′)  $(\forall w)(Yw \rightarrow \neg Sw) \wedge (\forall w)(Rw \rightarrow \neg Sw)$

---

10. Indeed, if we tried to apply the strategy of (6′), we would get $(7^X)$ $(\forall y)((Yy \wedge Ry) \rightarrow Sy)$. The conjunction in the antecedent narrows the category so that this reads: 'Every marble that is both yellow and red is scratched'. But (7) is about the broader category. Hence the disjunction in the antecedent of (7‴).

(10″)  $(\forall w)((Yw \lor Rw) \to \neg Sw)$

(11)  None of the yellow marbles other than the cracked ones are scratched

(11′)  $(\forall x)((Yx \land \neg Cx) \to \neg Sx)$

Of course, the consequent category of **A** and **E** claims can be broadened or narrowed. But this is generally more straightforward than what we have just been considering, since the choice of conjunction versus disjunction is obvious from the English.

(12)  All yellow marbles are cracked and scratched

(12′)  $(\forall z)(Yz \to (Cz \land Sz))$

(13)  No yellow marbles are both cracked and scratched

(13′)  $(\forall z)(Yz \to \neg(Cz \land Sz))$

(14)  Each yellow marble is either cracked or scratched

(14′)  $(\forall z)(Yz \to (Cz \lor Sz))$

(15)  None of the yellow marbles is either cracked or scratched

(15′)  $(\forall z)(Yz \to \neg(Cz \lor Sz))$

(16)  The yellow marbles are cracked but not scratched

(16′)  $(\forall z)(Yz \to (Cz \land \neg Sz))$

(17)  No yellow marbles are cracked but not scratched

(17′)  $(\forall z)(Yz \to \neg(Cz \land \neg Sz))$

The use of 'only' and 'only if' in categorical claims adds a slight complication to **A** and **E** claims. If, however, you remember the pattern for 'only if' (see Box 2.6), you should have little problem here.

(18)  Marbles are bigger than Big Red only if they are not smaller than Big Red

(18′)  $(\forall x)(Bxr \to \neg Sxr)$

Clearly, the claim refers to all marbles. Equally clearly, it is asserting a conditional claim. We simply need to be sure we locate the antecedent and consequent correctly. Since what follows the 'only if' is the consequent, we get (18′). Since the consequent happens to be negated, this turns out to be an **E** claim.

(19)  Only blue marbles are bigger than Big Red

(19′)  $(\forall y)(Byr \to By)$

It may be a little less obvious that (19) is an **A** form. But think of it this way, if only blue marbles are bigger than Big Red, what do we know about all marbles that are bigger than Big Red? They are blue. Hence the paraphrase, 'for all marbles y, if y is bigger than Big Red, then y is blue', and (19′) is the appropriate translation. Another way to look at it is that 'only', like 'only if', introduces the consequent of a conditional, since this consequent is not negated, it results in an **A** claim. Note that neither (19) nor (19′) claims that *all*

---

### Box 5.8: Variations on A and O Forms

| | |
|---|---|
| All $\mathbb{F}$s (that are) $\mathbb{G}$ are $\mathbb{H}$ | $(\forall x)((\mathbb{F}x \wedge \mathbb{G}x) \rightarrow \mathbb{H}x)$ |
| All $\mathbb{F}$s except/but the $\mathbb{G}$s are $\mathbb{H}$ | $(\forall x)((\mathbb{F}x \wedge \neg \mathbb{G}x) \rightarrow \mathbb{H}x)$ |
| Only $\mathbb{F}$s are $\mathbb{G}$ | $(\forall x)(\mathbb{G}x \rightarrow \mathbb{F}x)$ |

$$\text{All } \mathbb{F}\text{s and all } \mathbb{G}\text{s are } \mathbb{H} \quad \begin{cases} (\forall x)(\mathbb{F}x \rightarrow \mathbb{H}x) \wedge (\forall x)(\mathbb{G}x \rightarrow \mathbb{H}x) \\ (\forall x)[(\mathbb{F}x \rightarrow \mathbb{H}x) \wedge (\mathbb{G}x \rightarrow \mathbb{H}x)] \\ (\forall x)((\mathbb{F}x \vee \mathbb{G}x) \rightarrow \mathbb{H}x) \end{cases}$$

---

blue marbles are bigger than Big Red—just that all marbles bigger than Big Red are blue. There might be some blue marble at are not bigger than Big Red. Box 5.8 schematizes the main points of this discussion.

The existential **I** and **O** claims are much more straightforward. Generally your symbolizations will involve an existentially quantified conjunction, where any complications or complexity carry over from the English into one of the conjuncts as you would expect.

(20)  Some blue marbles are scratched
(20′)  $(\exists x)(Bx \wedge Sx)$

This is just your garden variety **I** claim. We can complicate it as follows (some of the complications result in **O** forms).

(21)  At least one cracked blue marble is scratched
(21′)  $(\exists x)((Cx \wedge Bx) \wedge Sx)$

(22)  At least one cracked blue marble is not scratched
(22′)  $(\exists x)((Cx \wedge Bx) \wedge \neg Sx)$

(23)  At least one cracked blue marble is scratched and smaller than The Green Giant
(23′)  $(\exists x)((Cx \wedge Bx) \wedge (Sx \wedge Sxg))$

(24)  At least one blue marble is scratched or cracked
(24′)  $(\exists x)(Bx \wedge (Sx \vee Cx))$

(25)  At least one blue marble is neither scratched nor cracked
(25′)  $(\exists x)(Bx \wedge \neg(Sx \vee Cx))$

(26)  At least one blue marble is not both scratched and cracked
(26′)  $(\exists x)(Bx \wedge \neg(Sx \wedge Cx))$

(27)  At least one blue marble is scratched but not cracked
(27′)  $(\exists x)(Bx \wedge (Sx \wedge \neg Cx))$

We can, of course, form truth-functional compounds of individual categorical claims. (We saw a number of such cases above). Here is one more example.

(28)  Some, but not all, blue marbles are bigger than Big Red
(28′)  $(\exists x)(Bx \wedge Bxr) \wedge \neg(\forall x)(Bx \rightarrow Bxr)$
(28″)  $(\exists x)(Bx \wedge Bxr) \wedge (\exists x)(Bx \wedge \neg Bxr)$

## 'Any'

So far I have mostly avoided the quantity term 'any', because it can be used in English as either a universal or an existential quantifier, depending on the context. Consider the following examples:

(29)  Any marble that is yellow is scratched
(30)  Every marble that is yellow is scratched

It is probably obvious to you that these both say the same thing. 'Any' in (29) is functioning as a universal quantifier, just as 'Every' is in (30). Thus, both can be translated as the same **A** claim:

(29′, 30′)  $(\forall x)(Yx \rightarrow Sx)$

But 'any' does not always function as a universal. Consider the contrast between the following two claims:

(31)  If every marble is yellow, then Old Yeller is
(32)  If any marble is yellow, then Old Yeller is

(31) should strike you as obvious. If *every* marble is yellow, then, of course, Old Yeller is—it is one of the marbles and all of them are yellow. It practically goes without saying. Indeed, (31) is a quantificationally true claim—because of its quantificational structure, it cannot be false (see Section 6.2). Translated into **P**, it is a conditional with a universally quantified antecedent:

(31′)  $(\forall x)Yx \rightarrow Yo$

(32), however, is different. Unlike (31), it actually tells you some useful information. It does not tell us whether there are any yellow marbles, but it does tell us that if *there are* any yellow marbles, then Old Yeller is one of them. Also unlike (31), (32) can be false—suppose Sky is yellow, but Old Yeller is not, this yields a true antecedent and a false consequent. Thus, (32) cannot be saying the same thing as (31). The phrase 'there are' (three sentences back) tips us off—'any' is functioning as an existential quantifier in (32). We can make this perfectly clear with a paraphrase, and the corresponding translation:

(32*)  If there is at least one marble that is yellow, then Old Yeller is yellow
(32′)  $(\exists x)Yx \rightarrow Yo$

---

**Box 5.9: The Vagaries of 'Any'**

---

$$\left.\begin{array}{l} \text{Any } \mathbb{F} \text{ is } \mathbb{G} \\ \text{If any thing is } \mathbb{F} \text{ then } it \text{ is } \mathbb{G} \end{array}\right\} \quad (\forall x)(\mathbb{F}x \to \mathbb{G}x)$$

with pronominal cross-reference

If any thing is $\mathbb{F}$ then $\mathbb{P}$    $(\exists x)\mathbb{F}x \to \mathbb{P}$

no pronominal cross-reference

$$\text{Not any thing is } \mathbb{F} \quad \begin{cases} \neg(\exists x)\mathbb{F}x \\ (\forall x)\neg\mathbb{F}x \end{cases}$$

$$\text{Not any } \mathbb{F} \text{ is } \mathbb{G} \quad \begin{cases} \neg(\exists x)(\mathbb{F}x \wedge \mathbb{G}x) \\ (\forall x)(\mathbb{F}x \to \neg\mathbb{G}x) \end{cases}$$

---

Now we have a conditional with an existentially quantified antecedent. It is the difference in the antecedents of (31') and (32') that reflects the difference between (31) and (32).

From the discussion so far you might guess that 'any' plays the role of an existential quantifier when it appears in the antecedent of a conditional (as the 'if any' in (32)). But, as the following shows, this is not quite right.

(33)  If any marble is yellow, then it is scratched

This is actually an **A** claim, equivalent to (29) and (30). Hence it receives an identical translation.

(33')  $(\forall x)(Yx \to Sx)$

Though (33') provides a counterexample to our first guess, it also suggests a modification of that guess. Note that (33), unlike (32), has a pronoun in its consequent that refers back to the object(s) specified in the antecedent. It is the cross-reference of the pronoun from the consequent to the antecedent that indicates the need for a universal quantifier.

Finally, 'any' should be treated as an existential quantifier when it is negated. That is, in contexts involving 'not any' we should use a negated existential. For instance:

(34)  Not any marbles are scratched
(34')  $\neg(\exists x)Sx$

(35)  Not any of the yellow marbles are scratched
(35')  $\neg(\exists x)(Yx \wedge Sx)$

Of course, there are equivalent wffs of *P* using universal quantifiers (see Boxes 5.4 and 5.7), but it is simplest to treat 'not any' as 'it's not the case that there exists'. Box 5.9 sums up these points.

## Notes on Interpretations

It is worth making a couple of points about interpretations before moving on to exercises. First, it was mentioned above that the horizontal and vertical relationships of the Simple Square of Opposition (Box 5.4), do not hold for the Modern Categorical Square (Box 5.7). While it seems intuitive that if an **A**-form claim (all $\mathbb{F}$ are $\mathbb{G}$) is true, then the corresponding **I**-form claim (some $\mathbb{F}$ are $\mathbb{G}$) must be true, this is not the case on our analysis of categorical claims. Similar remarks apply to **E** and **O** claims. Moreover, it may seem strange that (contrary to what you might expect from the Simple Square) corresponding **A** and **E** forms could both be true, or corresponding **I** and **O** claims could both be false. But this is possible on the Modern Categorical Square as a result of our allowing the use of predicate letters with empty extensions—i.e., predicate letters that are not true of any member of the **UD**.

Given the same interpretation we've been working with, let's suppose that *there are no blue marbles* in Fred's collection. How are we to interpret the following wffs of **P**?

(36) $(\forall x)(Bx \rightarrow Cx)$
(37) $(\forall x)(Bx \rightarrow \neg Cx)$
(38) $(\exists x)(Bx \wedge Cx)$
(39) $(\exists x)(Bx \wedge \neg Cx)$

(36) says that all blue marbles are cracked. You are likely tempted to interpret this as false, since (as we have stipulated) there are no blue marbles. But this is incorrect. Recall that we interpret a universally quantified claim as true iff the open wff that results from removing the quantifier ('$Bx \rightarrow Cx$' in this case) is satisfied by (true of) every member of the **UD**. Recall also that the material conditional is true whenever the antecedent is false. Note that '$Bx$' fails to be satisfied by (is false of) every member of the **UD**, hence '$Bx \rightarrow Cx$' *is* satisfied by (true of) every member of the **UD**. As a result (36) is true. Similar reasoning applies to (37). They are both true, so they are not contraries. The reason is that there are no blue marbles at all—so they are said to be vacuously true. Note that (38) and (39) are both false. There must be at least one object of the **UD** that satisfies '$Bx \wedge Cx$' or '$Bx \wedge \neg Cx$' for (38) or (39) (respectively) to be true. But there are no members of the **UD** that satisfy '$Bx$', so no members that satisfy either of the two relevant conjunctions. Hence both (38) and (39) are false. So they are not subcontraries. Nor are they implied by (36) or (37) (respectively), for those are both true on this interpretation but (38) and (39) are false.[11]

The second point worth making is that, exactly how we choose our **UD** will affect how the same English sentences are translated. For instance, suppose we modify our working interpretation to include not just Fred's marbles, but also his Superball collection, Fred himself, his friend Susie, and her marble and Superball collections. We also add a few new predicates and constants:

---

11. In classical Aristotelian logic empty predicates are not allowed, hence the above situation never arises, and the horizontal (contrary and subcontrary) and vertical (implication downward) relationships do hold in9 Aristotelian logic. .

**UD**: Fred, Susie, Fred's marbles and Superballs, Susie's marbles and Superballs

M①: ① is a marble                      B₂①②: ① belongs to ②
B①: ① is blue                          S①②: ① is smaller than ②
G①: ① is green
R①: ① is red                           f:  Fred
Y①: ① is yellow                        g:  The Green Giant
C①: ① is cracked                       o:  Old Yeller
S₁①: ① is a Superball                  r:  Big Red
S₂①: ① is scratched                    s₁: Susie
B₁①②: ① is bigger than ②               s₂: Sky

All of the translations we have so far performed become more complex. Earlier we treated claims about marbles as unrestricted quantifications, because there were only marbles in the **UD**. Now we have marbles, Superballs, and a couple of people. Thus, to translate

(40)  All the marbles are blue

we cannot write the simple universal:

(40$^X$)  $(\forall x)Bx$                          **INCORRECT**

This symbolization says that everything in the **UD** is blue, but on the new interpretation that includes not only the marbles, but the Superballs, as well as Fred and Susie. Hence (40$^X$) is incorrect. What we need to do is translate (40) as an **A** claim, since now the **UD** contains more than just marbles. Thus:

(40′)  $(\forall x)(Mx \rightarrow Bx)$

All the translations of this and the previous section would have to become correspondingly more complex under this new interpretation. We would have to take care to make clear that we were singling out the marbles. But this interpretation is correspondingly more powerful. For we can talk about more things, and subdivide them in more and different ways. Here is just a single example:

(41)  All the cracked blue Superballs bigger than Big Red are scratched or belong to Susie
(41′)  $(\forall z)([(S_1z \wedge Cz) \wedge (Bz \wedge B_1zr)] \rightarrow [S_2z \vee B_2zs_1])$

## 5.3.4 **Exercises**

Using the interpretation given, translate each of the following into *P*.

**UD**: People in Kate's office

A①: ① is an accountant
C①: ① is a cashier
D①: ① is diligent
L①: ① is lazy
U①: ① is upbeat

L①②: ① likes ②
W①②: ① works for ②

f: Fred
k: Kate

1. Everyone is an accountant

2. Everyone likes Kate

3. All the accountants are diligent

4. No accountants are diligent

5. Some accountants are diligent

6. Some accountants are not diligent

7. Not every accountant is diligent

8. Cashiers are not diligent

9. Some cashiers are not diligent

10. Everyone who works for Kate likes her

11. Everyone who works for Kate or does not like Fred likes Kate

12. Not everyone who likes Kate works for her

13. No one who works for Fred likes him

14. Some accountant likes both Fred and Kate

15. Some diligent accountants like Fred but not Kate

16. Some accountants and cashiers work for Fred

17. Someone who is both an accountant and a cashier works for Fred

18. Kate does not like everyone who works for her

19. Kate does not like anyone who works for her

20. Kate likes only those who are both upbeat and diligent

21. Only those who work for Kate like her

22. All the accountants and cashiers like Kate

**23.** All the diligent accountants work for Kate, but none of the lazy ones

**24.** Any accountant who is upbeat but not diligent either works for Kate or is liked by Fred

**25.** Accountants who are not diligent do not work for Kate

**26.** Accountants and cashiers who are not diligent do not work for Kate

**27.** Accountants and cashiers who are neither diligent nor upbeat do not work for Kate

**28.** Accountants and cashiers who are neither diligent nor upbeat either do not work for Kate or are not liked by Kate

**29.** If Fred is a lazy cashier and he likes himself, then there is a lazy cashier whom Fred likes

**30.** If Fred works for Kate, then anyone who works for Fred works for Kate

**31.** Everyone who works for Kate likes her, and everyone who likes Kate works for her

**32.** All and only those who work for Kate like her

**33.** Everyone is such that they like Kate if and only if they work for her

## 5.4 **Complex Symbolizations**

So far we have seen only rather simple use of the quantifiers—cases that involve only a single quantifier or truth-functional combinations thereof. But quantifiers can be used in combination to produce wffs of greater expressive power.

### 5.4.1 **Basics of Overlapping Quantifiers**

For most of this section we will use the following interpretation:

**UD**: People, living and dead

L①②: ① loves ②  
M①②: ① is the mother of ②

f: Fred  
k: Kate

Consider the following progression of claims and their translations:

(1) Fred loves Kate
(1′) Lfk

(2) Fred loves someone
(2′) (∃y)Lfy

(3) Someone loves someone

(3') (∃x)(∃y)Lxy

(1') and (2') should be no problem for you. (3') is produced from (2') by existentially quantifying in to the first predicate position previously occupied by 'f'. In a sense, we remove reference to Fred, and, via the existential x-quantifier, refer instead to at least one person. We get 'there is at least one person who loves at least one person'. Note that, because of how we interpret the existential quantifier, the values of x and y (the lover and the beloved) need not be distinct, they may be one and the same person. For instance, suppose Fred is the only one who loves anyone and (contrary to (1)) the only person he loves is himself. (3) and (3') are true under this supposition.

A similar progression illustrates the use of two universal quantifiers:

(4) Fred loves Kate
(4') Lfk

(5) Fred loves everyone
(5') (∀y)Lfy

(6) Everyone loves everyone
(6') (∀x)(∀y)Lxy

With (6') we wind up with, 'everyone is such that he or she loves everyone'. This, of course, implies that everyone loves himself or herself. As above, the values of x and y need not be distinct. So, according to (6'), not only does each person love everyone else, each person loves himself or herself.

Since, in (3') and (6'), we are not mixing existential and universal quantifiers, neither the order of the quantifiers nor the order of the variables in the predicate make any difference to the interpretation. In each of the following two columns the wffs are equivalent:

| | |
|---|---|
| (3') (∃x)(∃y)Lxy | (6') (∀x)(∀y)Lxy |
| (3'') (∃y)(∃x)Lxy | (6'') (∀y)(∀x)Lxy |
| (3''') (∃x)(∃y)Lyx | (6''') (∀x)(∀y)Lyx |
| (3'''') (∃y)(∃x)Lyx | (6'''') (∀y)(∀x)Lyx |

There are syntactical differences here. In each column, each of the four wffs has a different combination of quantifier order and variable order in the predicate. If we were strict about reading these wffs, we would come up with different readings for each: 'there is some x such that there is some y such that x loves y', 'there is some y such that there is some x such that...' and so on. But there are some shared patterns. The first and last member of each column has the first quantifier linked to the first predicate position and the second quantifier linked to the second position (because the variable in the first predicate position matches that of the first quantifier, and similarly for the second). The second and third of each column has the first quantifier linked to the second predicate position and the second quantifier linked to the first position. Thus the first and the fourth, as well as the second and third in each column are identical, except that the 'x's are switched for 'y's.

It is important to see that when we are not mixing universal and existential quantifiers in the same wff, these syntactical differences are irrelevant to the interpretation. The existentially quantified wffs all say 'someone loves someone' while the universally quantified wffs all say 'everyone loves everyone'. The existential wffs each basically say 'there is some pair of people (not necessarily distinct) such that one of them loves the other'. The universal wffs say 'for every pair of people (not necessarily distinct) one of them loves the other'.

For example, suppose Fred loves Kate, but nobody else loves anybody (in particular, Kate does not love Fred). In this situation, each of (3′)–(3‴‴) will turn out true. For (3′) and (3‴‴) we just need to find at least one pair the first member of which loves the second. The pair consisting of Fred and Kate (in that order) suffices. For (3″) and (3‴) we just need to find a pair the second member of which loves the first. Fred and Kate will not do, for we have stipulated that Kate does not love Fred, but Kate and Fred (in that order) suffices. Though we have not proved it (we have merely illustrated a single case), (3′)–(3‴) will be true in all and only the same situations, even despite the syntactical differences. Similar remarks apply to (6′)–(6‴‴).

But what happens when we mix universal and existential quantifiers in the same wff? In that case the order of the quantifiers and the variables in the predicates makes all the difference in the world. Consider the following pair:

(7′)  $(\forall x)(\exists y)Myx$

(7*)  For every x there is at least one y such that y is the mother of x

(7)  Everyone has a mother

(8′)  $(\exists y)(\forall x)Myx$

(8*)  There is at least one y such that for every x, y is the mother of x

(8)  Someone is the mother of everyone

Ignoring issues about the first people,[12] (7) and (7′) are true. Each of us has (or had) a mother. Your siblings, if you have some, have the same mother as you, but that is neither required nor barred by (7) or (7′). All that is required is that for each object in the **UD**, each person in this case, there is at least one object from the **UD** that is its mother. (8) and (8′) say something quite different. They claim that there is at least one person who (single-handedly!?) is the mother of us all, including herself. Clearly this is false.

The change in quantifier order is crucial here, and it is a question of the kinds of pairs being posited by the wff. (7′) has each of us related to our own mother, whereas (8′) has one cosmic mother related to all of us. Note that if (8′) were true, (7′) would be as well—if one person is the mother of all, then everyone has a mother (we would all have the same one in this case). Thus, (8′) implies (7′), but not vice versa.

This is not the only difference we need to keep in mind. Suppose we take the two previous wffs, and change the order of the variables in the predicate:

(9′)  $(\forall x)(\exists y)Mxy$

---

12. Actually (7′) requires either an infinite regress or some kind of loop (e.g., that some person be the mother of herself or one of her ancestors). Closer to the truth is that every person *alive today* has or had a mother (not necessarily alive today).

---

**Box 5.10: Reading Overlapping Quantifiers**

$(\exists x)(\exists y) \ldots$ $\begin{cases} \text{there is an x such that there is a y such that} \ldots \\ \text{there is at least one pair x,y such that} \ldots \end{cases}$

$(\forall x)(\forall y) \ldots$ $\begin{cases} \text{for all x and for all y} \ldots \\ \text{for all pairs x,y} \ldots \end{cases}$

$(\forall x)(\exists y) \ldots$ $\quad$ for each x, there is at least one y such that $\ldots$

$(\exists y)(\forall x) \ldots$ $\quad$ there is at least one y such that for each x $\ldots$

---

(9*)  For every x there is at least one y such that x is the mother of y

(9)  Everyone is a mother (of at least one child)

(10′)  $(\exists y)(\forall x)Mxy$

(10*)  There is at least one y such that for every x, x is the mother of y

(10)  There is someone for whom every person is his/her mother

First, consider the contrasts with (7′) and (8′). Rather than everyone having a mother (as in (7′)), (9′) claims that everyone is a mother. Though the order of the quantifiers is the same in (7′) and (9′), the order of the variables in the predicate has changed, and this accounts for the difference in the interpretations of the wffs. Similar remarks apply to (8′) and (10′).

Next, contrast (9′) with (10′). Again it is a question of how quantifier order affects what is being claimed. (9′) claims that everyone is a mother—everyone has his/her own child, possibly, but not necessarily, distinct from others' children. (10′), on the other hand, claims that there is one child (possibly more) of whom we are all mothers. Both of these are false, of course. But note that (10′) implies (9′), but not vice versa.

Boxes 5.10 and 5.11 sum up some of the points made in this discussion.

## Negation and Overlapping

Adding negation into the mix only slightly complicates matters. If you are familiar with the Simple Square of Opposition (Box 5.4, p. 163), you should not have too much trouble with the following examples. Given the equivalences in the corners of that square, there will always be more than one brief way of translating the following. I show only one or two for each. Also, given the flexibility of English quantifiers and negations, there will often be equivalent, but very different sounding, English phrasings.

(11)  Not everybody loves somebody

(11′)  $\neg(\forall x)(\exists y)Lxy$

(11″)  $(\exists x)\neg(\exists y)Lxy$

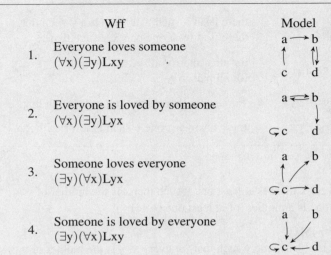

**Box 5.11: Quantifiers, Order, and Love**

| | Wff | Model |
|---|---|---|
| 1. | Everyone loves someone<br>$(\forall x)(\exists y)Lxy$ | |
| 2. | Everyone is loved by someone<br>$(\forall x)(\exists y)Lyx$ | |
| 3. | Someone loves everyone<br>$(\exists y)(\forall x)Lyx$ | |
| 4. | Someone is loved by everyone<br>$(\exists y)(\forall x)Lxy$ | |

Each model satisfies the wff to its left. In addition, since 4. ⊨ 1., model 4. also satisfies wff 1. Of course, 1. ⊭ 4., and we can see that model 1. does not satisfy wff 4. Similarly, since 3. ⊨ 2., model 3. also satisfies wff 2. But 2. ⊭ 3., and we see that model 2. does not satisfy wff 3. Indeed, with the exception of the entailments just noted, no model satisfies any of the other wffs. See if you can explain this. Note: These are not the only models which satisfy their respective wffs. They were chosen to illustrate the above points.

(12)  Everyone has someone they don't love
(12′)  $(\forall x)(\exists y)\neg Lxy$
(12″)  $\neg(\exists x)(\forall y)Lxy$

(13)  Nobody loves nobody
(13′)  $\neg(\exists x)\neg(\exists y)Lxy$
(13″)  $(\forall x)(\exists y)Lxy$

(14)  Not everyone is loved by someone
(14′)  $\neg(\forall x)(\exists y)Lyx$
(14″)  $(\exists x)\neg(\exists y)Lyx$

(15)  Nobody is unloved by everybody
(15′)  $\neg(\exists x)(\forall y)\neg Lyx$
(15″)  $(\forall x)(\exists y)Lyx$

If you find these tricky, a good approach is to build them up step by step from a wff with fewer, or no, quantifiers, as we did at the start of this section. Take (11), for example:

(11) Not everybody loves somebody

Suppose we just start with 'Fred loves Kate':

Lfk

We can easily change this to 'Fred loves somebody':

(∃y)Lfy

Now we just need to take 'Fred' out and replace it with 'not everybody'. The result is one of the correct translations of (11):

(11′) ¬(∀x)(∃y)Lxy

Here is an example involving (15)

(15) Nobody is unloved by everybody

Kate is unloved by Fred
¬Lfk

Kate is unloved by everybody
(∀y)¬Lyk

(15) Nobody is unloved by everybody
(15′) ¬(∃x)(∀y)¬Lyx

## Categorical Claims and Overlapping

Overlapping quantifiers can come into play when dealing with complex categorical claims. In this situation, the strategy will also be to try to construct the complex claims by successive modifications of less complex claims, though this can sometimes mislead. Take the following examples:

(16) Every mother loves all her children
(17) Every mother is loved by all her children
(18) At least one mother loves some of her children
(19) At least one mother loves none of her children

These could be treated as simple categorical claims if we were willing to define predicate letters for such predicates as 'is a mother', 'loves all her children', 'is loved by all her children', 'loves some of her children'. But this strategy is highly undesirable, because it leaves too much logical structure hidden within the defined predicates (namely the quantity terms). Moreover, such a strategy would proliferate predicate letters while simultaneously obscuring the common structure of our sentences. A better strategy is to work with the interpretation we have. To deal with (16), we can start with a straightforward **A** claim:

Kate loves all her children

$(\forall y)(Mky \rightarrow Lky)$

For all y, if Kate is the mother of y, then Kate loves y

We wind up with a translation of the candidate sentence, even though we do not have a predicate letter for 'is a child of'. For by saying that ① is the mother of ② we are essentially saying that ② is the child of ①. To translate (16) we need to generalize by removing 'Kate':

(16)  Every mother loves all her children

(16*)  For all x and for all y, if x is the mother of y, then x loves y

(16′)  $(\forall x)(\forall y)(Mxy \rightarrow Lxy)$

To translate (17) we simply reverse the order of the variables in the consequent of (16′):

(17′)  $(\forall x)(\forall y)(Mxy \rightarrow Lyx)$

The following progressions lead us to (18) and (19):

Kate loves some of her children

$(\exists y)(Mky \wedge Lky)$

(18)  At least one mother loves some of her children

(18′)  $(\exists x)(\exists y)(Mxy \wedge Lxy)$

Kate loves none of her children

$(\forall y)(Mky \rightarrow \neg Lky)$

(19)  At least one mother loves none of her children

(19′)  $(\exists x)(\forall y)(Mxy \rightarrow \neg Lxy)$

This strategy can lead astray when we start with a claim of one form (an **E** claim, for instance), but our result should be of another form (an **A** claim, perhaps).

(20)  Every mother loves some of her children

Kate loves some of her children

$(\exists y)(Mky \wedge Lky)$

Now we might be tempted to replace the constant 'k' with an 'x' and append a universal x-quantifier, yielding:

(20$^X$)  $(\forall x)(\exists y)(Mxy \wedge Lxy)$          **INCORRECT**

(20$^X$), however, is not a correct translation of (20). (20$^X$) is a translation of 'Everyone is a mother of a child he or she loves', which is not the same as (20). The problem is that (20) is basically a complex **A** claim, but we started with the **E** claim, 'Kate loves some of her children'. We simply need to change the conjunction in (20$^X$) into a material conditional, yielding:

$(20')$ $(\forall x)(\exists y)(Mxy \rightarrow Lxy)$

The potential variations are endless. You will need to apply your knowledge of the basics in order to generate a candidate translation, which you then compare against the original to see if you have been successful.

## 5.4.2 Exercises

Using the interpretation given, translate each of the following into *P*.

**UD**: People in Kate's office

A①: ① is an accountant
C①: ① is a cashier
D①: ① is diligent
L①: ① is lazy
U①: ① is upbeat

L①②: ① likes ②
W①②: ① works for ②

f: Fred
k: Kate

**1.** Everyone likes everyone

**2.** Someone likes someone

**3.** Everyone likes someone

**4.** Everyone is liked by someone

**5.** Someone likes everyone

**6.** Someone is liked by everyone

**7.** Someone is liked by no one

**8.** It's not the case that there's someone who likes no one

**9.** Someone likes nobody

**10.** Nobody likes everybody

**11.** There is someone nobody likes

**12.** Nobody is liked by everybody

**13.** Somebody is not liked by somebody

## 5.4.3   Identity, Numerical Quantification, and Definite Descriptions

While we have been dealing with some rather complex quantificational claims, the quantity terms we are able to translate remain nonspecific (except, of course, for 'none'). Neither 'at least one' nor 'all' tells us about the number of objects involved. What are we to do if we want to be more specific about the number of objects involved? For example, how can we translate the following? (We'll use the interpretation from the previous exercises (page 187).)

(1)  There are at least two cashiers

Clearly the following is incorrect,

$(1^X)$  $(\exists x)Cx$                                   **INCORRECT**

for $(1^X)$ will be true if there is only one cashier, whereas (1) will be false. Perhaps more tempting are these,

$(1^{XX})$  $(\exists x)Cx \land (\exists y)Cy$                 **INCORRECT**
$(1^{XXX})$  $(\exists x)(\exists y)(Cx \land Cy)$              **INCORRECT**

Each of these uses two existential quantifiers, so you might think that you get 'at least two' by combining them in one or both of these ways. But this is a misinterpretation of how quantifiers and variables work. Remember that, even though we use different variables 'x' and 'y' in $(1^{XX})$ and $(1^{XXX})$, the objects picked out by those variables need not be distinct. As a result, $(1^{XX})$ just says twice over that there is at least one cashier—changing the variable in one conjunct doesn't say anything new. So $(1^{XX})$ is equivalent to $(1^X)$, and will be true if there is just one cashier. Using overlapping quantifiers, as in $(1^{XXX})$, doesn't help either. $(1^{XXX})$ is also true if there is only one cashier. It says that there is some x and there is some (not necessarily distinct) y, such that x is a cashier, and y is a cashier. Because of the '(not necessarily distinct)' part (which we usually leave implicit), $(1^{XXX})$ is also equivalent to $(1^X)$. Hence, it does not translate (1).

Clearly, then, what we need to do is make clear that we mean to pick out an x and a y that are different. One way to do this would be to introduce a new predicate for '① is different from ②'. Then we could say 'There is some x and there is some y, they are different, and each is a cashier'. Indeed, it will be more convenient to introduce a predicate for sameness, and use its denial to assert difference.

Before doing so, let me introduce a technical term:

**Identity:**
>  By *identity* we mean numerical identity—that what appear to be distinct objects are actually *one and the same thing*. Object x is *identical* to object y iff x is y.

So by 'identical' we do not mean exactly similar, as two cars from the same production line are similar, or as identical twins are similar. If there are *two* cars, *two* twins, then they are not identical in our sense.[13] Identity applies only to one object—every object is identical to itself and to no others. Fred is identical to Fred and to no one else. While Fred's self-identity may seem obvious, and stating it may seem pointless, assertions of identity for objects that have more than one name, or in combination with definite descriptions, are quite useful and informative. Consider:

> Edda Kathleen van Heemstra Hepburn-Ruston is (*identical to*) Audrey Hepburn
> Barack Obama is (*identical to*) the 44th President of the USA

To accommodate identity claims in **P**, we now modify our language by introducing the *identity sign*,

$$=$$

as a new two-place predicate letter. This predicate letter will be somewhat different from the others.

First, unlike other two-place predicate letters, the identity sign is written between the terms that fill its places. For example, we write,

(i) $a = b$
(ii) $a = x$
(iii) $x_1 = z$

Second, rather than using a hook to negate the identity sign, we put a slash though it,

(iv) $a \neq b$
(v) $a \neq x$
(vi) $x_1 \neq z$

Third, unlike any other predicate letter, the identity sign will receive a fixed interpretation:

**Interpretation of '=':**

①$=$②:  ① is identical to ②

Thus, the identity sign is always available for use with its fixed interpretation and need not be mentioned in specific interpretations. For the sake of absolute clarity, here are the appropriate interpretations of wffs (i)–(vi):

(i) a is identical to b
(ii) a is identical to x
(iii) $x_1$ is identical to z

---

13. What *is* identical (or nearly so) in the case of the cars is the design and method of production. In the case of identical twins it is the *sequence* of genes (not the actual genetic material itself, but its chemical configuration) that is identical, and which explains their near exact similarity.

  (iv)  a is not identical to b

  (v)  a is not identical to x

  (vi)  $x_1$ is not identical to z

To return to the sentence under consideration:

  (1)  There are at least two cashiers

We said that in translating (1) we need to make clear that x and y are to pick out distinct objects. This is now easily done with '='.

  (1′)  $(\exists x)(\exists y)(x \neq y \wedge (Cx \wedge Cy))$

Here are a strict and more colloquial reading of (1′):

  (1*)  there is at least one x and at least one y, such that x and y are not identical and x is a cashier and y is a cashier

  (1**)  there are at least two objects, they are distinct objects, and each is a cashier

We can, of course, introduce variations on this theme:

  (2)  There are at least two cashiers working for Fred

Here the only complication occurs in the ensuing description. We will start out essentially as we did in (1′), and then assert of each x and y both that it is a cashier and that it works for Fred.

  (2′)  $(\exists x)(\exists y)(x \neq y \wedge [(Cx \wedge Wxf) \wedge (Cy \wedge Wyf)])$

  (2*)  there is at least one x and at least one y, such that x and y are not identical and both x is a cashier and x works for Fred and y is a cashier and y works for Fred

  (2**)  there are at least two objects, they are distinct objects, and each is a cashier who works for Fred

Finally, consider the following:

  (3)  There are at least three cashiers

All we need do is use three quantifiers, deny identity between each relevant pair, and apply the appropriate predicates to each variable; thus,

  (3′)  $(\exists x)(\exists y)(\exists z)([(x \neq y \wedge x \neq z) \wedge y \neq z] \wedge [(Cx \wedge Cy) \wedge Cz])$

  (3*)  there is at least one x and at least one y and at least one z, such that x and y are not identical and x and z are not identical and y and z are not identical and x is a cashier and y is a cashier and z is a cashier

  (3**)  there are at least three objects, they are distinct objects, and each is a cashier

As you may imagine, these can get rather long and cumbersome. You should be able to extrapolate how to translate 'at least four', 'at least five', and so on.

To this point, there is still a lack of specificity in our existential quantification. We had 'at least one', now we have 'at least $n$' for any positive integer $n$. However, we may well like to be numerically exact in stating how many things meet a certain description. This is easily achieved with the identity symbol. Take the following:

(4) There is (exactly) one cashier

Clearly none of the wffs we've seen so far will be adequate to translate (4). We can easily say 'there is at least one...', but we need to say more than just that. In fact we need say that there is at least one cashier and no more than one. Consider the following wff with a strict and brief interpretation:

(4′) $(\exists x)(Cx \land (\forall y)(Cy \to x = y))$

(4*) there is at least one x, such that x is a cashier and for any y, if y is a cashier then y is identical to (the same object as) x

(4**) there is at least one cashier and at most one cashier

(4′) is the appropriate translation of (4). The construction consists of a standard existential portion (up to the '$\land$') followed by a uniqueness portion (following the '$\land$'). We are basically asserting that there is at least one cashier (existence), and anything in the **UD** that is a cashier just is that self-same cashier (uniqueness). In other words, there is at least one, and no more than one.

(4′) will be true if and only if there is exactly one cashier in the **UD**. If there are none, then there will be no object to satisfy 'Cx'. So the existence portion of the wff will be false, making the whole claim false. If there is more than one, then there will be a pair of objects (two distinct cashiers) that satisfies the antecedent of '$(\forall y)(Cy \to x = y)$' but that fails to satisfy the consequent (since they are distinct, not identical). That makes the universally quantified conditional false. That is, the uniqueness portion of the claim is false, so the whole wff would be false. If there is exactly one cashier, however, both the existence and the uniqueness conditions will be satisfied, making the whole true.

There are two additional, quantificationally equivalent, ways of stating the uniqueness condition that might be more intuitive to you. I produce them here with their strict and abbreviated readings:

(4″) $(\exists x)(Cx \land (\forall y)(x \neq y \to \neg Cy))$

(4†) There is at least one x, such that x is a cashier, and for any y, if x is not identical to (or is distinct from) y, then y is not a cashier

(4††) There is an x that is a cashier and everything other than x is not a cashier

(4‴) $(\exists x)(Cx \land \neg(\exists y)(x \neq y \land Cy))$

(4‡) There is at least one x, such that x is a cashier, and there does not exist a y, such that both x is distinct from y, and y is a cashier

(4‡‡) There is an x that is a cashier, and nothing is both a cashier and distinct from x

(4′), (4″), and (4‴) are all quantificationally equivalent, and each is an appropriate translation of (4). I prefer the style of (4′), as it is the most economical in terms of symbols on the page.

It is, of course, possible to symbolize the following:

(5)  There are exactly two cashiers

Parallel to the construction of 'exactly one', we will have an existential portion ('there are at least two') and a uniqueness portion ('there are no more than two'). We know from above how to do the 'there are at least two' portion. The uniqueness portion will be a bit more complicated than in (4′). Here is an example with the familiar portion of the reading slightly compressed:

(5′)  $(\exists x)(\exists y)((x \neq y \land [Cx \land Cy]) \land (\forall z)(Cz \rightarrow [x = z \lor y = z]))$

(5*)  There are at least two cashiers x and y, and for any z, if z is a cashier, then it is either identical to x or identical to y

Again, the higher the number we wish to specify, the longer and more complex the construction becomes. If we were planning on using such constructions frequently for higher integers, it would be useful to introduce some abbreviations. As things stand I'll simply illustrate the next iteration.

(6)  There are exactly three cashiers

(6′)  $(\exists x)(\exists y)(\exists z)[([(x \neq y \land x \neq z) \land y \neq z] \land [(Cx \land Cy) \land Cz]) \land$
$(\forall w)(Cw \rightarrow ([x = w \lor y = w] \lor z = w))]$

Try to read this out (in one breath!) and see if you understand it. Box 5.12 sums up this discussion.

The introduction of the identity sign also allows construction of definite descriptions. Definite descriptions were mentioned at the start of this chapter as a kind of singular term in natural language (page 144ff.). Definite descriptions were not, however, symbolized by individual constants—I promised a more complex treatment, to which I now turn. I start with a definition.

**Definite Description:**

A *definite description* is a phrase that is supposed to designate an object via a unique description of it—i.e., a description that, if satisfied, is satisfied by one and only one object.

Consider the following sentence, which contains a definite description:

(7)  The cashier who works for Kate is diligent

The definite article 'the' is supposed to imply the uniqueness of the thing described. One way to achieve this would be to assign an individual constant, say 'c', to 'the cashier who works for Kate'. This would give us 'Dc' as a translation for (7). But this would obscure

---

**Box 5.12: 'At least' and 'Exactly'**

There are at least two $\mathbb{F}$s     $(\exists x)(\exists y)(x \neq y \land (\mathbb{F}x \land \mathbb{F}y))$

There are at least three $\mathbb{F}$s     $(\exists x)(\exists y)(\exists z)([(x \neq y \land x \neq z) \land y \neq z] \land$
$[(\mathbb{F}x \land \mathbb{F}y) \land \mathbb{F}z])$

There is exactly one $\mathbb{F}$
$\begin{cases} (\exists x)(\mathbb{F}x \land (\forall y)(\mathbb{F}y \to x = y)) \ * \\ (\exists x)(\mathbb{F}x \land (\forall y)(x \neq y \to \neg \mathbb{F}y)) \\ (\exists x)(\mathbb{F}x \land \neg(\exists y)(x \neq y \land \mathbb{F}y)) \end{cases}$

There are exactly two $\mathbb{F}$s     $(\exists x)(\exists y)([x \neq y \land (\mathbb{F}x \land \mathbb{F}y)] \land$
$(\forall z)(\mathbb{F}z \to (x = z \lor y = z)))$

*Preferred, though not required. For 'exactly two' I show only this form.

---

important features of the logical structure of the definite description. In particular, the definite description involves a number of the predicates of our interpretation, and it would be undesirable to hide this fact through the use of 'c', especially if we want to assess inferences made with wffs involving the definite description.

So we need to display the internal structure of the description 'cashier who works for Kate'. Thus, we might be tempted to symbolize (7) as follows:

$(7^X)$  $(\exists x)((Cx \land Wxk) \land Dx)$       **INCORRECT**

This, however, is inadequate. We said that the 'the' implied uniqueness of the object described. But $(7^X)$ simply says there is at least one cashier who works for Kate and is diligent. We need to add a uniqueness condition into our symbolization. As you might guess, we do it in essentially the same way as above. Here it is with a compressed paraphrase:

$(7')$  $(\exists x)([(Cx \land Wxk) \land (\forall y)((Cy \land Wyk) \to x = y)] \land Dx)$
$(7^*)$  There is at least one cashier who works for Kate, he's the only cashier who works for Kate, and he's diligent

Essentially what we are doing with $(7')$ is saying 'there is exactly one cashier who works for Kate, and he is diligent'. Note that the only predicates appearing in the statement of uniqueness are those involved in the definite description ('C' and 'W'). The predicate ascribed ('D') to the thing identified by the description appears outside of the uniqueness portion of the wff. I put it at the very end to set it off from the rest, but since the main sub-wff is a long conjunction, it could go just about anywhere, except within the uniqueness condition.

Consider the following wffs and their interpretations to see why placement of the various predicates is important (I reiterate (7)):

---

**Box 5.13: Definite Descriptions and 'The'**

The $\mathbb{F}$ is $\mathbb{G}$    $(\exists x)([\mathbb{F}x \land (\forall y)(\mathbb{F}y \to x = y)] \land \mathbb{G}x)$

The two $\mathbb{F}$s are $\mathbb{G}$    $(\exists x)(\exists y)([(x \neq y \land [\mathbb{F}x \land \mathbb{F}y]) \land$
$(\forall z)(\mathbb{F}z \to [x = z \lor y = z])] \land [\mathbb{G}x \land \mathbb{G}y])$

---

(7′)  $(\exists x)([(Cx \land Wxk) \land (\forall y)((Cy \land Wyk) \to x = y)] \land Dx)$

(7*)  There is at least one cashier who works for Kate, he's the only cashier who works for Kate, and he's diligent

(7)  The cashier who works for Kate is diligent

(8′)  $(\exists x)([Cx \land (\forall y)(Cy \to x = y)] \land (Wxk \land Dx))$

(8*)  There is at least one cashier, he's the only cashier, and he works for Kate and is diligent

(8)  The cashier works for Kate and is diligent

(9′)  $(\exists x)([[(Cx \land Wxk) \land Dx] \land (\forall y)([(Cy \land Wyk) \land Dx] \to x = y))$

(9*)  There is at least one cashier who works for Kate and is diligent, he's the only cashier who works for Kate and is diligent

(9)  There is exactly one cashier who works for Kate and is diligent

The differences among these involve what they claim or allow regarding how many cashiers there are in the office, how many cashiers there are who work for Kate, and what other descriptions those cashiers satisfy. In particular, (7′) has exactly one cashier working for Kate (and he's diligent)—there may well be other cashiers in the office (whether diligent or not), but Kate has only one. (8′), on the other hand, says there is exactly one cashier in the whole office, he happens to work for Kate and is diligent. (9′), finally, claims that there is exactly one cashier who both works for Kate and is diligent. In contrast to (7′) Kate may have other cashiers (but they are not diligent). In contrast to (8′) there may also be other cashiers in the office (whether diligent or not) who do not work for Kate.

Another way to make these points is with the following schema:

The $\mathbb{F}$ is $\mathbb{G}$    $(\exists x)([\mathbb{F}x \land (\forall y)(\mathbb{F}y \to x = y)] \land \mathbb{G}x)$

(7′) has 'Cx $\land$ Wxk' in the $\mathbb{F}$ portion and 'Dx' in the $\mathbb{G}$ portion. In contrast (8′) has 'Cx' in the $\mathbb{F}$ portion and 'Wxk $\land$ Dx' in the $\mathbb{G}$ portion. It is this difference that dictates the different interpretations. (9′) is not even an instance of the above form. Rather (9′) has the form of 'exactly one' as in (4′). Its form is $(\exists x)(\mathbb{F}x \land (\forall y)(\mathbb{F}y \to x = y))$ with '(Cx $\land$ Wxk) $\land$ Dx' in the $\mathbb{F}$ portion.

Finally, just as we progressed from 'exactly one' to 'exactly two', and so on, we can progress from 'the (one)' to 'the two' and so on.

(10) The two cashiers working for Kate are diligent

(10\*) There are at least two cashiers working for Kate, no more than two cashiers working for Kate, and they are diligent

(10′) $(\exists x)(\exists y)[(x \neq y \wedge [([Cx \wedge Wxk] \wedge [Cy \wedge Wyk]) \wedge$
$(\forall z)([Cz \wedge Wzk] \rightarrow [x = z \vee y = z])]) \wedge (Dx \wedge Dy)]$

Note again the contrasts with the following:

(11) The two cashiers work for Kate and are diligent

(11\*) There are at least two cashiers, no more that two cashiers, and they work for Kate and are diligent

(11′) $(\exists x)(\exists y)[(x \neq y \wedge [(Cx \wedge Cy) \wedge (\forall z)(Cz \rightarrow [x = z \vee y = z])]) \wedge$
$([Dx \wedge Wxk] \wedge [Dy \wedge Wxk])]$

(12) There are exactly two cashiers who work for Kate and are diligent

(12\*) There are at least two cashiers who work for Kate and are diligent, and no more than two cashiers who work for Kate and are diligent

(12′) $(\exists x)(\exists y)[(x \neq y \wedge ([(Cx \wedge Wxk) \wedge Dx] \wedge [(Cy \wedge Wyk) \wedge Dy])) \wedge$
$(\forall z)([(Cz \wedge Wzk) \wedge Dz] \rightarrow [x = z \vee y = z])]$

As you can see this gets rather long and complex. Box 5.13 gives the basic schemas from which these are built.

## 5.4.4 Exercises

Using the interpretation given, translate each of the following into **P**.

**UD**: People in Kate's office

A①: ① is an accountant      L①②: ① likes ②
C①: ① is a cashier           W①②: ① works for ②
D①: ① is diligent
L①: ① is lazy                 f: Fred
U①: ① is upbeat            k: Kate

1. There are at least two cashiers

2. There are at least two cashiers working for Kate

3. There is exactly one cashier

4. There is exactly one cashier working for Kate

5. There is exactly one upbeat cashier working for Kate

6. There is exactly one upbeat cashier and he works for Kate

7. There are exactly two accountants

8. Exactly two accountants work for Kate

9. Everyone but Fred works for Kate

10. The cashier is lazy

11. The cashier working for Fred is lazy

12. The lazy cashier who works for Fred is not upbeat

13. Fred is the only diligent accountant who works for Kate

14. Only one diligent accountant works for Kate

15. Some accountants work for Kate

16. At least two accountants work for Kate

17. Exactly one accountant works for Kate

18. Some diligent accountants work for Kate

19. At least two diligent accountants work for Kate

20. Exactly one diligent accountant works for Kate

21. There is exactly one diligent accountant and he works for Kate

22. The diligent accountant works for Kate

23. The diligent accountant who works for Kate is upbeat

24. The diligent accountant neither works for Kate nor for Fred

## 5.5  **Chapter Glossary**

**Atomic Formula, Molecular Formula:**
Any wff that qualifies simply in virtue of clause (1) of the definition of a wff (that is, any wff that just is some predicate letter with the appropriate number of terms), is called an *atomic* formula. By analogy, all other wffs are *molecular*. (157)

**Bound Variable, Free Variable:**
An occurrence of a variable x in a wff $\mathbb{P}$ is *bound* iff it is within the scope of an x-quantifier. An occurrence of a variable is *free* iff it is not bound. (158)

**Closed Wff, Sentence of *P*:**
A wff of *P* is *closed* iff it contains no free occurrences of variables. We also call such wffs *sentences* of *P*. (159)

**Definite Description:**

A *definite description* is a phrase that is supposed to designate an object via a unique description of it—i.e., a description that, if satisfied, is satisfied by one and only one object. (192)

**Expression of *P*:**

An *expression of P* is any finite sequence of the symbols of *P*. (155)

**Free Variable:**

See Bound Variable. (158)

**Identity:**

By *identity* we mean numerical identity—that what appear to be distinct objects are actually *one and the same thing*. Object x is *identical* to object y iff x is y. The two place predicate '①$=$②' is always interpreted as ① is identical to ②. (188, 189)

**Individual Terms:**

**Individual Constants:**

$a, b, \ldots, v, a_1, b_1, \ldots, v_1, a_2, \ldots$

**Individual Variables:**

$w, x, y, z, w_1, x_1, y_1, z_1, w_2, \ldots$     (154)

**Interpretation of *P*:**

An *interpretation of P* consists of 3 components:

(1) a non-empty universe of discourse, **UD**

(2) an assignment of truth values or natural language statements to statement letters and of natural language predicates to predicate letters via a translation key

(3) an assignment of objects from the **UD** to constants via a translation key such that

(a) every individual constant is assigned to an object, and

(b) no individual constant is assigned to more than one object     (146)

**Main Operator, Well-Formed Components:**

Atomic wffs have no main operator. The *main operator* of a molecular wff $\mathbb{R}$ is the operator appearing in the clause of the definition of a wff cited last in showing $\mathbb{R}$ to be a wff. The *immediate well-formed components* of a molecular wff are the values of $\mathbb{P}$ and $\mathbb{Q}$ (or simply $\mathbb{P}$) in the last-cited clause of the definition of a wff. The *well-formed components* of a wff are the wff itself, its immediate well-formed components, and the well-formed components of its immediate well-formed components. The *atomic components* of a wff are the well-formed components that are atomic wffs. (158)

**Metavariables:**

$\mathbb{A}, \mathbb{B}, \mathbb{C}, \ldots, \mathbb{Z}, \mathbb{A}_1, \ldots$

$$\mathbb{A}_k^n, \mathbb{B}_k^n, \ldots, \mathbb{Z}_k^n$$
$$\mathrm{a}, \mathrm{b}, \mathrm{c}, \ldots, \mathrm{z}, \mathrm{a}_1, \ldots \quad (155)$$

**Molecular Formula:**
   See Atomic Formula. (157)

**Open Wff:**
   A wff of *P* is *open* iff it contains at least one free occurrence of a variable. Otherwise it is a closed wff. (158) See also **Closed Wff, Sentence of *P*.**

**Operator:**
   The truth-functional connectives and quantifiers are *operators*. (157)

**Predicate:**
   A *predicate* is a series of words with one or more blanks that yields a sentence when all its blanks are filled with singular terms. Conversely, we could think of a predicate as what remains after removing one or more singular terms from a sentence. (145)

**Predicate Letters:**
$$A_1^0,\ B_1^0, \ldots, Z_1^0,\ A_2^0,\ B_2^0, \ldots, Z_2^0,\ A_3^0, \ldots$$
$$A_1^1,\ B_1^1, \ldots, Z_1^1,\ A_2^1,\ B_2^1, \ldots, Z_2^1,\ A_3^1, \ldots$$
$$A_1^2,\ B_1^2, \ldots, Z_1^2,\ A_2^2,\ B_2^2, \ldots, Z_2^2,\ A_3^2, \ldots$$
$$A_1^3,\ B_1^3, \ldots, Z_1^3,\ A_2^3,\ B_2^3, \ldots, Z_2^3,\ A_3^3, \ldots$$
$$\vdots \qquad\qquad\qquad \ddots$$

   That is, any uppercase letter with zero or positive integer superscript, $n$, indicating the number of places, and positive integer subscript, $k$, to give us an infinite (denumerable) supply. (154)

**Punctuation Marks:**
   ( )    (154)

**Quantificational Logic:**
   *Quantificational Logic* is the logic of sentences involving quantifiers, predicates, and names. It investigates the properties that arguments, sentences, and sets of sentences have in virtue of their quantificational structure. (144)

**Quantifier of *P*:**
   Where x ranges over individual variables, expressions of the form $(\forall x)$ are called *universal quantifiers*, while expressions of the form $(\exists x)$ are called *existential quantifiers*. We may also refer to a quantifier by the particular variable it contains— e.g., '$(\forall y_3)$' is a universal $y_3$-quantifier, while '$(\exists x)$' is an existential x-quantifier. (155)

**Quantifier Symbols:**
   $\forall$  $\exists$    (154)

**Satisfaction and Truth in *P* (Informal):**

Given an interpretation, where $\mathbb{F}x$ is a wff with only instances of the variable x free, and a is a constant:

(1) a *satisfies* $\mathbb{F}x$ (or $\mathbb{F}x$ is *true of* a) iff $\mathbb{F}a$ is true

(2) A universally quantified wff $(\forall x)\mathbb{F}x$ is true iff the condition expressed by the immediate subcomponent $\mathbb{F}x$ is satisfied by *every* object in the **UD**

(3) An existentially quantified wff $(\exists x)\mathbb{F}x$ is true iff the condition expressed by the immediate subcomponent $\mathbb{F}x$ is satisfied by *at least one* object in the **UD** (161)

**Scope:**

The *scope* of an operator is that portion of the wff containing its immediate sentential component(s). (158)

**Singular Term:**

A *singular term* is a word or phrase that designates or is supposed to designate some individual object. Natural language singular terms are either proper nouns or definite descriptions (a phrase that is supposed to designate an object via a unique description of it). (144)

**Substitution Instance:**

Let $\mathbb{Q}(a/x)$ indicate the wff that is just like $\mathbb{Q}$ except for having the constant a in every position where the variable x appears in $\mathbb{Q}$. Where $\mathbb{P}$ is a closed wff of the form $(\forall x)\mathbb{Q}$ or $(\exists x)\mathbb{Q}$, then $\mathbb{Q}(a/x)$ is a *substitution instance* of $\mathbb{P}$, with a as the *instantiating constant*. (159)

**Symbols of *P*:**

See Predicate Letters (154), Individual Terms (154), Individual Constants (154), Individual Variables (154), Truth-Functional Connectives (154), Quantifier Symbols (154), Punctuation Marks (154).

**Truth-Functional Connectives:**

$\neg \wedge \vee \rightarrow \leftrightarrow$     (154)

**Well-Formed Components:**

See Main Operator, Well-Formed Components. (158)

**Well-Formed Formula, Wff, of *P*:**

A *well-formed formula* or *wff* of *S* is a grammatical formula of the language *S*. Wffs of *P* may be open or closed. See full definition in text. (155)

# 6 Formal Semantics for *P*

## 6.1 Semantics and Interpretations

### 6.1.1 Basics of Interpretations

The previous chapter focused on translation between *P* and English. Our definition of interpretation was tailored to that task, centering around translation keys:

**Interpretation of *P* (Informal):**
    An *interpretation of* **P** consists of 3 components:[1]

> (1) a non-empty universe of discourse, **UD**, specifying the things about which we'll be talking
> (2) an assignment of truth values or natural language statements to statement letters and of natural language predicates or extensions to predicate letters
> (3) an assignment of objects from the **UD** to constants via a translation key such that
>> (a) every individual constant is assigned to an object, and
>> (b) no individual constant is assigned to more than one object

Since truth values are directly assigned only to statement letters (zero-place predicate letters), we needed the following rough definition to determine the truth values of more complex wffs:

**Satisfaction and Truth in *P* (Informal):**
    Given an interpretation, where $\mathbb{F}$x is a wff with only instances of the variable x free, and a is a constant:[2]

> (1) The denotation of a *satisfies* $\mathbb{F}$x (or $\mathbb{F}$x is *true of* the object named by a) iff $\mathbb{F}$a is T
>
> (2) A universally quantified wff $(\forall x)\mathbb{F}$x is true iff the condition expressed by the immediate subcomponent $\mathbb{F}$x is satisfied by *every* object in the **UD**
>
> (3) An existentially quantified wff $(\exists x)\mathbb{F}$x is true iff the condition expressed by the immediate subcomponent $\mathbb{F}$x is satisfied by *at least one* object in the **UD**

---

1. See p. 146 for original context.
2. See p. 161 for original context.

Here is a simple example.

**M1.**    **UD**: Ancient Greek Philosophers

|  |  |
|---|---|
| B①: ① is bald | a: Aristotle |
| H①: ① is human | p: Plato |
| T①②: ① teaches ② | s: Socrates |

Interpretation **M1** allows us to translate between English and *P* aided by the preceding definition of truth conditions for the predicates and quantifiers. That definition relies on an intuitive notion of satisfaction—the idea that an object *satisfies* a predicate iff that predicate is *true of* that object. So, for example, Socrates satisfies the predicate '① is bald' because it is true of Socrates that he is (was) bald. Consider these wffs:

(1)  Bs                           (4)  (∃y)By
(2)  Bp                           (5)  (∀z)Bz
(3)  (∀x)Hx

When working with the interpretation above, we say that (1) is T because the object in the **UD** denoted by 's' (Socrates) satisfies the open wff 'Bx'. Plato, however, does not satisfy the predicate '① is bald' because it is false of Plato that he is (was) bald. Thus, on the interpretation above, we say that (2) is F because the object in the **UD** denoted by 'p' does not satisfy 'Bx'.

Further, wff (3) '(∀x)Hx' is T on the interpretation because every object in the **UD** (even those not named by the three constants) satisfies the open wff 'Hx'. Similarly, (4) '(∃y)By' is T on this interpretation since there is at least one object in the **UD** that satisfies the open wff 'By'—Socrates, for one, and likely a few others in the **UD** but not named in the interpretation.

Finally, (5) '(∀z)Bz' is F on this interpretation, since it is not the case that every object in the **UD** satisfies 'Bz'.

It is important to see clearly that if we change the interpretation in any one of three ways, the truth values of the formula might change. We can change the **UD**. We can change the interpretation of one or more of the predicate letters. We can change the interpretation of one or more of the constants.

Suppose we change the **UD**. This produces a different interpretation:

**M2.**    **UD**: Prime Numbers and Ancient Greek Philosophers

|  |  |
|---|---|
| B①: ① is bald | a: Aristotle |
| H①: ① is human | p: Plato |
| T①②: ① teaches ② | s: Socrates |

Because the **UD** in **M2** is different from that of **M1**, the wff '(∀x)Hx' is F on interpretation **M2**. Prime numbers are not human, so not everything in the **UD** is human, so the formula is false on interpretation **M2**.

Let's go back to the **UD** of **M1** and now change the interpretation of one of the predicate letters:

**M3.** **UD**: Ancient Greek Philosophers

      B①: ① is bigger than a city bus      a: Aristotle
      H①: ① is human      p: Plato
      T①②: ① teaches ②      s: Socrates

On interpretation **M3**, both 'Bs' and '(∃y)By' are F, since neither Socrates, nor any other member of the **UD** is bigger than a city bus.

The last variation is to take **M1** and change the interpretation of at least one of the constants:

**M4.** **UD**: Ancient Greek Philosophers

      B①: ① is bald      a: Aristotle
      H①: ① is human      p: Plato
      T①②: ① teaches ②      s: Zeno

Here, while '(∃y)By' is still T, 'Bs' is F. On interpretation **M4** the constant 's' denotes Zeno, not Socrates, and Zeno was not bald.[3]

The upshot here is that, whenever we are discussing the truth value of a formula of **P**, that truth value is relative to some interpretation or other. As illustrated above, the truth value of the formula might change depending on the interpretation. So, to be strict, we should always mention the interpretation when citing a truth value. E.g.: 'Bs' is T on **M1** and **M2**, but F on **M3** and **M4**. Frequently—when doing translation exercises, for instance—we work within a single interpretation for an extended period. So, once the interpretation is set, we can often leave reference to it implicit. Otherwise it becomes unwieldy to always say "on interpretation **M1**...".

Two shortcomings of our current approach need to be addressed. First, our rough definitions of interpretation, satisfaction, and truth really only handle one-place predicate letters, and don't fully explain how to extrapolate from satisfaction of simple predicate letters to satisfaction of truth-functional combinations or satisfaction for quantifier-variable-constant combinations. In the previous chapter we relied on our intuitive understanding of English to ignore this issue.

Second, to investigate the quantificational logical properties of wffs of **P**, we need to focus not on any particular natural language translation, but on what happens with the truth value of a wff or wffs on *all interpretations*. For example, recall from Chapter 3 that a truth-functionally true wff of **S** is one that is true on every truth value assignment. A parallel, though more complex, concept is that of a quantificationally-true wff of **P**—true, not on every truth-value assignment, but on every interpretation of **P**. To investigate such logical properties, we need to go beyond particular translation keys.

---

3. Zeno of Elea, famous for the paradoxes of motion, is the Zeno I have in mind here.

However, there are an infinity of interpretations of **P**. Not only are there all the permutations of translation keys for the over 6,000 natural languages in the world, but there are all the permutations to be had when taking an arbitrary universe of discourse and assigning arbitrary extensions and objects to the predicate letters and constants. Remember that there are an infinite supply of each of these, and our **UD** can be infinitely large. To make the case a little more clearly, consider that one possible **UD** is the set of natural numbers $\mathbb{N}$,

**UD**: $\mathbb{N} = \{0, 1, 2, 3, \ldots\}$

Now consider the interpretation that assigns the number 0 to *every* constant in **P**. That is one interpretation (though maybe not a very useful one!). Next, consider the interpretation that assigns 1 to every constant, then the interpretation that assigns 2 to every constant, and so on... There is clearly an infinite number of interpretations of **P**.

How are we to examine or account for all the infinite interpretations? When dealing with wffs of *S*, the truth table method gave us a decision procedure for determining the logical properties of wffs. We have no such method here. Instead of a mechanical decision procedure, we will have to reason generally about all interpretations based on our understanding of quantificational semantics. Sometimes, we will have to show that an interpretation of a certain sort does, in fact, exist. In these cases, we simply need to produce an example interpretation. In other cases, we will have to produce semantic meta-proofs of the sort we used in Chapter 3.

To do any of this it is important to have a clear understanding of interpretations, satisfaction, and truth. And for this, it will help to understand a bit about sets and ordered *n*-tuples. Those familiar with set theory may skip the following subsection.

## 6.1.2   Interlude: A Little Bit of Set Theory

We briefly mentioned sets in previous chapters.[4] We have talked about consistency and implication of sets of wffs and we have described the universe of discourse in an interpretation as a set of objects. Put most simply, sets can be understood as collections of objects. The objects—the *members* or *elements* of the set—need not actually be near each other in time or space. That is, they need not be physically collected together. They can be any sort of object—concrete objects, such a chairs or people, or abstract objects, such as numbers or other sets. As long as we can specify the members of a set clearly, either by listing them or by giving some sort of rule or clear description for determining membership, then the sets will be well defined. Take the following example of a set with just three elements; I'll call the set '$\Gamma$':

$\Gamma = \{$Socrates, Plato, Aristotle$\}$

Here we specify the set $\Gamma$ by listing (or *enumerating*) its elements or members. Note that the order or method of specification does not matter. The identity of a set is completely determined by its members, no matter how they are specified. So consider:

---

4. See Chapter 8 for a full introduction to set theory.

$\Gamma' = \{\text{Plato, Aristotle, Socrates}\}$
$\Gamma'' = \{\text{the three objects assigned constants in interpretation } \mathbf{M1}\}$

Because the membership of $\Gamma$, $\Gamma'$, and $\Gamma''$, is exactly the same,

$$\Gamma = \Gamma' = \Gamma''$$

We use the epsilon symbol '$\in$' and its negation '$\notin$' to express or deny membership. For example:

(6)  Plato $\in \Gamma$       (7)  Zeno $\notin \Gamma$

Each statement may be read in any of the following corresponding ways:

(6′)  Plato is a member of $\Gamma$     (7′)  Zeno is not a member of $\Gamma$
(6″)  Plato is an element of $\Gamma$    (7″)  Zeno is not an element of $\Gamma$
(6‴)  Plato is in $\Gamma$       (7‴)  Zeno is not in $\Gamma$

Note that $\Gamma$ is not the same set as the **UD** of interpretation **M1** above. $\Gamma$ has exactly three members, no more, no less. The **UD** of **M1** contains *all* the ancient Greek philosophers, not just the three for whom we have specified constants. Indeed, we know of over 300 ancient Greek philosophers, and there may well be some about whom we do not know— all of them are in the **UD**. So, every member of $\Gamma$ is also a member of the **UD**, but not every member of the **UD** is a member of $\Gamma$. The **UD** of **M1** is larger than $\Gamma$. In particular, we can see that Zeno is in the **UD**, but not in $\Gamma$. So we say that $\Gamma$ is a *proper subset* of the **UD** above, and we use the proper subset symbol '$\subset$' to express this. Here is a sketch of the reasoning I just employed:

(1)  Zeno $\notin \Gamma$
(2)  Zeno $\in \mathbf{UD}$
(3)  for all x, if $x \in \Gamma$, then $x \in \mathbf{UD}$
(4)  some x is such that $x \in \mathbf{UD}$ but $x \notin \Gamma$      from (1) and (2)
(5)  $\Gamma \subset \mathbf{UD}$              from (3) and (4)[5]

As mentioned above, we can specify a set via a rule or description determining the members. Here is an example:

$$\Delta = \{\text{all x such that x is an even perfect square}\}$$

Or, more briefly:

---

5. Proper subset, '$\subset$', is distinguished from subset, '$\subseteq$', in the following way:

$$\alpha \subseteq \beta \leftrightarrow (\forall x)(x \in \alpha \rightarrow x \in \beta)$$
$$\alpha \subset \beta \leftrightarrow ((\forall x)(x \in \alpha \rightarrow x \in \beta) \wedge \alpha \neq \beta)$$

note that $\alpha \subset \beta \rightarrow (\exists y)(y \in \beta \wedge y \notin \alpha)$.

$$\Delta = \{x \mid x \text{ is an even perfect square}\}$$

Both specifications are read as "Delta is the set of all x, such that x is an even perfect square". A perfect square is a number made by squaring a whole number. An even number is divisible by 2 without remainder. Thus, the following statements are all true:

| | |
|---|---|
| $4 \in \Delta$ | $9 \notin \Delta$ |
| $36 \in \Delta$ | $8 \notin \Delta$ |
| $64 \in \Delta$ | Socrates $\notin \Delta$ |

Note that the set $\Delta$ is infinitely large, so we could not write out the whole list of members. However, whether any given object is a member is clearly determinable from the description. Infinitely large sets are discussed further in Chapter 8.

An important aspect of sets is that sets are not ordered objects. As mentioned, the identity of a set is determined solely by the members, regardless of how they might be ordered or presented. Thus, the following all specify the same set; call it '$\Theta$'

$$\Theta = \{\text{Socrates, Plato}\} = \{\text{Plato, Socrates}\} = \{\text{Plato, Socrates, Plato}\}$$

It makes no difference that we named Plato twice in the third specification of $\Theta$. It still contains only two members and is thus still the same set. Note as well that $\Theta \subset \Gamma$.

Sometimes, however, we do want to take order into account, and we do so by using a special kind of set called an ordered *n*-tuple—that is, an ordered sequence of *n* elements, such as an ordered pair, ordered triple, ordered quadruple, and so on. We indicate ordered *n*-tuples by using parentheses '(' and ')' instead of curly brackets. Sticking to ordered pairs for a moment, here are some examples:

(Socrates, Plato)                    (Plato, Socrates)

Order matters here:

$$(\text{Socrates, Plato}) \neq (\text{Plato, Socrates})$$

Ordered pairs (and ordered *n*-tuples generally) are not determined simply by their members, but by their members and their order. For pairs, we can express this as follows:

$$(x, y) = (v, w) \text{ iff both } x = v \text{ and } y = w$$

Moreover, while:

$$\{\text{Plato, Socrates}\} = \{\text{Plato, Socrates, Plato}\}$$

The same does not hold for *n*-tuples:

$$(\text{Plato, Socrates}) \neq (\text{Plato, Socrates, Plato})$$

Throwing Plato in on the right yields an ordered triple where on the left we have a pair. So we have two different $n$-tuples there.

Note that we can have 1-tuples. These are different from both the ordered pair of one object and the one-membered or singleton set:

$$(Plato) \neq \{Plato\}$$
$$(Plato) \neq (Plato, Plato)$$

It should be noted that neither the 1-tuple, nor the singleton are identical to the object itself (in this case, the man, Plato). That is:

$$(Plato) \neq Plato$$
$$\{Plato\} \neq Plato$$

We can even have a 0-tuple. This is just the empty set:

$$() = \{\} = \varnothing$$

It is worth having a rough understanding of functions. Intuitively a function takes inputs called arguments and for each argument yields a unique output value or image, but two different arguments may yield the same image. Think of the mathematical function of squaring. If we take the arguments to be positive and negative integers, then each input yields a unique output (its square), but both $n$ and $-n$ yield $n^2$. Where x is the argument and y is the image, we write:

$$f(x) = y$$

Here, x is the argument and y is the image (or value) of x under f.

This is sufficient for now. Chapter 8 contains a more thorough discussion of set theory.

## 6.1.3 Formal Interpretation of $P$

The elements of interpretations have not changed, though we will augment them somewhat and present them in a more technical guise. If the level of technicality in this section is intimidating, feel free to skim it to get the gist, then move on to the next section, 6.1.4, Constructing Interpretations.

As before, we have a Universe of Discourse and we interpret the predicate letters and the constants. Further, we will need to specify how this all comes together with variables, quantifiers, and the truth-functional connectives. To facilitate this we add the set of all variable assignments and a denotation function for singular terms of $P$.

**Interpretation of** $P$**:**

An *interpretation* $M$ *of* $P$ consists of:

(1) a non-empty universe of discourse, **UD**

(2) a function, **E**, that assigns extensions to predicate letters and objects to constants, such that

    (a) if $\mathbb{F}_k^0$ is a 0-place predicate, then $\mathbf{E}(\mathbb{F}_k^0) \in \{\mathsf{T}, \mathsf{F}\}$

    (b) if $\mathbb{F}_k^n$ is an $n$-place predicate where $n \geq 1$, then
$$\mathbf{E}(\mathbb{F}_k^n) = \{(\mathbf{o}_{1i}, \ldots, \mathbf{o}_{ni}), (\mathbf{o}_{1j}, \ldots, \mathbf{o}_{nj}), \ldots\}, \text{ where every } \mathbf{o}_m \in \mathbf{UD}$$

    (c) if $\mathbf{c}$ is an individual constant, then $\mathbf{E}(\mathbf{c}) = \text{some } \mathbf{o} \in \mathbf{UD}$

(3) the set **A** of all variable assignments, **a**, in which every individual variable, $\mathbf{x}_i$, of $P$ is assigned an object in **UD**, in other words:
$$\mathbf{A} = \{\mathbf{a} \mid \mathbf{a}(\mathbf{x}_i) = \mathbf{o}_j \in \mathbf{UD}\}$$

(4) for an interpretation **M** and each $\mathbf{a} \in \mathbf{A}$ we define a denotation function $\mathbf{d}_\mathbf{a}^\mathbf{M}$ that assigns objects from the **UD** to individual terms of $P$ such that

    (a) if $\mathfrak{t}$ is an individual constant $\mathbf{c}$, then $\mathbf{d}_\mathbf{a}^\mathbf{M}(\mathfrak{t}) = \mathbf{E}(\mathbf{c})$

    (b) if $\mathfrak{t}$ is an individual variable $\mathbf{x}$, then $\mathbf{d}_\mathbf{a}^\mathbf{M}(\mathfrak{t}) = \mathbf{a}(\mathbf{x})$

Before we define satisfaction and truth, let's take it piece by piece.

The Universe of Discourse, or **UD**, is any non-empty set—that is, the **UD** must have at least one member. We can specify the **UD** by listing the members or by giving a rule or definition for determining membership. The following specify the same **UD** in two different ways:

    **UD**: $\{11, 12, 13, 14, 15, 16, 17, 18, 19, 20\}$
    **UD**: $\{x \mid x \text{ is a positive integer and } 11 \leq x \leq 20\}$

The first one lists the members while the second one specifies them by mathematical definition. Let's turn to predicate letters.

Zero-place predicate letters of $P$ are treated just like statement letters of $S$. They just need to be assigned a $\mathsf{T}$ or $\mathsf{F}$, and not both. For example:

    $A^0$: $\mathsf{T}$
    $B^0$: $\mathsf{F}$

Note that the zero superscript indicates that they are each zero-place predicates. Normally we will do without the superscript, unless clarity demands it. So it could look like this:

    A: $\mathsf{T}$
    B: $\mathsf{F}$

To interpret greater-than-zero-place predicate letters of $P$ we need to specify the *extension* of the predicate letter. The *extension* is just the set of all $n$-tuples of objects from the **UD** that satisfy the predicate letter. For a one-place predicate letter, it will be a set of ordered singles. For a two-place predicate letter, it will be a set of ordered pairs. For a three-place predicate, it will be a set of ordered triples, and so on.

When our interest is just translation to and from English, we can specify the extension of predicate letters of *P* by matching them up with actual English predicates. It works well enough because English predicates have extensions themselves as a result of the way we use the words and the way the world is. So, when we assign an English predicate to a predicate letter of *P*, we are specifying an extension.[6]

Now, however, we want to be more abstract and more precise than natural language might allow. So we specify the extension as a set, either by listing the members of the set or by giving a rule or definition for determining membership. Here are some examples, assuming the **UD** above:

$$E①: \{12, 14, 16, 18, 20\}$$
$$O①: \{x \mid x \text{ is odd}\}$$
$$K①: \{\}$$
$$G①②: \{(x, y) \mid x > y\}$$
$$M①②③: \{(16, 13, 12), (11, 19, 14), (17, 16, 17), (11, 11, 11)\}$$

Here the one place predicate letter 'E' equates to '① is even' (in the current **UD**), and we could easily have specified its extension in a fashion similar to 'O'.[7] The extension of 'K' is the empty set. This means there is nothing in the **UD** that satisfies 'K'. 'K' is false of everything in this **UD**. The two-place predicate letter 'G' has as its extension the set of all ordered pairs $(x, y)$ such that $x > y$. Since our **UD** is finite, we could have listed all these pairs, but it would have been a little unwieldy. The three-place predicate 'M' doesn't have any obvious natural interpretation—I chose the triples with no pattern in mind. It is, however, an entirely legitimate three-place predicate letter, interpreted as a set of triples of members of the **UD**.

Finally, we have the constants:

| | | | |
|---|---|---|---|
| l: | 12 | q: | 17 |
| m: | 13 | r: | 18 |
| n: | 14 | s: | 19 |
| o: | 15 | t: | 20 |
| p: | 16 | u: | 20 |

Note that 20 is the denotation of two constants, 't' and 'u'. There is nothing technically wrong with this, though it is redundant. I did it here simply to illustrate the possibility. Also note that 11, while in the **UD**, is not the denotation of any constant on this interpretation. This, too, is mainly a point of illustration—not every object in the **UD** must be the denotation of a constant. What is required is that every constant be assigned one and only

---

6. Of course, the extensions of natural language predicates are not always precisely defined. Moreover, exactly how linguistic behavior of a community generates meanings and extensions is a wonderfully complex philosophical issue, best treated in a Philosophy of Language course.

7. Note that when dealing with ordered singles in the extension of a one-place predicate, we drop the parentheses: $\{12, 14, 16, 18, 20\}$ instead of $\{(12), (14), (16), (18), (20)\}$.

one object from the **UD**. Each of the constants above is assigned one and only one object. So we are fine.[8]

So here is what we have in our interpretation so far:

**M5.**    **UD**: $\{x \mid x \text{ is a positive integer and } 11 \leq x \leq 20\}$

| | | |
|---|---|---|
| A: T | k: 11 | p: 16 |
| B: F | l: 12 | q: 17 |
| E①: $\{12, 14, 16, 18, 20\}$ | m: 13 | r: 18 |
| O①: $\{x \mid x \text{ is odd}\}$ | n: 14 | s: 19 |
| K①: $\{\}$ | o: 15 | t: 20 |
| G①②: $\{(x, y) \mid x > y\}$ | | |
| M①②③: $\{(16, 13, 12), (11, 19, 14), (17, 16, 17), (11, 11, 11)\}$ | | |

Dealing with the **UD**, predicate letters, and constants is familiar from before, but presented a little more precisely and technically. Elements (3) and (4) are new.

The set **A** is defined in clause (3). This is just the set of all variable assignments, **a**. Remember that *P* has an infinite number of variables:

$$v, w, x, y, z, v_1, w_1, x_1, y_1, z_1, v_2, \ldots$$

So each **a** assigns one and only one object, **o**, to each variable of *P*. We can think of these variable assignments as infinite sequences of objects from the **UD**, where the first object in a particular sequence is assigned to 'v', the second is assigned to 'w', the sixth is assigned to '$v_1$', and so on. The general form looks like this:

$$
\begin{array}{ccccccc}
x_i & = & v & w & x & y & z & v_1 & \cdots \\
 & & \downarrow & \downarrow & \downarrow & \downarrow & \downarrow & \downarrow & \cdots \\
\mathbf{a}_j(x_i) & = & \mathbf{o}_1 & \mathbf{o}_2 & \mathbf{o}_3 & \mathbf{o}_4 & \mathbf{o}_5 & \mathbf{o}_6 & \cdots
\end{array}
$$

This depicts a particular variable assignment, $\mathbf{a}_j$. Each individual variable, $x_i$, is assigned some object, $\mathbf{a}_j(x_i) = \mathbf{o}_k$. In particular, here $\mathbf{a}_j(v) = \mathbf{o}_1$, $\mathbf{a}_j(w) = \mathbf{o}_2$, and so on. It is important to understand that the notation '$\mathbf{a}_j(w)$' refers, not to the variable 'w', but to the object $\mathbf{o}_2$ that assignment $\mathbf{a}_j$ assigns to 'w'.

This $\mathbf{a}_j$ is just one particular variable assignment in the set **A** of all variable assignments. A different one assigns $\mathbf{o}_1$ to every variable, or $\mathbf{o}_7$ to every variable, or $\mathbf{o}_{15}$ to the first ten variables and $\mathbf{o}_7$ to the rest—the combinations are endless. As long as there are 2 or more members of the **UD**, there will be an infinite number of variable assignments in the set **A**. And we will have to reason about all of these.[9]

---

8. But recall that—by use of subscripting—there is an infinite supply of predicate letters and constants in *P*. Thus, what we have specified above leaves an infinite number of predicate letters and constants uninterpreted. Theoretically, we get around this in the same way we get around the parallel issue with truth value assignments: this interpretation actually represents the infinite variety of interpretations which vary with respect to all predicate letters and constants *other than those specified*. In practice we do not worry about this at all—we make sure we interpret the predicate letters and constants we plan to use.

9. What if the UD only has one member? I leave this as an exercise for the reader.

Given the **UD** of **M5**, one of the variable assignments is the following:

$$\begin{array}{ccccccc}
x_i & = & v & w & x & y & z & v_1 & \cdots \\
& & \downarrow & \downarrow & \downarrow & \downarrow & \downarrow & \downarrow & \cdots \\
a_j(x_i) & = & 11 & 11 & 16 & 11 & 13 & 18 & \cdots
\end{array}$$

Another is:

$$\begin{array}{ccccccc}
x_i & = & v & w & x & y & z & v_1 & \cdots \\
& & \downarrow & \downarrow & \downarrow & \downarrow & \downarrow & \downarrow & \cdots \\
a_k(x_i) & = & 12 & 11 & 15 & 18 & 20 & 20 & \cdots
\end{array}$$

Part (4) defines a set of denotation functions that take us from ($n$-tuples of) individual terms of $P$ to objects in the **UD**. Remember that a term is either a constant or a variable. So denotation functions give us a way of combining the assignments of objects to constants in (2c) and the variable assignments in (3). For a given interpretation, **M**, there is a denotation function, **d**, for each variable assignment, **a**. That is why we have the superscripted '**M**' and subscripted '**a**'. Because the denotation function is relative to a variable assignment, **a**, there will be the same number of each in a given interpretation—one denotation function per variable assignment.

Let's look at the definition:

4a. if t is an individual constant $c$, then $d_a^M(t) = E(c)$
4b. if t is an individual variable x, then $d_a^M(t) = a(x)$

What this is telling us is that when the term is a constant, $c$, its denotation is exactly that object the function **E** assigns it—that is, the thing it names. Next, it tells us that if the term is a variable, x, its denotation is exactly what that particular variable assignment **a** assigns that variable. This may seem an unnecessary complication, but it allows us to talk about variable assignments satisfying wffs when the wff in question contains a mix of variables and constants. Specifying the **d** functions relative to the variable assignments, **a**, allows the denotation of the variables to vary. Identifying the denotation of the constants with the object assigned by **E** allows the denotation of constants to remain fixed no matter the variable assignment. This is exactly what we need. To illustrate, consider fragments of two different variable assignments:

$$\begin{array}{ccccc}
x_i & = & x & y & z & \cdots \\
& & \downarrow & \downarrow & \downarrow & \cdots \\
a_j(x_i) & = & 16 & 11 & 13 & \cdots
\end{array}
\qquad
\begin{array}{ccccc}
x_i & = & x & y & z & \cdots \\
& & \downarrow & \downarrow & \downarrow & \cdots \\
a_k(x_i) & = & 15 & 18 & 20 & \cdots
\end{array}$$

Next, a fragment of the assignment of objects to constants from **M5** above:

$$\begin{array}{ccccc}
c_i & = & q & r & s & \cdots \\
& & \downarrow & \downarrow & \downarrow & \cdots \\
E(c_i) & = & 17 & 18 & 19 & \cdots
\end{array}$$

Next, fragments of two denotation functions for **M5**, corresponding to the variable assignments above:

$$\begin{array}{ccccc}
t_i & = & s & z & r & \cdots \\
& & \downarrow & \downarrow & \downarrow & \cdots \\
d_{a_j}^{M5}(t_i) & = & 19 & 13 & 18 & \cdots
\end{array}
\qquad
\begin{array}{ccccc}
t_i & = & s & z & r & \cdots \\
& & \downarrow & \downarrow & \downarrow & \cdots \\
d_{a_k}^{M5}(t_i) & = & 19 & 20 & 18 & \cdots
\end{array}$$

Note how, because they are constants, the denotations of 'r' and 's' are fixed by **M5**, while, because it is a variable, that of 'z' varies depending on the particular variable assignment **a**.

## Satisfaction and Truth in $P$

We are now in a position to define satisfaction and truth.

**Satisfaction in $P$:**

Given an interpretation **M**, satisfaction is a relation between a variable assignment, **a**, and $\mathbb{P}$, a wff of the language $P$. Thus, for all $\mathbf{a} \in \mathbf{A}$ and all wffs $\mathbb{P}$ of $P$:

(1) If $\mathbb{P}$ is $\mathbb{F}_k^0$, then **a** satisfies $\mathbb{P}$ iff $\mathbf{E}(\mathbb{F}_k^0) = \mathsf{T}$

(2) If $\mathbb{P}$ is $\mathbb{F}_k^n \mathfrak{t}_1 \ldots \mathfrak{t}_n$, then **a** satisfies $\mathbb{P}$ iff $(\mathbf{d_a^M}(\mathfrak{t}_1), \ldots, \mathbf{d_a^M}(\mathfrak{t}_n)) \in \mathbf{E}(\mathbb{F}_k^n)$, for $n \geq 1$

(3) If $\mathbb{P}$ and $\mathbb{Q}$ are wffs of $P$, then

    (a) **a** satisfies $\neg\mathbb{P}$ iff **a** does not satisfy $\mathbb{P}$

    (b) **a** satisfies $\mathbb{P} \wedge \mathbb{Q}$ iff both **a** satisfies $\mathbb{P}$ and **a** satisfies $\mathbb{Q}$

    (c) **a** satisfies $\mathbb{P} \vee \mathbb{Q}$ iff either **a** satisfies $\mathbb{P}$ or **a** satisfies $\mathbb{Q}$

    (d) **a** satisfies $\mathbb{P} \rightarrow \mathbb{Q}$ iff either **a** does not satisfy $\mathbb{P}$ or **a** satisfies $\mathbb{Q}$

    (e) **a** satisfies $\mathbb{P} \leftrightarrow \mathbb{Q}$ iff either **a** satisfies both $\mathbb{P}$ and $\mathbb{Q}$, or **a** satisfies neither $\mathbb{P}$ nor $\mathbb{Q}$

(4) If $\mathbb{P}$ is $(\forall x_i)\mathbb{F}_k^n \mathfrak{t}_1 \ldots x_i \ldots \mathfrak{t}_n$, then **a** satisfies $\mathbb{P}$ iff every $\mathbf{a}'$ that differs from **a** in at most $\mathbf{a}'(x_i)$ satisfies $\mathbb{F}_k^n \mathfrak{t}_1 \ldots x_i \ldots \mathfrak{t}_n$

(5) If $\mathbb{P}$ is $(\exists x_i)\mathbb{F}_k^n \mathfrak{t}_1 \ldots x_i \ldots \mathfrak{t}_n$, then **a** satisfies $\mathbb{P}$ iff at least one $\mathbf{a}'$ that differs from **a** in at most $\mathbf{a}'(x_i)$ satisfies $\mathbb{F}_k^n \mathfrak{t}_1 \ldots x_i \ldots \mathfrak{t}_n$

Clause (1) deals with 0-place predicates (statement letters). Since variable assignments are irrelevant to the interpretation of these, we simply stipulate that every **a** satisfies the statement letter iff it is assigned T by **E**. This also implies that $\mathbf{E}(\mathbb{F}_k^0) = \mathsf{F}$ iff no **a** satisfies it.

Clause (2) tells us that when we have an $n$-place predicate letter followed by $n$ terms (any mixture of constants or variables), then an assignment **a** satisfies that wff iff the ordered $n$-tuple consisting of the denotations of those terms is in the extension of the predicate. Remember that, according to our definition of $\mathbf{d_a^M}$, the denotation of constants will be the object assigned to them by **E**, while the denotation of the variables will differ depending on which assignment **a** we consider. So, some open wffs will be satisfied by all **a**s, some by some **a**s but not others, and some by none.

Clause (3) accommodates satisfaction involving the truth-functional connectives.

Clauses (4) and (5) capture with precision what we have been after. A universally quantified wff with instances of $x_i$ as the quantified variable is satisfied by a variable assignment, **a**, iff *every* variable assignment $\mathbf{a}'$ which differs from **a** only with respect to the assignment to $x_i$, satisfies the immediate sub-wff. In contrast, an existentially quantified wff with instances of $x_i$ as the quantified variable is satisfied by a variable assignment, **a**, iff *at least one* variable assignment $\mathbf{a}'$ which differs from **a** only with respect to the assignment to $x_i$, satisfies the immediate sub-wff.

Note, first, that this is a recursive definition. This allows us to iteratively apply the clauses in order to determine whether a complex wff is satisfied on the basis of the satisfaction (or not) of its sub-wffs. Notice also that in the case of both universally and existentially quantified wffs, either all variable assignments satisfy the wff, or none do. Pay close attention to how clauses (4) and (5) are worded. This is important for the following definition.

**Truth in *P*:**

> A closed wff, or sentence, $\mathbb{P}$ of *P* is T on interpretation **M** iff every variable assignment **a** satisfies $\mathbb{P}$ on **M**. A closed wff of $\mathbb{P}$ is F on **M** iff no **a** satisfies $\mathbb{P}$ on **M**.

This should be clear for zero-place predicates. It is stipulated in clause (1) of the definition of satisfaction that every **a** satisfies $\mathbb{F}_k^0$ iff it is T. For an atomic closed wff, $\mathbb{F}_k^n \mathbf{c}_1 \dots \mathbf{c}_n$—i.e., an *n*-place predicate followed by *n* constants—a given variable assignment **a** satisfies it iff the *n*-tuple of the denotations of the constants is in the extension. If it is, then every **a** satisfies it, and it is T. If it is not, then no **a** satisfies it, and it is F. In the case of a closed quantified wff, either every **a** satisfies it, making it T, or none do, making it F. Truth-functional combinations of closed wffs are treated as usual. As for open wffs, while we do define satisfaction, we do not define truth for open wffs.

## 6.1.4 **Constructing Interpretations**

When reasoning about wffs of *P* we will often need to show that there is an interpretation which makes a certain wff (or all members of a set of wffs) T or F. We can do this in any number of ways. One is to provide a familiar **UD** and a natural language translation key that does the trick. Take, for example, the following wffs:

(1)  $(\exists y)(Ry \wedge Gy)$
(2)  $\neg Rb \wedge Gb$

We can give any number of natural language interpretations on which (1) and (2) are T. Here are two:

> **M6.**    **UD**: Presidents of the United States of America
> G①: ① served a term as a state governor
> R①: ① is a Republican
> b: Bill Clinton

> **M7.**    **UD**: Hollywood Celebrities
> G①: ① is good-looking
> R①: ① is an Emmy Award winner
> b: Jon Hamm

Wff (1) is T on interpretation **M6**, because there are a number of members of the **UD** that satisfy both 'Ry' and 'Gy'. Ronald Reagan, for instance, is among the Presidents, served as Governor of California, and was a Republican. (2) is also T on **M6**, since Bill Clinton is in the **UD**, not a Republican (making '¬Rb' T), and was Governor of Arkansas (making 'Gb' T). (1) and (2) are also both T on interpretation **M7**. For (1), there are any number of members of the **UD** that satisfy 'Ry' and 'Gy' on that interpretation, making the existential claim T. And for (2), Hamm, though good-looking, has not won an Emmy Award.

A problem with **M7** is that 'good-looking' is not well-defined here. You and I might disagree about which Emmy Award winners are good-looking. Another issue is that Hamm may one day win an Emmy, which is not logically problematic, but it would render this particular interpretation a poor example of one on which wff (2) is T. These are weaknesses of the natural language translation key approach.[10]

A better approach is to use the natural numbers, $\mathbb{N} = \{0, 1, 2, 3, \ldots\}$, as the **UD**. Any consistent set of wffs of *P* can be given an interpretation with the natural numbers as **UD**. All we have to do is come up with the appropriate extensions of predicates and constants. (Often easier said than done of course!) Here are two simple ones:

| **M8.** | **UD:** $\mathbb{N}$ | **M9.** | **UD:** $\mathbb{N}$ |
|---|---|---|---|
| | G①: ① is even | | G: $\{31, 9\}$ |
| | R①: ① is prime | | R: $\{9\}$ |
| | b: 4 | | b: 31 |

On **M8**, both (1) and (2) are T. 2 is the only even prime, so there is at least one even prime, as (1) claims. 4 is even, but not prime, as (2) claims. But (1) and (2) are both T on **M9** as well. Note that with **M9**, rather then giving a mathematical translation of the predicate letters, we simply specify their extensions as sets of objects from the **UD**. Here, (1) is T because there is at least one object from the **UD**, 9, that satisfies (is in the extension of)[11] both 'Ry' and 'Gy'. Moreover, (2) is T since the denotation of 'b', 31, satisfies 'Gy' and does not satisfy 'Ry', thereby satisfying '¬Ry'.

Interpretation **M9** gives us a very sparse and simple way of showing that (1) and (2) are T on some interpretation. As long as the wff in question does not require a very large or infinitely large **UD**, we can be even more economical with our interpretations. We can construct a small **UD** with just a few elements, and construct extensions for the predicates to get the result we want. Here is an example:

| **M10.** | **UD:** $\{a, b\}$ |
|---|---|
| | G: $\{a, b\}$ |
| | R: $\{a\}$ |
| | b: b |

---

10. Indeed, Hamm won his first Emmy in 2015, not long after the initial draft of this chapter.
11. When clarity does not require otherwise, I will use 'satisfies', 'makes true', and 'is in the extension of' interchangeably.

On interpretation **M10**, both (1) and (2) are T. There is at least one object that is both in the extension of 'R'[12] and in the extension of 'G', namely, a. So (1) is T. Moreover, b satisfies 'G', but is not in the extension of 'R', so (2) is also T.

What are a and b? It doesn't really matter what they are, we can think of them as arbitrary objects, and think of the extensions of 'R' and 'G' as arbitrarily chosen so as to make (1) and (2) T on **M10**. In most circumstances, if you like, you can think of a and b, as constants that name themselves on these minimalist interpretations. In fact, when using these minimalist interpretations we often won't bother to show the assignment of constants to objects. Further, typically we will be able to read off the number of places of a predicate letter by looking at the members of its extension. Guidelines for minimalist interpretations appear below.

Here is a minimalist interpretation that makes (1) and (2) both F:

**M11.**    UD: $\{a, b\}$
       G: $\{a\}$
       R: $\{b\}$

On **M11**, there is no object that makes both 'R' and 'G' T, so (1) is F. For (2), b fails to satisfy both '¬R' and 'G', making it F on **M11**.

We will use this minimalist approach to constructing interpretations whenever possible. We will depart from it only when we must, or when it is more instructive to use a different approach. There is no rule that we must construct the most minimal interpretation that meets our needs. In fact, to make (1) and (2) both F, we could have gotten rid of object a in **M11**, making the **UD** even smaller. On the other hand, we could have thrown more objects into the **UD**s above, or made the extensions more complicated. But that gives us more to keep track of. Better to keep it simple when we are able, and find an interpretation that works the way we want it.

## Guidelines for Constructing Minimalist Interpretations

1. Keep the **UD** and extensions of predicate letters simple. You need not find the simplest possible interpretation, but, in general, simpler is better. Some exercises require you to use a **UD** of a particular size, so be sure to pay attention to that.

2. For ease of reading, we drop the superscripts for predicate letters and drop the parentheses for 1-tuples. We do still need the parentheses for $n$-tuples when $n \geq 2$.

       B: $\{a, b, c\}$                    B: $\{(a, b), (b, b), (c, d)\}$
         instead of                      instead of
      $B^1$: $\{(a), (b), (c)\}$              $B^2$: $\{(a, b), (b, b), (c, d)\}$

3. The **UD** is specified using the relevant constants, where those constants are taken to denote arbitrary objects.

   **a.** Every constant that appears in the relevant wffs must appear in the **UD**.

---

12. For brevity, I will sometimes use "in 'R'" for "in the extension of 'R'".

    **b.** We forego explicitly interpreting constants—they denote whatever they denote. So, 'b' denotes whatever b is.

    **c.** Unless explicitly stated (e.g., 'a = b'), each constant denotes a unique member of the **UD**.[13]

One final pair of examples here:

(3) $(\exists y)(\forall x)Lyx$
(4) $(\exists y)(\forall x)Lxy$

**M12.**   **UD:** $\{a, b, c, d\}$         **M13.**   **UD:** $\{a, b, c, d\}$
        **L:** $\{(c,a), (c,b), (c,c), (c,d)\}$         **L:** $\{(a,c), (b,c), (c,c), (d,c)\}$

On **M12**, (3) is T, but (4) is F. On **M13**, (4) is T and (3) is F. See if you can explain why. If it helps, glance back at Box 5.11 in the previous chapter. (3) and (4) here correspond to (3) and (4) there, and the illustrations there correspond to the two interpretations here.

# 6.2 **Semantic Properties of Individual Wffs**

Just as with wffs of *S*, we can investigate individual wffs of *P* to determine their quantificational semantic properties. Of course, rather than investigating all truth value assignments, we must deal with all interpretations of the relevant portions of *P*. Consider the following definitions of semantic properties:

**Quantificationally True Wff:**
    A wff $\mathbb{P}$ of *P* is *quantificationally true* iff $\mathbb{P}$ is true on every interpretation (every interpretation is a model).

**Quantificationally False Wff:**
    A wff $\mathbb{P}$ of *P* is *quantificationally false* iff $\mathbb{P}$ is false on every interpretation (no interpretation is a model).

**Quantificationally Contingent Wff:**
    A wff $\mathbb{P}$ of *P* is *quantificationally contingent* iff $\mathbb{P}$ is neither quantificationally true nor quantificationally false; i.e., iff it is false on at least one interpretation and true on at least one interpretation (at least one interpretation is model, but not all are).

**Quantificationally Satisfiable Wff:**
    A wff $\mathbb{P}$ of *P* is *quantificationally satisfiable* iff $\mathbb{P}$ is not quantificationally false; i.e., it is true on at least one interpretation. We also say that the wff has a *model* or is *modeled*.

---

13. Note that when we do have identity claims in the object language, we should not take constants to denote themselves, as that would cause trouble: 'a = b' would imply that 'a' = 'b', which is false.

### Quantificationally Falsifiable Wff:

A wff $\mathbb{P}$ of $\boldsymbol{P}$ is *quantificationally falsifiable* iff $\mathbb{P}$ is not quantificationally true; i.e., it is false on at least one interpretation (at least one interpretation is not a model).

These definitions parallel those of Section 3.2. But, while truth-table tests give us a decision procedure for truth-functional properties, there is in general no decision procedure for the quantificational properties. For cases such as quantificational contingency, satisfiability, and falsifiability, we simply need to produce the relevant interpretations. For cases of quantificational truth and quantificational falsehood, we will have to show via a semantic proof that the wff in question is true on every interpretation, or false on every interpretation.

Take the following three wffs as examples:

(1) $(\exists v)Jv$
(2) $(\forall x)Fx \rightarrow (\exists y)Fy$
(3) $\neg(\exists z)(\neg Bz \lor Bz)$

Wff (1) is quantificationally contingent, and therefore both quantificationally satisfiable and quantificationally falsifiable. We can show this by producing two interpretations, one on which (1) is T and one on which it is F. Since (1) is an existential claim, for it to be true on an interpretation there must be at least one object **o** in the **UD** which is also in the extension of 'J'. For it to be F on an interpretation there must be no object **o** in the **UD** that is in the extension of 'J'. There are any number of each kind of interpretation, but we need only show one of each, so let's keep it nice and simple. Here are two minimalist interpretations:

**M14.**   **UD**: $\{b\}$          **M15.**   **UD**: $\{b\}$
        J: $\{b\}$                      J: $\{\}$

In both **M14** and **M15** the **UD** consists of just one object, b. It doesn't really matter who or what b is. What does matter is that, on interpretation **M14**, b is in the extension of 'J', so there is at least one object in the **UD** which satisfies 'Jv', so '$(\exists v)Jv$' is T on **M14**. Thus (1) is quantificationally satisfiable. Note that on interpretation **M15** the extension of 'J' is the empty set. So on **M15** there is no object in the **UD** which satisfies 'Jv', so '$(\exists v)Jv$' is F on **M15**. Thus (1) is quantificationally falsifiable. Of course, the combination of **M14** and **M15** shows that (1) is quantificationally contingent.

Now consider (2). We can give an interpretation to show it is satisfiable—indeed something similar to **M14** would work, giving 'F' the same extension as 'J'. But we cannot give an interpretation to show it is falsifiable, because it is quantificationally true—true on every interpretation. How can we show this? There are an infinite number of interpretations, we cannot produce them all! We must show by reasoning about quantificational semantics that (2) is true on every interpretation. That is, we must give a semantic meta-proof similar to the ones we did in Section 3.4 with truth-functional semantics. Let's give it a try:

'$(\forall x)Fx \rightarrow (\exists y)Fy$' is quantificationally true. Suppose it were not quantificationally true. Then there would be an interpretation, **N**, on which it is F. This would require that on **N** the antecedent '($\forall$x)Fx' is T while the consequent '($\exists$y)Fy' is F. If '($\forall$x)Fx' is T on **N**, then every object, **o**, in the **UD** is in the extension of 'F'. If '($\exists$y)Fy' is F on **N**, then *no* object in the **UD** is in the extension of 'F'. So we have both all objects in the extension of 'F' and all objects not in the extension of 'F'. This is a contradiction, so there is no interpretation on which '($\forall$x)Fx → ($\exists$y)Fy' is F. So it is quantificationally true.

This proof shows by contradiction that (2) is true on every interpretation. It is quantificationally true (and therefore quantificationally satisfiable as well).

Let's consider (3). It is quantificationally false. Again, showing this requires a meta-proof, rather than a display of one or two interpretations. Here is such a proof:

'¬($\exists$z)(¬Bz $\lor$ Bz)' is quantificationally false. On every interpretation, there is at least one object **o** in the **UD**. That object **o** is either not in the extension of 'B' or in the extension of 'B'. In either case the wff '¬Bz $\lor$ Bz' is satisfied by **o**, since it satisfies either the left disjunct or the right disjunct. So on every interpretation at least one object **o** satisfies '¬Bz $\lor$ Bz'. Thus, on every interpretation '($\exists$z)(¬Bz $\lor$ Bz)' is T. But this means that on every interpretation, its negation, '¬($\exists$z)(¬Bz $\lor$ Bz)' is F. So '¬($\exists$z)(¬Bz $\lor$ Bz)' is quantificationally false.

Since it is quantificationally false, (3) is also quantificationally falsifiable.

## 6.2.1  **Exercises**

1. Construct a minimalist interpretation to show that the following wff is not quantificationally true:
   [($\forall$x)Kx → ($\forall$y)Hy] → ($\forall$z)(Kz → Hz)

2. Construct a minimalist interpretation to show that the following wff is not quantificationally false:
   [($\exists$x)Bx $\land$ ($\exists$x)Cx] $\land$ ¬($\exists$x)(Bx $\land$ Cx)

3. Construct a minimalist interpretation to show that the following wff is not quantificationally true:
   ($\forall$z)(Fz → Gz) → ($\forall$z)Gz

4. Construct a minimalist interpretation to show that the following wff is not quantificationally true:
   ($\forall$x)($\exists$y)Lxy → ($\exists$y)($\forall$x)Lxy

5. Construct two minimalist interpretations to show that the following wff is quantificationally contingent:
   ($\forall$x)Bax → ($\forall$x)¬Bax

6. The following wff is quantificationally true, prove it:
   [($\forall$x)Dx $\lor$ ($\forall$x)Ex] → ($\forall$x)(Dx $\lor$ Ex)

**7.** The following wff is quantificationally true, prove it:
$(\exists x)(Dx \wedge Ex) \rightarrow (\exists x)Dx \wedge (\exists x)Ex$

# 6.3  **Semantic Properties of Sets of Wffs**

Here we define semantic properties that apply, not to individual wffs, but to sets of one, two, or more wffs.

**Quantificationally Equivalent Pair of Wffs:**
> Wffs $\mathbb{P}$ and $\mathbb{Q}$ of $P$ are *quantificationally equivalent* iff there is no interpretation on which they differ in truth value (every model of $\mathbb{P}$ is a model of $\mathbb{Q}$, and every model of $\mathbb{Q}$ is a model of $\mathbb{P}$).

**Quantificationally Contradictory Pair of Wffs:**
> Wffs $\mathbb{P}$ and $\mathbb{Q}$ of $P$ are *quantificationally contradictory* iff there is no interpretation on which they have the same truth value (no model of $\mathbb{P}$ is a model of $\mathbb{Q}$, and no model of $\mathbb{Q}$ is a model of $\mathbb{P}$).

Consider the following wffs:

(1)  $(\exists x)(Gx \wedge Hx)$
(2)  $(\forall x)(Gx \wedge Hx)$
(3)  $(\forall x)Gx \wedge (\forall x)Hx$
(4)  $(\forall x)(Gx \rightarrow \neg Hx)$

Wffs (1) and (2) are neither quantificationally equivalent nor quantificationally contradictory. We can demonstrate this with two interpretations—one on which they have different truth values, and one on which they have the same truth value:

| **M16.** | UD: $\{b,c\}$ | **M17.** | UD: $\{b,c\}$ |
|---|---|---|---|
| | G: $\{b\}$ | | G: $\{b,c\}$ |
| | H: $\{b,c\}$ | | H: $\{b,c\}$ |

On **M16** (1) and (2) have different truth values. The object b is in the extension of both 'G' and 'H' so at least one object satisfies 'Gx ∧ Hx', so '$(\exists x)(Gx \wedge Hx)$' is T on **M16**. But object c is not in the extension of 'G', so at least one object, c, does not satisfy 'Gx ∧ Hx', so '$(\forall x)(Gx \wedge Hx)$' is F on **M16**. In contrast, on **M17** (1) and (2) have the same truth value. The same reasoning regarding object b shows that '$(\exists x)(Gx \wedge Hx)$' is T on **M17**. Now, since on **M17** b and c are the only members of the **UD**, and both b and c are in the extension of both 'G' and 'H', then every member of the **UD** satisfies 'Gx ∧ Hx'. So '$(\forall x)(Gx \wedge Hx)$' is T on **M17**. We have shown that (1) and (2) are neither equivalent nor contradictory.

You may recall from the previous chapter that (2) and (3) are equivalent. They have the same truth value on every interpretation. We can reason as follows:

Take any interpretation where '$(\forall x)(Gx \land Hx)$' is T. On any such interpretation, every **o** in the **UD** satisfies '$Gx \land Hx$', which is equivalent to saying that both every **o** in the **UD** satisfies '$Gx$' and every **o** in the **UD** satisfies '$Hx$'. But this is equivalent to saying that on any such interpretation, both '$(\forall x)Gx$' and '$(\forall x)Hx$' are T. And this is equivalent to '$(\forall x)Gx \land (\forall x)Hx$' being T. So (2) and (3) will have the same truth value on every interpretation, and are quantificationally equivalent.

Note that because we reasoned through a number of equivalences, we do not need to consider what happens when (2) is F. Though we could do so.

Consider now (1) and (4). They are quantificationally contradictory.

Take any interpretation where '$(\forall x)(Gx \rightarrow \neg Hx)$' is T. On any such interpretation, every object **o** satisfies '$Gx \rightarrow \neg Hx$'. This equates to saying that every object either fails to satisfy '$Gx$' or satisfies '$\neg Hx$'. And that equates to: every object either fails to satisfy '$Gx$' or fails to satisfy '$Hx$'. But this is equivalent to saying that no object satisfies both '$Gx$' and '$Hx$', so no object satisfies '$Gx \land Hx$'. So '$(\exists x)(Gx \land Hx)$' is F on any such interpretation. Thus (1) and (4) have the opposite truth values on every interpretation.

Again, because we reasoned via equivalences, we do not need to consider what happens when (4) is F.

As you can see, what we need to do in these cases is reason generally about interpretations by making use of our definitions of satisfaction and truth. It both requires and enhances our understanding of quantificational semantics.

Three definitions remain. First, consistency:

### Quantificationally Consistent Set of Wffs:

A set $\Gamma$ of wffs of *P* is *quantificationally consistent* iff there is at least one interpretation on which all members of $\Gamma$ are true. We also say that $\Gamma$ has a *model* or is *modeled*. $\Gamma$ is *quantificationally inconsistent* iff it is not quantificationally consistent.

Consider two relatively simple sets:

(5) $\{(\exists y)(Dy \land Gy), (\exists w)(Dw \land \neg Gx), (\exists z)(\neg Dz \land Gz), (\exists z)(\neg Dz \land \neg Gz)\}$
(6) $\{(\exists x)(Mx \land Ox), (\forall y)(My \rightarrow Ny), (\forall v)(\neg Ov \leftrightarrow Nv)\}$

Set (5) is quantificationally consistent. To show this we need an interpretation which makes each of its members T. Clearly, we need an interpretation with one object that satisfies both 'D' and 'G', one object that satisfies 'D' but not 'G', one that satisfies 'G' but not 'D', and one that satisfies neither. This fits the bill:

**M18.**    UD: $\{b, c, d, e\}$
        D: $\{b, c\}$
        G: $\{b, d\}$

Every member of (5) is T on interpretation **M18**.

Set (6) is quantificationally inconsistent. There is no interpretation on which every member is T. To show this we need to show that on every interpretation at least one member of the set is F. The easiest way to do this is to suppose the set is quantificationally consistent, and show that this leads to a contradiction.

> Suppose set (6) is quantificationally consistent. So there is some interpretation **L** on which every member of the set is T. Since, '$(\exists x)(Mx \land Ox)$' is T, then there is at least one object in the **UD**, call it e, which satisfies '$Mx \land Ox$'. So, in particular, e satisfies 'M' and e satisfies 'O'. Since '$(\forall y)(My \rightarrow Ny)$' is T, any object which satisfies 'M' must also satisfy 'N', so e satisfies 'N'. But since '$(\forall v)(\neg Ov \leftrightarrow Nv)$' is also T, we find that every object satisfies 'N' iff it satisfies '$\neg O$'; that is, iff it fails to satisfy 'O'. Since e satisfies 'N', it does not satisfy 'O'. So e does and does not satisfy 'O'. This is a contradiction, so there can be no such interpretation **L**. Set (6) is quantificationally inconsistent.

Finally, we have entailment and validity. As mentioned in Section 1.5, the definitions of these two properties are very similar, differing mainly in how the question is posed and in that entailment allows for infinitely large sets. Here are the definitions:

**Quantificationally Entails:**

> A set $\Gamma$ of wffs of **P** *quantificationally entails* a wff $\mathbb{P}$ iff there is no interpretation on which all the members of $\Gamma$ are true and $\mathbb{P}$ is false. (Every model of $\Gamma$ is a model of $\mathbb{P}$.)

**Argument of *P*:**

> An *argument* of **P** is a finite set of two or more wffs of **P**, one of which is the conclusion, while the others are premises.

**Quantificationally Valid Argument:**

> An argument of **P** is *quantificationally valid* iff there is no interpretation on which all the premises are true and the conclusion is false. An argument of **P** is *quantificationally invalid* iff it is not quantificationally valid.

Consider two claims about sets and what they do or do not entail:

(7) $\{(\exists x)(Ax \land Dx), (\exists x)(Ax \land Kx)\} \nvDash (\exists x)(Dx \land Kx)$
(8) $\{(\forall x)(Hx \rightarrow Mx), Hs\} \vDash Ms$

Next, consider the corresponding arguments:

(7') $(\exists x)(Ax \land Dx)$
$\underline{(\exists x)(Ax \land Kx)}$
$(\exists x)(Dx \land Kx)$

(8') $(\forall x)(Hx \rightarrow Mx)$
$\underline{Hs}$
$Ms$

As indicated, the set in (7) does not entail the target. Thus the argument (7') is invalid. We can show both results by producing an interpretation on which the members of the set (the premises of (7')) are all T and the target (conclusion) is F. Here is one:

**M19.**    UD: {b,c}
        A: {b,c}
        D: {b}
        K: {c}

Clearly, '$(\exists x)(Ax \land Dx)$' is T on **M19** since b satisfies both 'A' and 'D'. Similarly for '$(\exists x)(Ax \land Kx)$' and c. But note that the target wff/conclusion, '$(\exists x)(Dx \land Kx)$' is F on this interpretation. There is no object in the **UD** which satisfies both 'D' and 'K', So the entailment in (7) does not hold, and argument (7′) is quantificationally invalid.

For (8) and (8′), we need to argue that there is no interpretation on which all the members of the set (premises) are true and the target (conclusion) is false.

> Any interpretation which makes all members of the set (all premises) T will have a **UD** containing the object denoted by 's'. That object, s, has to be in the extension of 'H' since we are assuming 'Hs' is T. But if, as assumed, '$(\forall x)(Hx \to Mx)$' is also T, then every object in the extension of 'H' must also be in the extension of 'M'. Thus, s is in the extension of 'M'. That is, 'Ms' is T on any interpretation where all members of the set (all premises) are T. So there is no interpretation on which all members of the set (all premises) are T and the target (conclusion) is F. So the set in (8) quantificationally entails its target, and argument (8′) is quantificationally valid.

## 6.3.1  Exercises

1. Construct a minimalist interpretation to show that the following pair of wffs is not quantificationally equivalent:
   $(\forall x)Dx \lor (\forall x)Ex$     $(\forall x)(Dx \lor Ex)$

2. Construct a minimalist interpretation to show that the following pair of wffs is not quantificationally equivalent:
   $(\exists x)Dx \land (\exists x)Ex$     $(\exists x)(Dx \land Ex)$

3. Construct a minimalist interpretation to show that the following pair of wffs is not quantificationally equivalent:
   $(\exists x)Bx \to Ca$     $(\exists x)(Bx \to Ca)$

4. Construct a minimalist interpretation to show that the following pair of wffs is not quantificationally equivalent:
   $(\forall x)Bx \to Ca$     $(\forall x)(Bx \to Ca)$

5. Construct a minimalist interpretation to show that the following set of wffs is quantificationally consistent:
   $\{(\exists x)Dx, (\exists x)Ex, \neg(\forall x)(Dx \lor Ex)\}$

6. Construct a minimalist interpretation to show that the following set of wffs is quantificationally consistent:
   $\{(\forall x)(Fx \to Gx), (\forall x)(Gx \to Hx), (\exists x)(Fx \land Hx)\}$

7. Construct a minimalist interpretation to show that the following set of wffs is quantificationally consistent:
$$\{(\forall x)(Fx \rightarrow Gx), (\forall x)(Gx \rightarrow Hx), (\exists x)(Hx \wedge \neg Fx)\}$$

8. Construct a minimalist interpretation to show that the following set of wffs is quantificationally consistent:
$$\{(\forall x)(Fx \rightarrow Gx), (\forall x)(Gx \rightarrow Hx), (\exists x)Fx \wedge (\exists x)\neg Hx\}$$

9. Construct a minimalist interpretation to show that the following set of wffs is quantificationally consistent:
$$\{(\forall x)(Dx \rightarrow Gx), (\forall x)(Fx \rightarrow Gx), (\exists x)Dx \wedge (\exists x)Fx, (\forall x)(Dx \rightarrow \neg Fx)\}$$

10. Construct a minimalist interpretation to show that the following set of wffs is quantificationally consistent:
$$\{(\forall x)(Fx \rightarrow Gx), \neg(\exists x)Fx, (\exists x)Gx\}$$

11. Construct a minimalist interpretation to show that the following set of wffs is quantificationally consistent:
$$\{(\forall x)(Fx \rightarrow Gx), \neg(\exists x)Gx\}$$

12. Construct a minimalist interpretation to show that the following failure of entailment is correct (i.e., the set does not entail the target):
$$\{(\forall x)(Fx \rightarrow Gx), (\exists x)Fx\} \nvDash (\forall x)Gx$$

13. The wffs in the following pair are quantificationally equivalent; prove it:
$$(\exists x)Bx \vee (\exists x)Cx \qquad (\exists x)(Bx \vee Cx)$$

14. The wffs in the following pair are quantificationally equivalent; prove it:
$$(\forall x)Bx \wedge Ca \qquad (\forall x)(Bx \wedge Ca)$$

15. The wffs in the following pair are quantificationally equivalent; prove it:
$$(\exists x)Bx \rightarrow Ca \qquad (\forall x)(Bx \rightarrow Ca)$$

16. The following entailment claim is correct; prove it:
$$(\exists y)(\forall x)Lxy \vDash (\forall x)(\exists y)Lxy$$

# 6.4 **Quantifier Scope and Distribution**

While learning to translate to $P$ we saw that the universal quantifier distributes across conjunction while maintaining quantificational equivalence, and the existential does the same across disjunction (see Box 5.6). That is:

(1)   $(\forall x)Fx \wedge (\forall x)Gx \iff (\forall x)(Fx \wedge Gx)$

(2)   $(\exists x)Fx \vee (\exists x)Gx \iff (\exists x)(Fx \vee Gx)$

We can argue clearly for these equivalences using our understanding of quantificational semantics. Here I will prove just (1), the equivalence involving the universal and conjunction.

Suppose the left hand side, $(\forall x)Fx \wedge (\forall x)Gx$, is T on some interpretation **M**. This means that both $(\forall x)Fx$ and $(\forall x)Gx$ are T on **M**. But this means that every variable assignment in **M** satisfies $Fx$ and every variable assignment in **M** satisfies $Gx$. And this is equivalent to: every variable assignment in **M** satisfies both $Fx$ and $Gx$. Which is equivalent to: every variable assignment in **M** satisfies $Fx \wedge Gx$. Which is equivalent to the right hand side, $(\forall x)(Fx \wedge Gx)$, being T on **M**. Since we proceeded by equivalences, we do not need to consider the other direction. The left hand side and right hand side are T in all and only the same interpretations. Hence the two forms are quantificationally equivalent.

The proof of (2) is in the exercises.

There are similar equivalences in cases where one side of the formula is a closed wff $P$ that does not contain $x$:

(3)  $(\forall x)Fx \wedge P \;\Leftrightarrow\; (\forall x)(Fx \wedge P)$

(4)  $(\exists x)Fx \wedge P \;\Leftrightarrow\; (\exists x)(Fx \wedge P)$

(5)  $(\forall x)Fx \vee P \;\Leftrightarrow\; (\forall x)(Fx \vee P)$

(6)  $(\exists x)Fx \vee P \;\Leftrightarrow\; (\exists x)(Fx \vee P)$

Here we see that both quantifiers distribute across conjunction and disjunction, so long as the other disjunct does not contain an instance of the same quantified variable. A proof of (6):

Suppose the left hand side, $(\exists x)Fx \vee P$, is T on some interpretation **M**. This means that either i) $(\exists x)Fx$ is T, or ii) $P$ is T. Take case i), this means that there is at least one object in the **UD** that satisfies $Fx$. But then there is at least one object in the **UD** that satisfies $Fx \vee P$, so $(\exists x)(Fx \vee P)$ is T on **M**. Take case ii), this means that every object in the **UD** satisfies $P$. But then at least one object satisfies $Fx \vee P$, so $(\exists x)(Fx \vee P)$ is T on **M**. In either case the right hand side is T on **M**.

Suppose the right hand side, $(\exists x)(Fx \vee P)$, is T on **M**. That means that there is at least one object that satisfies $Fx \vee P$. That is, there is either i) at least one object that satisfies $Fx$ or ii) at least one object that satisfies $P$. In case i) $(\exists x)Fx$ is T, making the left hand side, $(\exists x)Fx \vee P$, T on **M**. Case ii) implies that every object satisfies $P$, which means that $P$ is T, making the left hand side, $(\exists x)Fx \vee P$, T on **M**. In either case the right hand side is T on **M**. So the two forms are quantificationally equivalent.

Proofs of (3)–(5) are in the exercises. Note that with (1)–(6), because of the commutativity of conjunction and disjunction, the order of the components around the '$\wedge$' and '$\vee$' doesn't matter.

When dealing with the conditional, things are a bit different.

(7)  $\mathbb{P} \to (\forall x)\mathbb{F}x \;\Leftrightarrow\; (\forall x)(\mathbb{P} \to \mathbb{F}x)$

(8)  $\mathbb{P} \to (\exists x)\mathbb{F}x \;\Leftrightarrow\; (\exists x)(\mathbb{P} \to \mathbb{F}x)$

(9)  $(\forall x)\mathbb{F}x \to \mathbb{P} \;\Leftrightarrow\; (\exists x)(\mathbb{F}x \to \mathbb{P})$

(10)  $(\exists x)\mathbb{F}x \to \mathbb{P} \;\Leftrightarrow\; (\forall x)(\mathbb{F}x \to \mathbb{P})$

In (7) and (8) the quantified variable x appears in the consequent. Only in these cases can we simply extend or retract the scope of the quantifier and maintain equivalence. In (9) and (10), when the quantified variable x is in the antecedent, the quantifier must change to maintain equivalence. Here I will prove (9).

> Suppose the left hand side, $(\forall x)\mathbb{F}x \to \mathbb{P}$, is T on some interpretation **M**. Then either i) $(\forall x)\mathbb{F}x$ is F, or ii) $\mathbb{P}$ is T. If i), then there is at least one object in the **UD** that does not satisfy $\mathbb{F}x$. But that implies that there is at least one object in the **UD** that does satisfy $\mathbb{F}x \to \mathbb{P}$. Thus, the right hand side, $(\exists x)(\mathbb{F}x \to \mathbb{P})$, is T. If ii), then every object in the **UD** satisfies $\mathbb{P}$, so at least one object in the **UD** satisfies $\mathbb{F}x \to \mathbb{P}$. Thus, the right hand side, $(\exists x)(\mathbb{F}x \to \mathbb{P})$, is T. In either case the right hand side is T on **M**.

> Suppose the right hand side, $(\exists x)(\mathbb{F}x \to \mathbb{P})$, is T on some interpretation **M**. So there is at least one object in the **UD** that satisfies $\mathbb{F}x \to \mathbb{P}$. So either i) there is some object in the **UD** that does not satisfy $\mathbb{F}x$, or ii) there is one object that does satisfy $\mathbb{P}$. If i), then $(\forall x)\mathbb{F}x$ is F, in which case the left hand side, $(\forall x)\mathbb{F}x \to \mathbb{P}$, is T on **M**. If ii), then every object satisfies $\mathbb{P}$, making $\mathbb{P}$ T. And this makes the left hand side, $(\forall x)\mathbb{F}x \to \mathbb{P}$, T on **M**. In either case the left hand side is T on **M**. So the two forms are quantificationally equivalent.

Proofs of (7), (8), and (10) are in the exercises.

It might be worth showing for one of the cases (9) or (10) that the equivalence fails if we do not change the quantifier sign. For instance, let's show that:

(11)  $(\exists x)\mathbb{F}x \to \mathbb{P} \;\not\Leftrightarrow\; (\exists x)(\mathbb{F}x \to \mathbb{P})$

To show this, we just need to give an interpretation on which a wff with the form of the left hand side differs in truth value from a wff with the form of the right hand side. We will call these 'LHW', for left hand wff, and 'RHW' for right hand wff:

LHW: $(\exists z)\mathbb{B}z \to \mathbb{D}$
RHW: $(\exists z)(\mathbb{B}z \to \mathbb{D})$

Here is an interpretation on which LHW is F and RHW is T:

**M20.**   UD:  $\{b, c\}$

         D: F

         B: $\{b\}$

The antecedent of LHW is T since b is in the extension of 'B'. But the consequent of LHW is F, so LHW is F on **M20**. Since c is not in the extension of 'B', it fails to satisfy 'Bz'. But this means that c does satisfy 'Bz → D', making RHW T on **M20**. The exercises contain further questions of this sort.[14]

## 6.4.1 Exercises

For each of the following, prove that the quantificational equivalence holds, assuming that $\mathbb{P}$ is a closed wff that does not contain x:

1. $(\exists x)\mathbb{F}x \vee (\exists x)\mathbb{G}x \;\Leftrightarrow\; (\exists x)(\mathbb{F}x \vee \mathbb{G}x)$

2. $(\forall x)\mathbb{F}x \wedge \mathbb{P} \;\Leftrightarrow\; (\forall x)(\mathbb{F}x \wedge \mathbb{P})$

3. $(\exists x)\mathbb{F}x \wedge \mathbb{P} \;\Leftrightarrow\; (\exists x)(\mathbb{F}x \wedge \mathbb{P})$

4. $(\forall x)\mathbb{F}x \vee \mathbb{P} \;\Leftrightarrow\; (\forall x)(\mathbb{F}x \vee \mathbb{P})$

5. $\mathbb{P} \to (\forall x)\mathbb{F}x \;\Leftrightarrow\; (\forall x)(\mathbb{P} \to \mathbb{F}x)$

6. $\mathbb{P} \to (\exists x)\mathbb{F}x \;\Leftrightarrow\; (\exists x)(\mathbb{P} \to \mathbb{F}x)$

7. $(\exists x)\mathbb{F}x \to \mathbb{P} \;\Leftrightarrow\; (\forall x)(\mathbb{F}x \to \mathbb{P})$

# 6.5 Properties of Relations

In principle, our predicate letters can have any finite number of places and can be used to express relations of any finite number of objects. In practice, we rarely use predicate letters beyond 3- or 4-places and 1- and 2-places are most common. Examples of 2-place predicates expressing 2-place or binary relations are:

L①②:  ① loves ②
T①②:  ① teaches ②
   G:  $\{(x,y)|x > y\}$
   S:  $\{(a,b),(b,b),(c,d)\}$

All that defines a binary relation is that its extension is a set of ordered pairs in some **UD** (note: I did not bother specifying **UD**s for the examples above).

Logicians and mathematicians are often interested in the different properties exhibited (or not) by binary relations. Here we will review some of the most straightforward.

**Reflexivity:**
   A two-place relation $\mathbb{R}$ is *reflexive* in the **UD** of an interpretation **M** iff every object **o** of the **UD** bears the relation $\mathbb{R}$ to itself. That is, iff:

   $(\forall x)\mathbb{R}xx$

14. See questions **3.** and **4.** in Exercises 6.3.1 above.

Intuitively, the relation less or equal, $\leq$, where $\mathbf{UD} = \mathbb{N}$, is reflexive. Every natural number, $n$ is less than or equal to itself (though it is also less or equal to infinitely many other numbers, of course). In the interpretation below, we interpret one predicate letter as reflexive and one as not reflexive:

**M21.** UD: $\{a,b,c\}$
R: $\{(a,a),(a,b),(b,b),(b,c),(c,c)\}$
H: $\{(a,a),(b,b),(a,c)\}$

Clearly, in this **UD** on interpretation **M21** 'R' expresses a reflexive relation, since a, b, and c exhaust the **UD**, and $(a,a)$, $(b,b)$, and $(c,c)$ are all in the extension of 'R'. That is, $(\forall x)Rxx$ is T on **M21**. There are other pairs in the extension of 'R' as well, but their presence (or absence) doesn't matter, I included them only to illustrate this very point.

Strictly speaking, the predicate letter 'R' is not itself reflexive. It is just a symbol of $P$. Rather, it is the *extension* of 'R' on **M21**, the set defined in **M21** as $\mathbf{E}(R)$, that is reflexive. For the sake of simplicity, however, I will usually speak more loosely of a predicate being reflexive, or expressing a reflexive relation, or simply refer to the relation by the predicate letter assigned to it in the current interpretation. Moreover, when no clarity is lost by doing so, I will leave reference to the **UD** and interpretation implicit. In doing this I will also drop the single quotation marks, which should help remind us that we really mean the extension, not the predicate letter. So, above, R is reflexive.

If you look at H, you'll see that it is *not* reflexive. We have $(a,a)$ and $(b,b)$, but not $(c,c)$, so $(\forall x)Hxx$ is F on **M21**.

**Symmetry:**
A two-place relation $\mathbb{R}$ is *symmetric* in the **UD** of an interpretation **M** iff for objects $\mathbf{o}_1$ and $\mathbf{o}_2$ in the **UD**, whenever $\mathbf{o}_1$ bears $\mathbb{R}$ to $\mathbf{o}_2$, it is also the case that $\mathbf{o}_2$ bears $\mathbb{R}$ to $\mathbf{o}_1$. That is, iff:

$$(\forall x)(\forall y)(\mathbb{R}xy \rightarrow \mathbb{R}yx)$$

For an intuitive example, consider the relation being a sibling of. For any two people, a and b, if a is the sibling of b, then b is the sibling of a. Here is a minimalist example of symmetry and its failure:

**M22.** UD: $\{a,b,c\}$
R: $\{(a,b),(b,a),(b,b),(b,c),(c,b)\}$
H: $\{(a,b),(b,c)\}$

Here R is symmetric, but H is not. H fails because it is missing $(b,a)$ and $(c,b)$.

**Transitivity:**
A two-place relation $\mathbb{R}$ is *transitive* in the **UD** of an interpretation **M** iff for objects $\mathbf{o}_1$, $\mathbf{o}_2$, and $\mathbf{o}_3$ in the **UD**, whenever $\mathbf{o}_1$ bears $\mathbb{R}$ to $\mathbf{o}_2$ and $\mathbf{o}_2$ bears $\mathbb{R}$ to $\mathbf{o}_3$, it is also the case that $\mathbf{o}_1$ bears $\mathbb{R}$ to $\mathbf{o}_3$ . That is, iff:

$$(\forall x)(\forall y)(\forall z)((\mathbb{R}xy \wedge \mathbb{R}yz) \rightarrow \mathbb{R}xz)$$

A very intuitive example in mathematics is greater than: for any three numbers, if the first is greater than the second, and the second greater than the third, then the first is greater than the third. Another intuitive example is ancestor of. Here is a minimalist example:

**M23.**    UD: $\{a,b,c\}$
            R: $\{(a,b),(b,c),(a,c)\}$
            H: $\{(a,b),(b,c)\}$

Here R is transitive, but, because $(a,c)$ is not in its extension H is not transitive.

Reflexivity, symmetry, and transitivity are three basic properties of relations. Before we add a few others, let's look at these more closely.

One relation that has all three properties is identity. Recall that identity is expressed by '$=$' on every interpretation. Intuitively, therefore, each of the following must be true on every interpretation:

(1) $(\forall x)x = x$
(2) $(\forall x)(\forall y)(x = y \rightarrow y = x)$
(3) $(\forall x)(\forall y)(\forall z)((x = y \wedge y = z) \rightarrow x = z)$

The principle of substitutivity tells us that where a and b are singular terms of *P*, if $a = b$ then we can substitute a for one or more instances of a in any wff of *P* and maintain equivalence.

**Principle of Substitutivity:**
    Where a and b are singular terms of *P*,

$$\text{if } a = b, \text{ then } \mathbb{F}b \Leftrightarrow \mathbb{F}(a/b)$$

Given this principle, we can show that (1)–(3) are quantificationally true.

Suppose (1) is F on some interpretation, **M**. So on **M** '$\neg(\forall x)x = x$' is T. Which means that '$(\exists x)x \neq x$' is T. Call that 'x' 'a'. So on **M** '$a \neq a$' is T. But this is not possible, so there can be no such interpretation **M**. So (1) is T on every interpretation. (Note, we did not need substitutivity in this proof.)

Suppose (2) is F on some interpretation, **M**. So on **M** '$\neg(\forall x)(\forall y)(x = y \rightarrow y = x)$' is T. That is, '$(\exists x)(\exists y)(x = y \wedge y \neq x)$' is T. Call that 'x' 'a', and that 'y' 'b'. So '$a = b \wedge b \neq a$' is T. Given the left conjunct we can substitute 'a' for 'b' in the right and get '$a \neq a$' is T on **M**. But this is not possible, so there can be no such interpretation **M**. So (2) is T on every interpretation.

I leave the proof of (3) for the exercises.

It is important to see that these properties are logically independent of one another. That is, there are relations that have one, but neither of the other two. In interpretation **M21**, R is reflexive, but it is not symmetric, $(a,b)$ is there, but $(b,a)$ is not. Nor is it transitive, $(a,b)$ and $(b,c)$ are there, but $(a,c)$ is not. In **M22**, R is symmetric, but it is not reflexive, $(a,a)$ is missing. Nor is it transitive, $(a,b)$ and $(b,c)$ are there, but $(a,c)$ is not. Consider R in interpretation **M23**. It is transitive but not reflexive, $(a,a)$ is missing. Nor is it symmetric; $(a,b)$ is there, but $(b,a)$ is not.

Relations may also have any two of reflexivity, symmetry, and transitivity, without having the third. I leave these demonstrations for the exercises.

There are many properties of relations. We could, for instance define the failures discussed two paragraphs above in the following way:

**Nonreflexivity:**

$\neg(\forall x)Rxx$
$(\exists x)\neg Rxx$

**Nonsymmetry:**

$\neg(\forall x)(\forall y)(\mathbb{R}xy \rightarrow \mathbb{R}yx)$
$(\exists x)(\exists y)(\mathbb{R}xy \wedge \neg\mathbb{R}yx)$

**Nontransitivity:**

$\neg(\forall x)(\forall y)(\forall z)((\mathbb{R}xy \wedge \mathbb{R}yz) \rightarrow \mathbb{R}xz)$
$(\exists x)(\exists y)(\exists z)((\mathbb{R}xy \wedge \mathbb{R}yz) \wedge \neg\mathbb{R}xz)$

These "non" properties are not terribly interesting, since each only requires one instance of the failure of the original property, as is illustrated by the existential versions above.

More interesting are the following:

**Seriality:**

A two-place relation $\mathbb{R}$ is *serial* in the **UD** of an interpretation **M** iff every object $o_1$ in the **UD** bears $\mathbb{R}$ to some $o_2$ in the **UD**. That is, iff:

$$(\forall x)(\exists y)\mathbb{R}xy$$

**Irreflexivity:**

A two-place relation $\mathbb{R}$ is *irreflexive* in the **UD** of an interpretation **M** iff no object $o$ in the **UD** bears $\mathbb{R}$ to itself. That is, iff:

$$(\forall x)\neg\mathbb{R}xx$$

**Asymmetry:**

A two-place relation $\mathbb{R}$ is *asymmetric* in the **UD** of an interpretation **M** iff for objects $o_1$ and $o_2$ in the **UD**, whenever $o_1$ bears $\mathbb{R}$ to $o_2$, it is not the case that $o_2$ bears $\mathbb{R}$ to $o_1$. That is, iff:

$$(\forall x)(\forall y)(\mathbb{R}xy \rightarrow \neg\mathbb{R}yx)$$

**Antisymmetry:**

A two-place relation $\mathbb{R}$ is *antisymmetric* in the **UD** of an interpretation **M** iff for objects $o_1$ and $o_2$ in the **UD**, if $o_1$ bears $\mathbb{R}$ to $o_2$ and $o_2$ bears $\mathbb{R}$ to $o_1$, then $o_1 = o_2$. That is, iff:

$$(\forall x)(\forall y)((\mathbb{R}xy \land \mathbb{R}yx) \to x = y)$$

**Antitransitivity:**

A two-place relation $\mathbb{R}$ is *antitransitive* in the **UD** of an interpretation **M** iff for objects $o_1$, $o_2$, and $o_3$ in the **UD**, whenever $o_1$ bears $\mathbb{R}$ to $o_2$ and $o_2$ bears $\mathbb{R}$ to $o_3$, it is not the case that $o_1$ bears $\mathbb{R}$ to $o_3$. That is, iff:

$$(\forall x)(\forall y)(\forall z)((\mathbb{R}xy \land \mathbb{R}yz) \to \neg \mathbb{R}xz)$$

With the exception of antisymmetry, these are pretty straightforward, and the exercises give an opportunity to play around with them.

Antisymmetry may require a bit of explanation. A relation $\mathbb{R}$ is antisymmetric iff for every pair, if they are symmetrically related by $\mathbb{R}$, then they are really the same object. Perhaps more clearly, $\mathbb{R}$ is antisymmetric iff whenever $x \neq y$, then it is not the case that both $\mathbb{R}xy$ and $\mathbb{R}yx$. Thus,

**Antisymmetry:**

$$(\forall x)(\forall y)((\mathbb{R}xy \land \mathbb{R}yx) \to x = y)$$
$$(\forall x)(\forall y)(x \neq y \to \neg(\mathbb{R}xy \land \mathbb{R}yx))$$
$$\neg(\exists x)(\exists y)(x \neq y \land (\mathbb{R}xy \land \mathbb{R}yx))$$

The third reformulation above tells us that in an antisymmetric relation there is no pair of objects in which the members are distinct and bear the relation to each other in both directions. A classic example antisymmetry is less or equal, $\leq$, on the natural numbers.

$$(\forall x)(\forall y)((x \leq y \land y \leq x) \to x = y)$$
$$(\forall x)(\forall y)(x \neq y \to \neg(x \leq y \land y \leq x))$$
$$\neg(\exists x)(\exists y)(x \neq y \land (x \leq y \land y \leq x))$$

The exercises contain further work with these properties.

## 6.5.1  Exercises

This summary of the properties we have been discussing will be useful for the exercises.

**Reflexivity:**

$$(\forall x)\mathbb{R}xx$$

**Symmetry:**

$$(\forall x)(\forall y)(\mathbb{R}xy \to \mathbb{R}yx)$$

**Transitivity:**

$$(\forall x)(\forall y)(\forall z)((\mathbb{R}xy \land \mathbb{R}yz) \to \mathbb{R}xz)$$

**Seriality:**

$$(\forall x)(\exists y)\mathbb{R}xy$$

**Irreflexivity:**
$$(\forall x)\neg Rxx \; .$$

**Asymmetry:**
$$(\forall x)(\forall y)(Rxy \rightarrow \neg Ryx)$$

**Antisymmetry:**
$$(\forall x)(\forall y)((Rxy \wedge Ryx) \rightarrow x = y)$$

**Antitransitivity:**
$$(\forall x)(\forall y)(\forall z)((Rxy \wedge Ryz) \rightarrow \neg Rxz)$$

**Nonreflexivity:**
$$\neg(\forall x)Rxx$$
$$(\exists x)\neg Rxx$$

**Nonsymmetry:**
$$\neg(\forall x)(\forall y)(Rxy \rightarrow Ryx)$$
$$(\exists x)(\exists y)(Rxy \wedge \neg Ryx)$$

**Nontransitivity:**
$$\neg(\forall x)(\forall y)(\forall z)((Rxy \wedge Ryz) \rightarrow Rxz)$$
$$(\exists x)(\exists y)(\exists z)((Rxy \wedge Ryz) \wedge \neg Rxz)$$

**A.** For each of the following, construct a minimalist interpretation containing a relation R exhibiting the properties described. The **UD** must contain at least 3 distinct members. You may not use the empty relation (the relation containing no ordered pairs, i.e., the empty set).

1. Reflexivity and symmetry but not transitivity
2. Reflexivity and transitivity but not symmetry
3. Symmetry and transitivity but not reflexivity
4. Nonreflexivity, nonsymmetry, and nontransitivity

**B.** For each of the following properties of relations, construct a minimalist interpretation containing a relation R exhibiting the property. The **UD** must contain at least 4 distinct members. You may not use identity or the empty relation. Your relation may also exhibit other properties. Thus, you may use the same relation in answer to more than one question.

1. Reflexivity
2. Symmetry
3. Transitivity
4. Seriality
5. Irreflexivity
6. Asymmetry
7. Antisymmetry

    **8.** Antitransitivity

    **9.** Nonreflexivity

   **10.** Nonsymmetry

   **11.** Nontransitivity

**C.** For each of the following properties of relations, construct a minimalist interpretation containing a relation R exhibiting the property, but *none of the others* (where possible). The **UD** must contain at least 4 distinct members. You may not use identity or the empty relation.

    **1.** Reflexivity

    **2.** Symmetry

    **3.** Transitivity

    **4.** Seriality

    **5.** Irreflexivity

    **6.** Asymmetry

    **7.** Antisymmetry

    **8.** Antitransitivity

**D.** Prove the following relationships:

    **1.** Reflexivity $\Rightarrow$ Seriality

    **2.** Asymmetry $\Rightarrow$ Irreflexivity

    **3.** Asymmetry $\Rightarrow$ Antisymmetry

    **4.** Antitransitivity $\Rightarrow$ Irreflexivity

    **5.** Transitivity and Irreflexivity $\Rightarrow$ Asymmetry

**E.**  **1.** Using the principle of substitutivity, prove

    $(\forall x)(\forall y)(\forall z)((x = y \land y = z) \to x = z)$

    **2.** Prove that a **UD** with only one member can have only one variable assignment **a**.

    **3.** Argue that a relation that is transitive, irreflexive, and serial requires an infinite **UD**.

# 6.6 **Chapter Glossary**

**Antisymmetry:**

    $(\forall x)(\forall y)((\mathbb{R}xy \land \mathbb{R}yx) \to x = y)$    (229)

**Antitransitivity:**

    $(\forall x)(\forall y)(\forall z)((\mathbb{R}xy \land \mathbb{R}yz) \to \neg \mathbb{R}xz)$    (230)

**Argument of *P*:**

    An *argument* of *P* is a finite set of two or more wffs of *P*, one of which is the conclusion, while the others are premises. (221)

**Asymmetry:**

$$(\forall x)(\forall y)(\mathbb{R}xy \rightarrow \neg \mathbb{R}yx) \qquad (229)$$

**Interpretation of *P* (Informal):**

An *interpretation of P* consists of 3 components:

(1) a non-empty universe of discourse, **UD**

(2) an assignment of truth values or natural language statements to statement letters and of natural language predicates or extensions to predicate letters

(3) an assignment of objects from the **UD** to constants via a translation key such that

    (a) every individual constant is assigned to an object, and

    (b) no individual constant is assigned to more than one object    (201)

**Interpretation of *P*:**

An *interpretation M of P* consists of:

(1) a non-empty universe of discourse, **UD**

(2) a function, **E**, that assigns extensions to predicate letters and objects to constants, such that

    (a) if $\mathbb{F}_k^0$ is a 0-place predicate, then $\mathbf{E}(\mathbb{F}_k^0) \in \{\mathsf{T}, \mathsf{F}\}$

    (b) if $\mathbb{F}_k^n$ is an *n*-place predicate where $n \geq 1$, then
$$\mathbf{E}(\mathbb{F}_k^n) = \{(\mathbf{o}_{1i}, \ldots, \mathbf{o}_{ni}), (\mathbf{o}_{1j}, \ldots, \mathbf{o}_{nj}), \ldots\}, \text{ where every } \mathbf{o}_m \in \mathbf{UD}$$

    (c) if $\mathbf{c}$ is an individual constant, then $\mathbf{E}(\mathbf{c}) = $ some $\mathbf{o} \in \mathbf{UD}$

(3) the set **A** of all variable assignments, **a**, in which every individual variable, $x_i$, of *P* is assigned an object in **UD**, in other words:
$$\mathbf{A} = \{\mathbf{a} \mid \mathbf{a}(x_i) = \mathbf{o}_j \in \mathbf{UD}\}$$

(4) for an interpretation **M** and each $\mathbf{a} \in \mathbf{A}$ we define a denotation function $\mathbf{d}_\mathbf{a}^\mathbf{M}$ that assigns objects from the **UD** to individual terms of *P* such that

    (a) if $\mathfrak{t}$ is an individual constant $\mathbf{c}$, then $\mathbf{d}_\mathbf{a}^\mathbf{M}(\mathfrak{t}) = \mathbf{E}(\mathbf{c})$

    (b) if $\mathfrak{t}$ is an individual variable x, then $\mathbf{d}_\mathbf{a}^\mathbf{M}(\mathfrak{t}) = \mathbf{a}(x)$    (208)

**Irreflexivity:**

$$(\forall x)\neg \mathbb{R}xx \qquad (229)$$

**Nonreflexivity:**

$$\neg(\forall x)\mathbb{R}xx$$
$$(\exists x)\neg \mathbb{R}xx \qquad (229)$$

**Nonsymmetry:**

$$\neg(\forall x)(\forall y)(\mathbb{R}xy \rightarrow \mathbb{R}yx)$$
$$(\exists x)(\exists y)(\mathbb{R}xy \wedge \neg \mathbb{R}yx) \qquad (229)$$

**Nontransitivity:**

$$\neg(\forall x)(\forall y)(\forall z)((\mathbb{R}xy \wedge \mathbb{R}yz) \rightarrow \mathbb{R}xz)$$
$$(\exists x)(\exists y)(\exists z)((\mathbb{R}xy \wedge \mathbb{R}yz) \wedge \neg \mathbb{R}xz) \qquad (229)$$

**Principle of Substitutivity:**

Where a and b are singular terms of *P*,

$$\text{if } a = b, \text{ then } \mathbb{F}b \Leftrightarrow \mathbb{F}(a/b) \qquad (228)$$

**Quantificationally Consistent Set of Wffs:**

A set $\Gamma$ of wffs of *P* is *quantificationally consistent* iff there is at least one interpretation on which all members of $\Gamma$ are true. We also say that $\Gamma$ has a *model* or is *modeled*. $\Gamma$ is *quantificationally inconsistent* iff it is not quantificationally consistent. (220)

**Quantificationally Contingent Wff:**

A wff $\mathbb{P}$ of *P* is *quantificationally contingent* iff $\mathbb{P}$ is neither quantificationally true nor quantificationally false; i.e., iff it is false on at least one interpretation and true on at least one interpretation (at least one interpretation is model, but not all are). (216)

**Quantificationally Contradictory Pair of Wffs:**

Wffs $\mathbb{P}$ and $\mathbb{Q}$ of *P* are *quantificationally contradictory* iff there is no interpretation on which they have the same truth value (no model of $\mathbb{P}$ is a model of $\mathbb{Q}$, and no model of $\mathbb{Q}$ is a model of $\mathbb{P}$). (219)

**Quantificationally Entails:**

A set $\Gamma$ of wffs of *P* *quantificationally entails* a wff $\mathbb{P}$ iff there is no interpretation on which all the members of $\Gamma$ are true and $\mathbb{P}$ is false. (Every model of $\Gamma$ is a model of $\mathbb{P}$.) (221)

**Quantificationally Equivalent Pair of Wffs:**

Wffs $\mathbb{P}$ and $\mathbb{Q}$ of *P* are *quantificationally equivalent* iff there is no interpretation on which they differ in truth value (every model of $\mathbb{P}$ is a model of $\mathbb{Q}$, and every model of $\mathbb{Q}$ is a model of $\mathbb{P}$). (219)

**Quantificationally False Wff:**

A wff $\mathbb{P}$ of *P* is *quantificationally false* iff $\mathbb{P}$ is false on every interpretation (no interpretation is a model). (216)

**Quantificationally Falsifiable Wff:**

A wff $\mathbb{P}$ of *P* is *quantificationally falsifiable* iff $\mathbb{P}$ is not quantificationally true; i.e., it is false on at least one interpretation (at least one interpretation is not a model). (217)

**Quantificationally Satisfiable Wff:**

A wff $\mathbb{P}$ of *P* is *quantificationally satisfiable* iff $\mathbb{P}$ is not quantificationally false; i.e., it is true on at least one interpretation. We also say that the wff has a *model* or is *modeled*. (216)

**Quantificationally True Wff:**

A wff $\mathbb{P}$ of $P$ is *quantificationally true* iff $\mathbb{P}$ is true on every interpretation (every interpretation is a model). (216)

**Quantificationally Valid Argument:**

An argument of $P$ is *quantificationally valid* iff there is no interpretation on which all the premises are true and the conclusion is false. An argument of $P$ is *quantificationally invalid* iff it is not quantificationally valid. (221)

**Reflexivity:**

$(\forall x)\mathbb{R}xx$    (226)

**Satisfaction and Truth in $P$ (Informal):**

Given an interpretation, where $\mathbb{F}x$ is a wff with only instances of the variable $x$ free, and a is a constant:

(1) The denotation of a *satisfies* $\mathbb{F}x$ (or $\mathbb{F}x$ is *true of* the object named by a) iff $\mathbb{F}a$ is T

(2) A universally quantified wff $(\forall x)\mathbb{F}x$ is true iff the condition expressed by the immediate subcomponent $\mathbb{F}x$ is satisfied by *every* object in the **UD**

(3) An existentially quantified wff $(\exists x)\mathbb{F}x$ is true iff the condition expressed by the immediate subcomponent $\mathbb{F}x$ is satisfied by *at least one* object in the **UD** (201)

**Satisfaction in $P$:**

Given an interpretation **M**, satisfaction is a relation between a variable assignment, **a**, and $\mathbb{P}$, a wff of the language $P$. Thus, for all $\mathbf{a} \in \mathbf{A}$ and all wffs $\mathbb{P}$ of $P$:

(1) If $\mathbb{P}$ is $\mathbb{F}_k^0$, then **a** satisfies $\mathbb{P}$ iff $\mathbf{E}(\mathbb{F}_k^0) = \mathsf{T}$

(2) If $\mathbb{P}$ is $\mathbb{F}_k^n t_1 \ldots t_n$, then **a** satisfies $\mathbb{P}$ iff $(\mathbf{d}_\mathbf{a}^\mathbf{M}(t_1), \ldots, \mathbf{d}_\mathbf{a}^\mathbf{M}(t_n)) \in \mathbf{E}(\mathbb{F}_k^n)$, for $n \geq 1$

(3) If $\mathbb{P}$ and $\mathbb{Q}$ are wffs of $P$, then

   (a) **a** satisfies $\neg\mathbb{P}$ iff **a** does not satisfy $\mathbb{P}$

   (b) **a** satisfies $\mathbb{P} \wedge \mathbb{Q}$ iff both **a** satisfies $\mathbb{P}$ and **a** satisfies $\mathbb{Q}$

   (c) **a** satisfies $\mathbb{P} \vee \mathbb{Q}$ iff either **a** satisfies $\mathbb{P}$ or **a** satisfies $\mathbb{Q}$

   (d) **a** satisfies $\mathbb{P} \rightarrow \mathbb{Q}$ iff either **a** does not satisfy $\mathbb{P}$ or **a** satisfies $\mathbb{Q}$

   (e) **a** satisfies $\mathbb{P} \leftrightarrow \mathbb{Q}$ iff either **a** satisfies both $\mathbb{P}$ and $\mathbb{Q}$, or **a** satisfies neither $\mathbb{P}$ nor $\mathbb{Q}$

(4) If $\mathbb{P}$ is $(\forall x_i)\mathbb{F}_k^n t_1 \ldots x_i \ldots t_n$, then **a** satisfies $\mathbb{P}$ iff every $\mathbf{a}'$ that differs from **a** in at most $\mathbf{a}'(x_i)$ satisfies $\mathbb{F}_k^n t_1 \ldots x_i \ldots t_n$

(5) If $\mathbb{P}$ is $(\exists x_i)\mathbb{F}_k^n t_1 \ldots x_i \ldots t_n$, then **a** satisfies $\mathbb{P}$ iff at least one $\mathbf{a}'$ that differs from **a** in at most $\mathbf{a}'(x_i)$ satisfies $\mathbb{F}_k^n t_1 \ldots x_i \ldots t_n$    (212)

**Seriality:**

$(\forall x)(\exists y)\mathbb{R}xy$    (229)

**Symmetry:**

$$(\forall x)(\forall y)(\mathbb{R}xy \rightarrow \mathbb{R}yx) \qquad (227)$$

**Transitivity:**

$$(\forall x)(\forall y)(\forall z)((\mathbb{R}xy \wedge \mathbb{R}yz) \rightarrow \mathbb{R}xz) \qquad (227)$$

**Truth in *P*:**

A closed wff, or sentence, $\mathbb{P}$ of *P* is T on interpretation **M** iff every variable assignment **a** satisfies $\mathbb{P}$ on **M**. A closed wff of $\mathbb{P}$ is F on **M** iff no **a** satisfies $\mathbb{P}$ on **M**. (213)

# 7 *PD*: **Natural Deduction in** *P*

## 7.1 **Derivation Rules for the Quantifiers**

Given the complexities of the semantics of *P*, and the lack of a truth-table-like decision procedure for semantic properties of wffs of *P*, we might like to have a systematic procedure for checking for such properties. It would also be nice if this procedure, to some extent at least, paralleled our natural pattern of drawing quantificational inferences. A natural deduction system for *P*, called *PD*, gives us such a procedure.[1]

*PD* shall consist of all the rules of *SDE*, plus the four rules to be introduced in this section.

The rules carried over from *SD* and *SDE* apply to closed wffs of *P* (sentences, whether quantified or not) exactly as they did to wffs of *S*.[2] So, for example, the following are perfectly legitimate applications of →E within *PD*:

$$
\begin{array}{lll}
1 & Da \to (Bc \wedge Cc) & P \\
2 & \underline{Da} & P \qquad \vdash Bc \wedge Cc \\
3 & Bc \wedge Cc & 1, 2 \to E
\end{array}
$$

$$
\begin{array}{lll}
1 & (\forall x)Dx \to (Bc \wedge Cc) & P \\
2 & \underline{(\forall x)Dx} & P \qquad \vdash Bc \wedge Cc \\
3 & Bc \wedge Cc & 1, 2 \to E
\end{array}
$$

$$
\begin{array}{lll}
1 & Da \to (\exists y)(By \wedge Cy) & P \\
2 & \underline{Da} & P \qquad \vdash (\exists y)(By \wedge Cy) \\
3 & (\exists y)(By \wedge Cy) & 1, 2 \to E
\end{array}
$$

$$
\begin{array}{lll}
1 & (\forall x)Dx \to (\exists y)(By \wedge Cy) & P \\
2 & \underline{(\forall x)Dx} & P \qquad \vdash (\exists y)(By \wedge Cy) \\
3 & (\exists y)(By \wedge Cy) & 1, 2 \to E
\end{array}
$$

Since in each case the '→' is the main operator of the first line, and the second line matches the antecedent, we can perform →E to derive the consequent. The following, however, would not be allowed:

---

1. As with *SD*, *PD* will give us only a search procedure, not a decision procedure.
2. Here, I mostly use *SD* plus the four new *PD* rules, just in case the reader did not cover *SDE*.

$$
\begin{array}{ll}
1 \mid (\forall x)(Dx \rightarrow (Bx \wedge Cx)) & \text{P} \\
2 \mid \underline{(\forall x)Dx} & \text{P} \\
3 \mid (\forall x)(Bx \wedge Cx) & \text{1, 2} \rightarrow\!\text{E} \quad \textbf{MISTAKE}
\end{array}
$$

The problem here is that the main operator of line 1 is not the '→', but the '(∀x)', hence →E cannot apply to line 1.

In general, then, the rules from *SD* and *SDE* can deal with wffs of *P* only if they are treated as atomic (ignoring all internal structure) or truth-functional compounds (where the main operator is a truth-functional connective). The rules from *SD* and *SDE* cannot be used to deal with the constants, variables, or quantifiers themselves. To do this, we need introduction and elimination rules for the quantifiers. This is what the additional rules will give us.

Before the introductions begin we should remind ourselves of an important syntactic concept from Section 5.2.2:

**Substitution Instance:**
> Let $Q(a/x)$ indicate the wff which is just like $Q$ except for having the constant a in every position where the variable x appears in $Q$. Where $\mathbb{P}$ is a closed wff of the form $(\forall x)Q$ or $(\exists x)Q$, then $Q(a/x)$ is a *substitution instance* of $\mathbb{P}$, with a as the *instantiating constant*. (See page 160 for some examples.)

The idea here is that of removing the leftmost quantifier from a closed wff and replacing each instance of the variable bound by that quantifier with the same constant, resulting in a new closed wff. This syntactic transformation will be important to understanding the statement of the rules below.

I will now begin introducing the four derivation rules. Two of them, ∀E and ∃I, are quite easy and intuitive. I like to call them the friendly rules, and I will start with these. The other two, the unfriendly rules ∀I and ∃E, will come later.

> **Note:** A complete table of derivation rules appears in Appendix C

## 7.1.1 Universal Elimination—∀E

Probably the simplest and most obvious quantificational inference is the move from a universal claim to one of its instances (sometimes called Universal Instantiation). To illustrate, consider the following argument:

> Everything is looking up
> ───────────────
> Fred is looking up

The conclusion is obvious. Assuming everything is looking up, it follows that Fred is. Note, also, that we could have concluded that this book is looking up, or that the Sun is looking up, or... From any universal claim each particular instance follows.

What we need, then, is a syntactic rule which mimics such inferences. Enter Universal Elimination:

∀E

$$
\begin{array}{c|l}
i & (\forall x)\mathbb{P} \\
\triangleright & \mathbb{P}(a/x) \qquad i\ \forall E
\end{array}
$$

According to the rule schema, given a universally quantified wff we can derive any substitution instance of it. There is no restriction on the instantiating constant, a. In the justification we list the line containing the universally quantified wff.

The following derivation, which parallels the simple argument above, shows ∀E in action:

$$
\begin{array}{c|l l l}
1 & (\forall z)Lz & P & \vdash Lf \\
2 & Lf & 1\ \forall E &
\end{array}
$$

Here the instantiating constant is 'f', for Fred, but it could as easily be 'b$_3$', 's', 'g', or whatever you want. Choice of the instantiating constant is unrestricted, and the (semantic) interpretation of the constant is irrelevant to the (syntactic) derivation rules. It is, however, important to choose a constant that will allow you to complete the derivation at hand. So 'b$_3$', 's', and 'g' won't do us much good in the derivation above.

It is helpful to see how ∀E works in a slightly more typical argument:

All geniuses are misunderstood
Einstein was a genius
Einstein was misunderstood

Most people naturally make the inference from the premises to the conclusion in a single step. But we might also draw it out a bit by inserting an intermediate step (line (3)):

| | | |
|---|---|---|
| (1) | All geniuses are misunderstood | premise |
| (2) | Einstein was a genius | premise |
| (3) | If Einstein was a genius, then Einstein was misunderstood | from (1) |
| (4) | So, Einstein was misunderstood | (2), (3) |

Line (3) is a particular instance of the claim in line (1). Thus, the step from line (1) to line (3) can be mimicked by ∀E. Here, then, is how the corresponding derivation in **PD** looks:

$$
\begin{array}{c|l l l}
1 & (\forall x)(Gx \rightarrow Mx) & P & \\
2 & Ge & P & \vdash Me \\
3 & Ge \rightarrow Me & 1\ \forall E & \\
4 & Me & 2,3 \rightarrow E &
\end{array}
$$

As you can see, we use ∀E to derive line 3, an instance of line 1. Then lines 2 and 3 allow a familiar →E in order to reach the goal wff.

A final example:

All geniuses are misunderstood
All misunderstood people suffer from loneliness
Einstein was a genius
_____
Einstein suffered from loneliness

Symbolically deriving the conclusion from the premises can be done via two applications each of ∀E and →E:

$$
\begin{array}{lll}
1 & (\forall x)(Gx \rightarrow Mx) & P \\
2 & (\forall y)(My \rightarrow Ly) & P \\
3 & Ge & P \qquad \vdash Le \\
\hline
4 & Ge \rightarrow Me & 1\ \forall E \\
5 & Me \rightarrow Le & 2\ \forall E \\
6 & Me & 3, 4 \rightarrow E \\
7 & Le & 5, 6 \rightarrow E \\
\end{array}
$$

Lines 4 and 5 are particular instances of lines 1 and 2, respectively, and they ultimately allow the derivation of the goal sentence. Note how ∀E allows us to move from quantified wffs to non-quantified wffs, whereupon we can apply the rules familiar from **SD**.

Keep in mind that when applying ∀E *all* instances of the formerly quantified variable x must be replaced by *the same instantiating constant*. The following two attempts at applying ∀E are both mistakes:

$$
\begin{array}{lll}
1 & (\forall x)(Gx \rightarrow Mx) & P \\
\hline
2 & Ge \rightarrow Mx & 1\ \forall E \qquad \textbf{MISTAKE} \\
\end{array}
$$

This is a mistake, because we have failed to replace the final 'x' with a constant. Hence, line 2 is neither a closed wff, nor a substitution instance of line 1. ∀E requires that the result be a substitution instance of the universally quantified wff, and, as such, the result will be a closed wff. This mistaken application of ∀E, if it admits a meaningful interpretation at all, would be something like moving from 'All geniuses are misunderstood' to 'If Einstein was a genius, then *it* was misunderstood', where the reference of the '*it*' is indeterminate.

Here is another potential mistake:

$$
\begin{array}{lll}
1 & (\forall x)(Gx \rightarrow Mx) & P \\
\hline
2 & Ge \rightarrow Mf & 1\ \forall E \qquad \textbf{MISTAKE} \\
\end{array}
$$

Here line 2 is a closed wff, but two different constants have been used to replace instances of the same variable. So, line 2 is not a substitution instance of line 1, in violation of the rule. Hence, line 2 is a mistake. It is as if, from 'All geniuses are misunderstood', we have inferred 'If Einstein was a genius, then Fred was misunderstood'.

Another mistake:

$$
\begin{array}{lll}
1 & Ma \rightarrow (\forall x)Gx & P \\
2 & Ma \rightarrow Ge & 1\ \forall E \qquad \textbf{MISTAKE} \\
\end{array}
$$

Here, since the arrow, not the universal, is the main operator of line 1, attempting ∀E on 1 is a mistake. If we could derive 'Ma'—which we cannot do here—and use →E to get '(∀x)Gx' on a line by itself, then we could apply ∀E to that. Similarly:

| 1 | (∀x)Gx → Ma | P | |
|---|---|---|---|
| 2 | Ga → Ma | 1 ∀E | **MISTAKE** |

Without any parentheses in line 1, the scope of the universal quantifier extends only to the 'Gx', and the arrow is again the main operator. Thus, attempting to apply ∀E to line 1 is a mistake. We would somehow have to have '(∀x)Gx' on a line by itself—which we cannot do here—in order to apply ∀E. Here is a similar situation which is *not* a mistake:

| 1 | (∀x)(Gx → Ma) | P | |
|---|---|---|---|
| 2 | Ga → Ma | 1 ∀E | Correct! |

Since now the parentheses are present in line 1, the universal quantifier *is* the main operator. Thus, it is entirely correct to apply ∀E. Note that the consequent in line 1 contains the constant 'a', rather than a variable. This is perfectly fine. There are no restrictions on the choice of constant when applying ∀E. So we can choose 'a' to create 'Ga → Ma', as above, or some other constant, but the 'a' in 'Ma' should not be changed.

One last potential mistake worth looking at:

| 1 | (∃x)(∀y)Lxy | P | |
|---|---|---|---|
| 2 | (∃x)Lxf | 1 ∀E | **MISTAKE** |

Here, again, line 2 is not a substitution instance of line 1, and the rule has been misapplied. Any substitution instance of line 1 must result from removing the *outermost* quantifier (in this case the '(∃x)'). Another way to see the mistake is that we are trying to apply ∀E to a wff in line 1 which does not have '(∀y)' as its main operator. That is like trying to apply ∧E to a wff with '→' as the main connective. It turns out that the wff in line 2 is, in fact, derivable from that in line 1, but it will require more maneuvering, including ∃E, which we have yet to introduce.

## 7.1.2 Existential Introduction—∃I

Another simple quantificational inference is the move from a particular claim to a corresponding existential (sometimes called Existential Generalization). For example:

> Fred is looking up
> _____
> Something is looking up

This is a fairly obvious inference. Assuming Fred is, indeed, looking up, then there exists at least one thing which is looking up. In general, given a particular claim, we can infer the corresponding existential.

So, we need a syntactic rule which mimics this inference. Hence, Existential Introduction:

∃I
_____

$i$ | $\mathbb{P}(a/x)$

▷ | $(\exists x)\mathbb{P}$     $i$ ∃I

This rule allows us to derive an existentially quantified wff from one of its substitution instances. In the justification we cite the line containing the substitution instance. In other words, given a wff containing one or more occurrences of constant a, you may replace one or more of those occurrences of a with a variable x and append an existential x-quantifier. That is, you may existentially quantify into or existentially generalize over one or more of those a positions. The following derivation corresponds to the previous argument:

1 | Lf                    P        ⊢ $(\exists x)Lx$
2 | $(\exists x)Lx$           1 ∃I

Where multiple instances of the same constant appear, the rule allows the choice of which and how many of the constants to replace. So each of the following is a correct application of the rule (with a corresponding English argument):

| | | | |
|---|---|---|---|
| 1 | Gff | P | ⊢ $(\exists x)Gfx$ |
| 2 | $(\exists x)Gfx$ | 1 ∃I | |

Fred is good to himself
_____
Fred is good to someone

In the derivation above we simply quantify into the second 'f' position to mimic the English inference.

| | | | |
|---|---|---|---|
| 1 | Gff | P | ⊢ $(\exists x)Gxf$ |
| 2 | $(\exists x)Gxf$ | 1 ∃I | |

Fred is good to himself
_____
Someone is good to Fred

Above we quantify into the first 'f' position.

| | | | |
|---|---|---|---|
| 1 | Gff | P | ⊢ $(\exists x)Gxx$ |
| 2 | $(\exists x)Gxx$ | 1 ∃I | |

Fred is good to himself
_____
Someone is good to himself

This time we quantify into both 'f' positions.

| | | | |
|---|---|---|---|
| 1 | Gff | P | ⊢ $(\exists y)(\exists x)Gxy$ |
| 2 | $(\exists x)Gxf$ | 1 ∃I | |
| 3 | $(\exists y)(\exists x)Gxy$ | 2 ∃I | |

Fred is good to himself
_____
Someone is good to Fred
Someone is good to someone

Here, in line 2 we quantify into the first 'f' position. Then, in a separate application of ∃I, using a new variable, we quantify into the remaining 'f' position. Note that, had we not used a new variable in line 3 the result, '$(\exists x)(\exists x)Gxx$', would not qualify as a wff and so would be a violation of the rule. (Recall, also, that '$(\exists x)Gxx$' and '$(\exists y)(\exists x)Gxy$' are not equivalent.)

Note, however, that we cannot replace different constants with the same variable. The following is a mistake:

| 1 | Gfk | P | Fred is good to Kate |
|---|-----|---|---------------------|
| 2 | $(\exists x)Gxx$ | 1 $\exists$I   $\Leftarrow$ **MISTAKE** $\Rightarrow$ | Someone is good to himself |

The conclusion of the English argument clearly does not follow from the premise. Accordingly, the parallel move in **PD** would be in violation of the rule.

Let's take a look at $\exists$I in conjunction with $\forall$E:

> All geniuses are misunderstood
> Einstein was a genius
> Someone was misunderstood

The following derivation mimics the argument:

| 1 | $(\forall x)(Gx \rightarrow Mx)$ | P | |
|---|----|---|---|
| 2 | Ge | P | $\vdash (\exists x)Mx$ |
| 3 | Ge $\rightarrow$ Me | 1 $\forall$E | |
| 4 | Me | 2, 3 $\rightarrow$E | |
| 5 | $(\exists x)Mx$ | 4 $\exists$I | |

It is useful to note that, without line 2 (the second premise of the argument) the goal sentence could not be derived. We could stop at line 3, or apply '$\exists$I' to 3 to get '$(\exists x)(Gx \rightarrow Mx)$', though this would not get us to the goal wff.

## 7.1.3  Universal Introduction — $\forall$I

Consider the following argument.

> All geniuses are misunderstood
> Everyone is a genius
> Everyone is misunderstood

It is not difficult to see that it is valid, and we can make the inference from premises to conclusion in a single step. But to make fully explicit what underlies such an inference we can expand the line of reasoning, thus:

| (1) | All geniuses are misunderstood | premise |
|-----|-------------------------------|---------|
| (2) | Everyone is a genius | premise |
| (3) | If, say, Susie is a genius, then Susie is misunderstood | from (1) |
| (4) | Susie is a genius | (2) |
| (5) | Susie is misunderstood | (3), (4) |
| (6) | Everyone is misunderstood | (5) |

Note, first, that the name 'Susie' was chosen completely arbitrarily—we could as well have chosen 'Fred' or 'Einstein', and the steps above would follow. Because of this, nothing in the line of reasoning depends on the particular choice of the name 'Susie'. That is, nothing above depends on any special knowledge, assumptions, or premises regarding Susie. Hence, the crucial step from (5) to (6) (a Universal Generalization), is justified. Given (1) and (2) we could assert (3)–(5) about *anyone*, hence, given the premises, (6) follows: *Everyone* is misunderstood. Note, finally, that if we took line (5), "Susie is misunderstood", as a sole premise, it would *not* follow that "Everyone is misunderstood"; for in that case the premise offers special knowledge about Susie.

What we need, then, is a rule of *PD* which allows us to move from a particular substitution instance to a universally quantified wff, but which allows this *only in the appropriate circumstances*. Universal Introduction does this:

∀I

| | |
|---|---|
| $i$ | $\mathbb{P}(a/x)$ |
| ▷ | $(\forall x)\mathbb{P}$    $i$ ∀I |

Provided:
   (i)  a does not occur in an
        undischarged assumption.
   (ii) a does not occur in $(\forall x)\mathbb{P}$.

What this rule tells us is that we may move to a universally quantified wff '$(\forall x)\mathbb{P}$' from an instance, '$\mathbb{P}(a/x)$', but only if the two conditions are met. Condition (i) ensures that the constant, a, is entirely arbitrary (i.e., no special assumptions about a are still in play). Condition (ii) requires that every instance of a be replaced by the variable, x. The justification cites the line number of the substitution instance. In other words, the rule allows us to universally quantify into or universally generalize over *all* those a positions (condition (ii)), given that condition (i) is also met.

So the preceding argument is paralleled by:

| 1 | $(\forall x)(Gx \rightarrow Mx)$ | P |
|---|---|---|
| 2 | $(\forall z)Gz$ | P      ⊢ $(\forall x)Mx$ |
| 3 | $Gs \rightarrow Ms$ | 1 ∀E |
| 4 | $Gs$ | 2 ∀E |
| 5 | $Ms$ | 3, 4 →E |
| 6 | $(\forall x)Mx$ | 5 ∀I |

Take a look at the step from line 5 to line 6. Line 5 is clearly a substitution instance of line 6, so the form of the rule is met. Both restrictions are met as well. (i) The relevant constant 's' does not appear in any undischarged assumptions (lines 1 and 2), and (ii) 's' does not appear in line 6.

As always, it is useful to look at some of the more common mistakes. Here is one way of making a mistake with ∀I:

| 1 | $Ms$ | P | | Susie is misunderstood |
|---|---|---|---|---|
| 2 | $(\forall x)Mx$ | 1 ∀I | ⇐ **MISTAKE** ⇒ | Everyone is misunderstood |

This fits the form of the rule: line 1 is a substitution instance of line 2. The constant 's', however, does appear in an undischarged assumption (line 1). Hence, we have violated restriction (i) on ∀I, and our attempt is a mistake (as can be seen in the obviously invalid parallel argument).

Here is another way of making a mistake with ∀I:

| | | | |
|---|---|---|---|
| 1 | (∀z)Gzz | P | Everyone is good to himself |
| 2 | Gff | 1 ∀E | Fred is good to himself |
| 3 | (∀x)Gxf | 2 ∀I  ⇐ **MISTAKE** ⇒ | Everyone is good to Fred |

The step from line 2 to line 3 again fits the form of the rule since line 2 is a substitution instance of line 3. But the constant 'f' appears in the result (line 3)—we did not replace both of them as we should. This violates restriction (ii), so we have made a mistake (as the invalid parallel argument shows).

Let's take a look at one more derivation involving ∀I. Consider the following argument:

> All geniuses are misunderstood
> All misunderstood people suffer from loneliness
> All geniuses suffer from loneliness

Here is how the derivation runs:

| | | | |
|---|---|---|---|
| 1 | (∀z)(Gz → Mz) | P | |
| 2 | (∀y)(My → Ly) | P | ⊢ (∀x)(Gx → Lx) |
| 3 | Gs → Ms | 1 ∀E | |
| 4 | Ms → Ls | 2 ∀E | |
| 5 | Gs | A | |
| 6 | Ms | 3, 5 →E | |
| 7 | Ls | 4, 6 →E | |
| 8 | Gs → Ls | 5–7 →I | |
| 9 | (∀x)(Gx → Lx) | 8 ∀I | |

Here the step of interest is from line 8 to line 9. It fits the form of ∀I. It meets restriction (i): 's' does not appear in any undischarged assumptions. The constant 's' does appear in the assumption of line 5, but with the application of →I in line 8, the line 5 assumption is discharged. The only remaining undischarged assumptions are lines 1 and 2, and 's' does not appear. (Keep in mind, since lines 1 and 2 are primary assumptions, they never get discharged.) Finally, it meets restriction (ii) since 's' does not appear in line 9. Hence, we have an appropriate application of ∀I.[3]

---

3. Of course, by using HS from *SDE* we could eliminate the subderivation and shorten the whole thing.

## 7.1.4  Existential Elimination—∃E

The following argument is valid:

> All geniuses are misunderstood
> Some geniuses are agoraphobes
> ――――――――――――――――――――――
> Some agoraphobes are misunderstood

For many of us, the validity of this argument is not as obvious as some of the previous ones, but we could reason through it as follows:

| | | |
|---|---|---|
| (1) | All geniuses are misunderstood | premise |
| (2) | Some geniuses are agoraphobes | premise |
| (3) | Suppose Fred is a genius and an agoraphobe | assumption |
| (4) | If Fred is a genius, then Fred is misunderstood | from (1) |
| (5) | Fred is a genius | (3) |
| (6) | Fred is misunderstood | (4), (5) |
| (7) | Fred is an agoraphobe | (3) |
| (8) | Fred is an agoraphobe and misunderstood | (6), (7) |
| (9) | Some agoraphobes are misunderstood | (8) |
| (10) | Some agoraphobes are misunderstood | (3)–(9) |

The second premise, line (2), is an existential generalization. There is at least one genius agoraphobe, but we are not told his/her name. To reason with this information we choose an arbitrary name—a name about which we do not have any special knowledge, assumptions, or premises. In this case we choose 'Fred' (we could as well have chosen 'Susie', 'Einstein', etc.), and suppose that Fred is this genius agoraphobe. Once we have a name, we can make familiar steps in lines (4)–(9). Many of these steps depend on the supposition in line (3) involving Fred. But by the time we arrive at line (9) all mention of Fred has disappeared. Indeed, had we chosen any other name in line (3), we could execute the same steps, ending up with the same claim in line (9)—the name 'Fred', or 'Susie', or whatever we choose, is just an arbitrary place-holder. Thus, line (9) does not actually depend on line (3), but only on lines (1) and (2). So we rewrite the claim of line (9) on line (10), no longer indented. We have reasoned through to our conclusion.

What we need in *PD*, then, is a rule which allows us to perform a similar procedure—the elimination of an existentially quantified claim in favor of a substitution instance, where the instantiating constant of that instance eventually disappears from the derivation. Let's look at Existential Elimination:

∃E
――――――――――――――――――

| | | |
|---|---|---|
| $i$ | $(\exists x)\mathbb{P}$ | |
| $j$ | $\mathbb{P}(a/x)$ | A |
| $k$ | $\mathbb{Q}$ | |
| ▷ | $\mathbb{Q}$ | $i, j\text{–}k$ ∃E |

Provided:

(i)  a does not occur in an undischarged assumption.

(ii)  a does not occur in $(\exists x)\mathbb{P}$.

(iii)  a does not occur in $\mathbb{Q}$.

Given an existentially quantified wff, $(\exists x)\mathbb{P}$, this rule allows us to start a subderivation by assuming a substitution instance of that existential, $\mathbb{P}(a/x)$, derive some result, $\mathbb{Q}$, and then discharge the auxiliary assumption by ending the subderivation and writing $\mathbb{Q}$ one scope line to the left. But all this, only if the instantiating constant a (i) does not occur in any undischarged assumption, (ii) does not occur in the original existential wff, and (iii) does not occur in the result, $\mathbb{Q}$. The justification cites the line containing the existential and the lines of the subderivation.

As with the restrictions on $\forall$I, the restrictions on $\exists$E ensure that the constant in question not be the subject of any special knowledge, assumptions, or premises—i.e., that it operates as a neutral place-holder. Let's see how this works for the previous argument:

| | | |
|---|---|---|
| 1 | $(\forall w)(Gw \rightarrow Mw)$ | P |
| 2 | $(\exists y)(Gy \wedge Ay)$ | P     $\vdash (\exists z)(Az \wedge Mz)$ |
| 3 | Gf $\wedge$ Af | A |
| 4 | Gf $\rightarrow$ Mf | 1 $\forall$E |
| 5 | Gf | 3 $\wedge$E |
| 6 | Mf | 4, 5 $\rightarrow$E |
| 7 | Af | 3 $\wedge$E |
| 8 | Af $\wedge$ Mf | 6, 7 $\wedge$I |
| 9 | $(\exists z)(Az \wedge Mz)$ | 8 $\exists$I |
| 10 | $(\exists z)(Az \wedge Mz)$ | 2, 3–9 $\exists$E |

In line 3 we assume an instance of the existential of line 2, with 'f' as the instantiating constant. This allows a number of steps leading to line 9. Here the instantiating constant disappears, and the result can be rewritten one scope line to the left. All three restrictions are met: (i) 'f' does not occur in any undischarged assumptions (line 3 is discharged in applying the rule), (ii) 'f' does not appear in the existential in question (line 2), and (iii) 'f' does not appear in the result (lines 9 and 10).

Despite the name of the rule, $\exists$E often, though not always, results in a new existential wff. This is because the most common way of meeting condition (iii) is by reintroducing an existential via $\exists$I prior to ending the subderivation, thereby eliminating the instantiating constant (see line 9 above). The name 'Existential Elimination' refers to the elimination of the existential of line 2 in the auxiliary assumption. Not every correct use of $\exists$E results in a new existential, however. Consider:

| | | |
|---|---|---|
| 1 | $(\exists x)Gx$ | P |
| 2 | $(\forall z)(Gz \rightarrow Mz)$ | P |
| 3 | $(\exists y)My \rightarrow Ge$ | P     $\vdash Ge$ |
| 4 | Gs $\rightarrow$ Ms | 2 $\forall$E |
| 5 | Gs | A |
| 6 | Ms | 4, 5 $\rightarrow$E |
| 7 | $(\exists y)My$ | 6 $\exists$I |
| 8 | Ge | 3, 7 $\rightarrow$E |
| 9 | Ge | 1, 5–8 $\exists$E |

There are a number of interesting things to say about this derivation. First, note that line 4, which occurs prior to the subderivation, could have occurred during the subderivation if we had so chosen. Second, lines 8 and 9, which play the role of Q in the application of ∃E, do not contain an existential. Third, we have, nonetheless, correctly applied the rule. In line 1 we have an existential wff, in line 5 we assume a substitution instance of it, and in lines 8 and 9 we arrive at an appropriate result. We meet all restrictions since the instantiating constant, 's', does not appear (i) in any undischarged assumptions, nor (ii) in the existential of line 1, nor (iii) in the result (lines 8 and 9). Fourth, the result contains the constant 'e', which does appear in an undischarged assumption (line 3), but this does not lead to a violation of any of the restrictions on ∃E. The constant relevant to the restrictions is 's', the one introduced in line 5. The only way the presence of 'e' in line 3 could be a problem is if we had chosen 'e' as our instantiating constant in the assumption of line 5. In that case, when trying to apply ∃E in the move from 8 to 9, we would have violated restrictions (i) and (iii). Thus:

| | | | |
|---|---|---|---|
| 1 | $(\exists x)Gx$ | P | |
| 2 | $(\forall z)(Gz \rightarrow Mz)$ | P | |
| 3 | $(\exists y)My \rightarrow Ge$ | P | ⊢ Ge |
| 4 | Ge → Me | 2 ∀E | |
| 5 | Ge | A | |
| 6 | Me | 4, 5 →E | |
| 7 | $(\exists y)My$ | 6 ∃I | |
| 8 | Ge | 3, 7 →E | |
| 9 | Ge | 1, 5–8 ∃E    **MISTAKE** | |

Lines 6 and 7 here are somewhat unnecessary, but that is not the point. Note that the instantiating constant, 'e', appears in an undischarged assumption (line 3), violating restriction (i). It also appears in the Q result of lines 8 and 9, violating restriction (iii). Thus the attempted ∃E is a mistake. The next section discusses some strategies regarding choice of instantiating constants.

Finally, notice that the whole derivation could have gone slightly differently. We could have done the ∀E later, as mentioned above. In addition, we could have ended the subderivation with ∃E a bit earlier as well. Hence:

| | | | |
|---|---|---|---|
| 1 | $(\exists x)Gx$ | P | |
| 2 | $(\forall z)(Gz \rightarrow Mz)$ | P | |
| 3 | $(\exists y)My \rightarrow Ge$ | P | ⊢ Ge |
| 4 | Gs | A | |
| 5 | Gs → Ms | 2 ∀E | |
| 6 | Ms | 4, 5 →E | |
| 7 | $(\exists y)My$ | 6 ∃I | |
| 8 | $(\exists y)My$ | 1, 4–7 ∃E | |
| 9 | Ge | 3, 8 →E | |

Here, unlike our first successful example above, the ∃E does result in a new existential. The point is that, in this derivation and many others, the steps could go either way.

## 7.1.5 **Exercises**

Complete each of the following derivations.

**1.**

| | | | |
|---|---|---|---|
| 1 | $(\forall z)(Kz \rightarrow Jz)$ | P | |
| 2 | Kn | P | ⊢ Jn |

**2.**

| | | | |
|---|---|---|---|
| 1 | $(\forall y)Gy$ | P | |
| 2 | $(\forall z)(Gz \rightarrow Hz)$ | P | ⊢ $(\exists x)Hx$ |

**3.**

| | | | |
|---|---|---|---|
| 1 | $(\forall w)(Dw \leftrightarrow Cw)$ | P | ⊢ $(\forall x)Cx \rightarrow (\exists x)Dx$ |

**4.**

| | | | |
|---|---|---|---|
| 1 | $(\forall x)(Mx \rightarrow Bx)$ | P | |
| 2 | ¬Bc | P | ⊢ ¬Mc |

**5.**

| | | | |
|---|---|---|---|
| 1 | Fa | P | |
| 2 | $(\forall x)(Mx \leftrightarrow Gx)$ | P | |
| 3 | $(\forall x)(Fx \rightarrow Gx)$ | P | ⊢ $(\exists z)(Mz \wedge Gz)$ |

**6.**

| | | | |
|---|---|---|---|
| 1 | $(\forall y)My$ | P | |
| 2 | $(\forall z)(Kz \leftrightarrow Mz)$ | P | ⊢ $(\forall z)Kz$ |

**7.**

| | | | |
|---|---|---|---|
| 1 | $(\exists x)Fx$ | P | |
| 2 | $(\forall x)(Fx \rightarrow Gx)$ | P | ⊢ $(\exists x)Gx$ |

**8.**   1 │ $(\forall x)(Ax \rightarrow Bx)$          P       $\vdash (\exists x)Ax \rightarrow (\exists x)Bx$

**9.**   1 │ $(\forall x)Cx$                    P
         2 │ $(\forall x)(Cx \rightarrow (Dx \wedge Ex))$   P       $\vdash (\exists x)Dx \wedge (\forall y)Ey$

**10.**  1 │ $(\exists x)Cx$                    P
         2 │ $(\forall x)(Cx \rightarrow (Dx \wedge Ex))$   P       $\vdash (\exists x)Dx \wedge (\exists x)Ex$

## 7.2 Derivations: Strategies and Notes

As with derivations in *SD* and *SDE*, a key to success with derivations in *PD* will be the ability to work backward by allowing the goal sentence to guide your strategy. All of the lessons learned in Chapter 4 still apply, but there are some particulars to add regarding the four new rules of *PD*.

### The Friendly Rules

$\forall$E and $\exists$I are completely unrestricted and do not use subderivations, hence they are very "friendly" rules. Thinking in terms of working forward: whenever you have a universally quantified claim, you may derive any instance via $\forall$E; whenever you have some particular claim (i.e., involving some individual constant), you may derive an appropriate existential quantification of it via $\exists$I. Thinking backward: one way of getting a particular claim is from a relevant universal (if present); one way of getting an existential is from a relevant particular. Here is a simple example:

   1 │ Fe                          P
   2 │ $(\forall z)(Fz \rightarrow Gz)$          P       $\vdash (\exists x)(Fx \wedge Gx)$

This is not a horribly complex derivation, but it will allow us to illustrate some simple points. First, note that the goal is an existential wff. Hence, the easiest way to get this would be through an $\exists$I. That means we need, as subgoal, a substitution instance of the goal. This is not guaranteed to work, but we'll give it a try. The next issue, then, is what constant that substitution instance should incorporate. Since 'e' appears in the primary assumptions, we might try that. Hence:

```
1 │ Fe                        P
2 │ (∀z)(Fz → Gz)             P        ⊢ (∃x)(Fx ∧ Gx)
  │
  │
i │ Fe ∧ Ge                   1, ?? ∧I
j │ (∃x)(Fx ∧ Gx)             i ∃I
```

So line $i$ is our new (sub)goal, and since it will pretty obviously come from 1 by ∧I, I have entered that as partial justification. Now clearly, we need a line containing just 'Ge', for this will allow the ∧I we want to produce line $i$. Thus:

```
1 │ Fe                        P
2 │ (∀z)(Fz → Gz)             P        ⊢ (∃x)(Fx ∧ Gx)
  │
h │ Ge                        ??
i │ Fe ∧ Ge                   1, h ∧I
j │ (∃x)(Fx ∧ Gx)             i ∃I
```

The only problem at this point, then, is how to get the 'Ge' of line $h$. The key, of course, is to see that line 2 is just a universally quantified conditional with 'Gz' as its consequent. If we could get 'Fe → Ge', we could easily get line $h$, and we'd be home free. But, of course, ∀E will give us what we need—the conditional in question is just a substitution instance of line 2. So:

```
1 │ Fe                        P
2 │ (∀z)(Fz → Gz)             P        ⊢ (∃x)(Fx ∧ Gx)
3 │ Fe → Ge                   2 ∀E
4 │ Ge                        1, 3 →E
5 │ Fe ∧ Ge                   1, 4 ∧I
6 │ (∃x)(Fx ∧ Gx)             5 ∃I
```

There it is. You should always keep in mind that you can apply ∀E and ∃I without any worries.

## The Not-So-Friendly Rules

∀I has two restrictions, and ∃E has three restrictions and involves a subderivation. Hence, these two rules are not so friendly. In this and the next section we'll review some of the basic twists and turns which need to be kept in mind.

Contrast the previous derivation with the following, where line 1 and the goal are universal wffs:

```
1 │ (∀x)Fx                    P
2 │ (∀z)(Fz → Gz)             P        ⊢ (∀x)(Fx ∧ Gx)
  │
```

Working backward, we will again need a particular claim 'Fe ∧ Ge', say, as our second to last line, *i*, so that our final line, *j*, can be derived from ∀I. At this point, we may not be sure whether or not we will avoid violating the conditions on ∀I, but let's proceed and see what happens. Further, it looks likely that 'Fe ∧ Ge' will come from ∧I. (Note that we need not choose 'e' as the constant here—I do so only to maintain the parallel with the above derivation.) So our position is as follows:

| | | | |
|---|---|---|---|
| 1 | (∀x)Fx | P | |
| 2 | (∀z)(Fz → Gz) | P | ⊢ (∀x)(Fx ∧ Gx) |
| | | | |
| *g* | Fe | ?? | |
| *h* | Ge | ?? | |
| *i* | Fe ∧ Ge | *g*, *h* ∧I | |
| *j* | (∀x)(Fx ∧ Gx) | *i* ∀I | |

Now, clearly, we can get 'Fe' in line *g* from line 1 via ∀E. Moreover, using another ∀E on line 2 will give us 'Fe → Ge'. Then an application of →E will give us 'Ge', thus:

| | | | |
|---|---|---|---|
| 1 | (∀x)Fx | P | |
| 2 | (∀z)(Fz → Gz) | P | ⊢ (∀x)(Fx ∧ Gx) |
| 3 | Fe | 1 ∀E | |
| 4 | Fe → Ge | 2 ∀E | |
| 5 | Ge | 3, 4 →E | |
| 6 | Fe ∧ Ge | 3, 5 ∧I | |
| 7 | (∀x)(Fx ∧ Gx) | 6 ∀I | |

The only remaining question, then, is whether we have abided by the restrictions on ∀I in our step from line 6 to line 7. The instantiating constant in question is 'e'. It does not appear in any undischarged assumptions, satisfying restriction (i). Nor does it appear in the result, line 7, satisfying restriction (ii). So we have successfully completed the derivation.

Note that we avoided violating restriction (i) on ∀I because the variable in question entered the derivation only through an application of ∀E, hence it did not appear in any undischarged assumptions. This is a general rule: if a constant, a, enters a derivation only through an application of ∀E, then ∀I can be safely applied to that constant, a.

Suppose, now, we have the same derivation, except that line 1 is 'Fe'. This is a derivation that cannot be completed without violating some rule or other, but this is what I want to illustrate:

| | | | |
|---|---|---|---|
| 1 | Fe | P | |
| 2 | (∀z)(Fz → Gz) | P | ⊢ (∀x)(Fx ∧ Gx)  ⊬ ! |
| 3 | Fe → Ge | 2 ∀E | |
| 4 | Ge | 1, 3 →E | |
| 5 | Fe ∧ Ge | 1, 4 ∧I | |
| 6 | (∀x)(Fx ∧ Gx) | 5 ∀I | **MISTAKE** |

The problem here is that in moving from line 5 to line 6, we violate restriction (i) on ∀I, for the relevant constant, 'e', appears in an undischarged assumption, line 1.

Let's look at one more variation on this derivation. Now let line 1 and the goal sentence be existential wffs, thus:

$$
\begin{array}{ll}
1 & (\exists x)Fx \\
2 & (\forall z)(Fz \rightarrow Gz)
\end{array}
\qquad
\begin{array}{l}
P \\
P \qquad \vdash (\exists x)(Fx \wedge Gx)
\end{array}
$$

We might think backward through this in much the same way as we did with the others. We seem to need 'Fe ∧ Ge' as our second to last line, so that we can apply ∃I to reach the goal. 'Fe ∧ Ge' will clearly come from ∧I, and 'Fe → Ge' will come from line 2 via ∀E, thus:

$$
\begin{array}{ll}
1 & (\exists x)Fx \\
2 & (\forall z)(Fz \rightarrow Gz) \\
\\
g & Fe \rightarrow Ge \\
h & Ge \\
i & Fe \wedge Ge \\
j & (\exists x)(Fx \wedge Gx)
\end{array}
\qquad
\begin{array}{l}
P \\
P \qquad \vdash (\exists x)(Fx \wedge Gx) \\
\\
2\ \forall E \\
??,\ g \rightarrow E \\
??,\ h \wedge I \\
i\ \exists I
\end{array}
$$

This looks pretty good, so far. All we need is 'Fe', and can't we just get that straight from line 1? We might try this:

$$
\begin{array}{ll}
1 & (\exists x)Fx \\
2 & (\forall z)(Fz \rightarrow Gz) \\
3 & Fe \\
4 & Fe \rightarrow Ge \\
5 & Ge \\
6 & Fe \wedge Ge \\
7 & (\exists x)(Fx \wedge Gx)
\end{array}
\qquad
\begin{array}{l}
P \\
P \qquad \vdash (\exists x)(Fx \wedge Gx) \\
1\ \exists E \qquad \textbf{MISTAKE} \\
2\ \forall E \\
3,\ 4 \rightarrow E \\
3,\ 5 \wedge I \\
6\ \exists I
\end{array}
$$

Of course, line 3 is a mistake. Unlike with a universal wff, we cannot move from an existential straight to a particular. ∃E involves a subderivation and three restrictions. But not all is lost, we simply need to move all of our work on lines 3-7 into a subderivation, with line 3 as the assumption. This will leave us in the following position:

```
1 │ (∃x)Fx                  P
2 │ (∀z)(Fz → Gz)           P         ⊢ (∃x)(Fx ∧ Gx)
  │
3 │ │ Fe                    A
  │ │
4 │ │ Fe → Ge               2 ∀E
5 │ │ Ge                    3, 4 →E
6 │ │ Fe ∧ Ge               3, 5 ∧I
7 │ │ (∃x)(Fx ∧ Gx)         6 ∃I
  │
```

We are not finished yet. Although we have the goal wff, it is still inside a subderivation. But now we are in a position to apply ∃E. We have an existential in line 1 and a subderivation which begins with an instance of that existential. Thus:

```
1 │ (∃x)Fx                  P
2 │ (∀z)(Fz → Gz)           P         ⊢ (∃x)(Fx ∧ Gx)
  │
3 │ │ Fe                    A
  │ │
4 │ │ Fe → Ge               2 ∀E
5 │ │ Ge .                  3, 4 →E
6 │ │ Fe ∧ Ge               3, 5 ∧I
7 │ │ (∃x)(Fx ∧ Gx)         6 ∃I
8 │ (∃x)(Fx ∧ Gx)           1, 3–7 ∃E
```

The only issue is whether we have met all the restrictions of ∃E. The constant in question is 'e', (i) it is not in any undischarged assumptions, (ii) it is not in line 1, and (iii) it is not in line 7/8. Hence, we have successfully completed the derivation.

Of course, we could have avoided this slight backtracking if we had noticed that all we had to deal with in the primary assumptions was an existential. Nearly always, if there is an existential in the primaries, you will need to do a subderivation for ∃E at some point in your derivation. It is useful to keep this in mind when strategizing.

## Choice of Constants

As we have seen, ∀I and ∃E have restrictions involving the constants used. Consequently, the choice of constants used in a derivation can be crucial. Sometimes we need to choose a constant which does not appear earlier in the derivation, sometimes we need to choose one which does appear earlier, and sometimes it does not matter.

Consider the following derivation:

```
1 │ (∀x)(∀y)Cxy          .  P         ⊢ (∀y)(∃x)Cxy
2 │ (∀y)Cby                 1 ∀E
3 │ Cbb                     2 ∀E
4 │ (∃x)Cxb                 3 ∃I
5 │ (∀y)(∃x)Cxy             4 ∀I
```

In line 3 we choose the same constant, 'b', used in line 2. But this creates no problem. Because ∃I, applied in line 4, does not require us to quantify into all instances of the same constant, we can leave the second instance of 'b' alone. In line 5 we universally quantify into the remaining 'b'. We could as well have chosen a different constant in line 4, thus:

| | | | |
|---|---|---|---|
| 1 | (∀x)(∀y)Cxy | P | ⊢ (∀y)(∃x)Cxy |
| 2 | (∀y)Cby | 1 ∀E | |
| 3 | Cbc | 2 ∀E | |
| 4 | (∃x)Cxc | 3 ∃I | |
| 5 | (∀y)(∃x)Cxy | 4 ∀I | |

It makes no difference.

But contrast the following, which has the same primary assumption and a subtly different goal:

| | | | |
|---|---|---|---|
| 1 | (∀x)(∀y)Cxy | P | ⊢ (∃x)(∀y)Cyx |
| 2 | (∀y)Cby | 1 ∀E | |
| 3 | Cbb | 2 ∀E | |
| 4 | (∀y)Cyb | 3 ∀I | **MISTAKE** |
| 5 | (∃x)(∀y)Cxy | 4 ∃I | |

Here, again, we choose the same constant in line 3 as in line 2. This is not misuse of any rule, but it does force us into a mistake in the next step when trying to quantify into only the first 'b' position in line 4. Recall that restriction (ii) on ∀I requires that the instantiating constant, 'b', not appear in the result, '(∀y)Cyb'. That is, in applying ∀I to line 3, we are required to replace both instances of 'b' with the variable 'y', resulting in '(∀y)Cyy'. This would, of course, prevent us from being able to reach the goal wff. Thus, we must use two different constants in order to complete this derivation:

| | | | |
|---|---|---|---|
| 1 | (∀x)(∀y)Cxy | P | ⊢ (∃x)(∀y)Cyx |
| 2 | (∀y)Cby | 1 ∀E | |
| 3 | Cbc | 2 ∀E | |
| 4 | (∀y)Cyc | 3 ∀I | |
| 5 | (∃x)(∀y)Cyx | 4 ∃I | |

Here, in applying ∀I to line 3, we are able to quantify into exactly (and only) the position we need in order to set ourselves up for the final step from 4 to 5.

Now, suppose the goal had been '(∀y)Cyy'. In that case we would be required to choose the same constant in lines 2 and 3:

| | | | |
|---|---|---|---|
| 1 | (∀x)(∀y)Cxy | P | ⊢ (∀y)Cyy |
| 2 | (∀y)Cby | 1 ∀E | |
| 3 | Cbb | 2 ∀E | |
| 4 | (∀y)Cyy | 3 ∃I | |

The upshot of this series of examples is that if we need to use ∀I to quantify into some, but not all, constants in a wff, then we must set up the derivation by strategically choosing different variables—we must have instances of the same variable in places we do want to quantify, and instances of different variables in those places we do not want to quantify.

As one last example in this vein, consider the following:

$$
\begin{array}{ll}
1 \mid (\forall w)(\forall x)(\forall y)(\forall z)Fwxyzxw & \text{P} \qquad \vdash (\forall x)(\forall y)Fxyxxyx \\
2 \mid
\end{array}
$$

This is not as tough as it looks. If we check out the goal we can see that we need a second to last line which looks like this: '(∀y)Fayaaya'. That will allow the final ∀I. Obviously, then, the line prior to that must be something like: 'Fabaaba'. The only task, then, is to see whether, by successive application of ∀E with the strategic choice of constants, we can produce 'Fabaaba'. Here it is:

$$
\begin{array}{lll}
1 \mid (\forall w)(\forall x)(\forall y)(\forall z)Fwxyzxw & \text{P} & \vdash (\forall x)(\forall y)Fxyxxyx \\
2 \mid (\forall x)(\forall y)(\forall z)Faxyzxa & 1\ \forall E \\
3 \mid (\forall y)(\forall z)Fabyzba & 2\ \forall E \\
4 \mid (\forall z)Fabazba & 3\ \forall E \\
5 \mid Fabaaba & 4\ \forall E \\
6 \mid (\forall y)Fayaaya & 5\ \forall I \\
7 \mid (\forall x)(\forall y)Fxyxxyx & 6\ \forall I
\end{array}
$$

Here is a slightly different problem one might run into with '∀I':

$$
\begin{array}{ll}
1 \mid (\forall y)Bay & \text{P} \qquad \vdash (\forall y)(\exists x)Bxy \\
2 \mid
\end{array}
$$

We might try the following:

$$
\begin{array}{lll}
1 \mid (\forall y)Bay & \text{P} & \vdash (\forall y)(\exists x)Bxy \\
2 \mid Baa & 1\ \forall E \\
3 \mid (\exists x)Bxa & 2\ \exists I \\
4 \mid (\forall y)(\exists x)Bxy & 3\ \forall I & \textbf{MISTAKE}
\end{array}
$$

The step from 3 to 4 fits the form of ∀I, but it violates restriction (i), since 'a' appears in line 1, an undischarged assumption. The remedy is simple. We should choose a constant other than 'a' when applying ∀E to produce line 2:

$$
\begin{array}{lll}
1 \mid (\forall y)Bay & \text{P} & \vdash (\forall y)(\exists x)Bxy \\
2 \mid Bab & 1\ \forall E \\
3 \mid (\exists x)Bxb & 2\ \exists I \\
4 \mid (\forall y)(\exists x)Bxy & 3\ \forall I
\end{array}
$$

The moral is, then, that if you plan to apply ∀I to a constant introduced by ∀E, be sure the constant you introduce does not already appear in an undischarged assumption.
  Though the variations are endless, I'll present just one more example.

| | | | |
|---|---|---|---|
| 1 | (∃y)Ray | P | ⊢ (∃y)(∃x)Rxy |

The way to make a mistake with this is as follows:

| | | | |
|---|---|---|---|
| 1 | (∃y)Ray | P | ⊢ (∃y)(∃x)Rxy |
| 2 | Raa | A | |
| 3 | (∃x)Rxa | 2 ∃I | |
| 4 | (∃y)(∃x)Rxy | 3 ∃I | |
| 5 | (∃y)(∃x)Rxy | 1, 2–4 ∃E | **MISTAKE** |

Similar to the previous example, the steps here fit the forms of the rules, but our attempt to execute a ∃E in line 5 winds up violating restriction (i). The constant 'a' appears in the undischarged assumption of line 1. Hence, we should not have chosen 'a' in the auxiliary assumption of line 2. Thus:

| | | | |
|---|---|---|---|
| 1 | (∃y)Ray | P | ⊢ (∃y)(∃x)Rxy |
| 2 | Rab | A | |
| 3 | (∃x)Rxb | 2 ∃I | |
| 4 | (∃y)(∃x)Rxy | 3 ∃I | |
| 5 | (∃y)(∃x)Rxy | 1, 2–4 ∃E | |

In general, when starting a subderivation for the purpose of eventually applying ∃E, we should choose a constant which is entirely new to the subderivation (or at least, which does not appear in any undischarged assumption).

## 7.3  **Proof Theory in** *PD*

As with derivations in *SD* and *SDE*, we can use *PD* to test wffs and sets of wffs for certain syntactic or proof-theoretic properties. The relevant definitions exactly parallel those of *SD* from Section 4.3, so we'll run through them quickly.

**Derivation in *PD*:**

  A *derivation in PD* is a finite sequence of wffs of *P* such that each wff is either an assumption with scope indicated or justified by one of the rules of *PD*.

**Derivable in *PD*, Γ ⊢ P:**

  A wff ℙ of *P* is *derivable in PD* from a set Γ of wffs of *P* iff there is a derivation in *PD* the primary assumptions of which are members of Γ and ℙ depends on only those assumptions.

As usual, we use the single turnstile to assert or deny derivability. To show that $\Gamma \vdash \mathbb{P}^4$ we produce a derivation, all the primary assumptions of which are members of $\Gamma$, and then derive $\mathbb{P}$. So, for example, in the previous section we showed (among other things) that

$$\{(\exists x)Gx, (\forall z)(Gz \rightarrow Mz), (\exists y)My \rightarrow Ge\} \vdash Ge \quad \text{(page 248)}$$

**Valid in** *PD*:

An argument of *P* is *valid in PD* iff the conclusion is derivable from the set consisting of only the premises, otherwise it is invalid in *PD*.

Hence, in the same derivation on page 248, we showed that

$$(\exists x)Gx$$
$$(\forall z)(Gz \rightarrow Mz)$$
$$\underline{(\exists y)My \rightarrow Ge}$$
$$Ge$$

is valid in *PD*.

**Theorem of** *PD*:

A wff $\mathbb{P}$ is a *theorem of PD* iff $\mathbb{P}$ is derivable from the empty set; i.e., iff $\vdash \mathbb{P}$.

As an example, we can show that:

$$\vdash (\forall x)Fx \rightarrow (\exists x)Fx$$

| | | | |
|---|---|---|---|
| | | | $\vdash (\forall x)Fx \rightarrow (\exists x)Fx$ |
| 1 | | $(\forall x)Fx$ | A |
| 2 | | Fa | 1 $\forall$E |
| 3 | | $(\exists x)Fx$ | 2 $\exists$I |
| 4 | | $(\forall x)Fx \rightarrow (\exists x)Fx$ | 1–3 $\rightarrow$I |

**Equivalent in** *PD*:

Two wffs $\mathbb{P}$ and $\mathbb{Q}$ are *equivalent in PD* iff they are interderivable in *PD*; i.e., iff both $\mathbb{P} \vdash \mathbb{Q}$ and $\mathbb{Q} \vdash \mathbb{P}$.

A brief example, '$(\forall x)Nx \land (\forall x)Px$' and '$(\forall x)(Nx \land Px)$' are equivalent:

| | | | |
|---|---|---|---|
| 1 | $(\forall x)Nx \land (\forall x)Px$ | P | $\vdash (\forall x)(Nx \land Px)$ |
| 2 | $(\forall x)Nx$ | 1 $\land$E | |
| 3 | $(\forall x)Px$ | 1 $\land$E | |
| 4 | Nj | 2 $\forall$E | |
| 5 | Pj | 3 $\forall$E | |
| 6 | $Nj \land Pj$ | 4, 5 $\land$I | |
| 7 | $(\forall x)(Nx \land Px)$ | 6 $\forall$I | |

---

4. Strictly speaking the single turnstile should be subscripted with an indication of the relevant derivation system—the above should be $\Gamma \vdash_{PD} \mathbb{P}$, etc. I shall omit such subscripts except where clarity is threatened.

| 1 | $(\forall x)(Nx \wedge Px)$ | P | $\vdash (\forall x)Nx \wedge (\forall x)Px$ |
|---|---|---|---|
| 2 | $Nj \wedge Pj$ | 1 $\forall$E | |
| 3 | $Nj$ | 2 $\wedge$E | |
| 4 | $Pj$ | 2 $\wedge$E | |
| 5 | $(\forall x)Nx$ | 3 $\forall$I | |
| 6 | $(\forall x)Px$ | 4 $\forall$I | |
| 7 | $(\forall x)Nx \wedge (\forall x)Px$ | 5, 6 $\wedge$I | |

**Inconsistent in *PD*:**

A set $\Gamma$ of wffs is *inconsistent in **PD*** iff, for some wff $\mathbb{P}$, both $\Gamma \vdash \mathbb{P}$ and $\Gamma \vdash \neg \mathbb{P}$.

We show that the following set is inconsistent in *PD*:

$$\{(\forall y)(Ly \rightarrow Ty), (\exists w)(\neg Tw \wedge Lw)\}$$

| 1 | $(\forall y)(Ly \rightarrow Ty)$ | P | |
| 2 | $(\exists w)(\neg Tw \wedge Lw)$ | P | $\vdash \mathbb{P}, \neg \mathbb{P}$ |
| 3 | $\neg Td \wedge Ld$ | A | |
| 4 | $(\forall y)(Ly \rightarrow Ty)$ | A | |
| 5 | $Ld \rightarrow Td$ | 1 $\forall$E | |
| 6 | $Ld$ | 3 $\wedge$E | |
| 7 | $\neg Td$ | 3 $\wedge$E | |
| 8 | $Td$ | 5, 6 $\rightarrow$E | |
| 9 | $\neg(\forall y)(Ly \rightarrow Ty)$ | 4–8 $\neg$I | |
| 10 | $\neg(\forall y)(Ly \rightarrow Ty)$ | 2, 3–9 $\exists$E | |
| 11 | $(\forall y)(Ly \rightarrow Ty)$ | 1 R | |

Lines 10 and 11 contradict one another. Note that while lines 7 and 8 contradict one another, they do not rest on the main scope line. Hence, stopping at line 8 would not be sufficient to show the set is inconsistent in *PD*. Note, as well, that in line 4 we assume the wff already appearing in line 1. This is in order to use the contradiction from lines 7 and 8 in a $\neg$I in order to produce line 9. Since line 9 meets all the restrictions on $\exists$E, we can execute that rule and get it out on the main scope line in line 10. Reiterating line 1 produces the contradiction.

## 7.3.1 Exercises

**A.** For each of the following, construct a derivation in *PD* to prove the derivability claim.

**1.** $(\forall x)Jx \vdash (\forall x)Jz$

**2.** $\{(\forall x)(Dx \rightarrow Gx), (\forall w)(Gw \rightarrow Mw), (\forall y)Dy\} \vdash (\forall z)Mz$

**3.** $\{(\exists z)(Az \wedge Kz), (\forall x)(Kx \rightarrow Cx)\} \vdash (\exists x)(Cx \wedge Ax)$

   **4.** $\{(\forall z)(Gzz \rightarrow Hfz),\ (\exists x)((\forall y)Fxy \wedge Gxx)\} \vdash (\exists x)((\forall y)Fxy \wedge Hfx)$
   **5.** $\{(\forall z)(Gzz \rightarrow Hfz),\ (\exists x)((\forall y)Fxy \wedge Gxx)\} \vdash (\exists x)(\forall y)(Fxy \wedge Hfx)$

**B.** For each of the following, construct a derivation to show the argument is valid in *PD*.

   **1.** $(\forall x)Mx$
   $\dfrac{(\forall y)\neg Dyy}{(\forall x)(\neg Dxx \wedge Mx)}$

   **2.** $(\forall x)Mx$
   $\dfrac{(\forall y)\neg Dyy}{(\exists x)(\forall y)(\neg Dxx \wedge My)}$

   **3.** $(\forall x)(Mx \rightarrow Wx)$
   $\dfrac{(\exists y)\neg Wy}{(\exists z)\neg Mz}$

   **4.** $\dfrac{(\exists y)(\forall x)Lxy}{(\forall x)(\exists y)Lxy}$

   **5.** $\dfrac{(\exists y)(\forall x)Lyx}{(\forall x)(\exists y)Lyx}$

**C.** For each of the following, construct a derivation to show the wff is a theorem of *PD*.

   **1.** $(\forall x)(Jx \rightarrow Kx) \rightarrow ((\forall x)Jx \rightarrow (\forall x)Kx)$
   **2.** $(\forall x)(Fx \rightarrow Gx) \rightarrow ((\exists x)Fx \rightarrow (\exists x)Gx)$
   **3.** $(\exists x)(Fx \wedge Gx) \rightarrow ((\exists x)Fx \wedge (\exists x)Gx)$
   **4.** $(\forall x)(\exists y)(Fy \rightarrow Fx)$

**D.** For each of the following, construct derivations to show the pair are equivalent in *PD*.

   **1.** $(\forall x)Fx \wedge (\forall x)Gx$ $\qquad (\forall x)(Fx \wedge Gx)$
   **2.** $(\exists x)Fx \vee (\exists x)Gx$ $\qquad (\exists x)(Fx \vee Gx)$
   **3.** $\neg(\exists x)Bx$ $\qquad\qquad\quad (\forall x)\neg Bx$

# 7.4  *PDE*, **an Extension to** *PD*

Officially, *PD* includes all the rules of *SDE* plus the four quantifier rules. This is enough to complete any derivation with quantifiers. Depending on what you or your instructor find desirable, the inference and replacement rules added in *SDE* could be omitted. This would result in a system we might call *PD*$^-$, containing just the 11 rules of *PD* and the four quantifier rules. This is the minimal system needed to do any derivation involving our truth-functional connectives and quantifiers. We can, of course, add extended quantifier rules to *PD* to make certain derivations easier, creating the system *PDE*.

## 7.4.1  **Quantifier Negation** — QN

Though strictly unnecessary, the following Quantifier Negation rule can make many derivations significantly simpler. We know from our knowledge of translations and quantificational semantics that a negated universal is equivalent to an existential negation, while a negated existential is equivalent to a universal negation. Thus:

$$\neg(\forall x)Fx \Leftrightarrow (\exists x)\neg Fx$$
$$\neg(\exists x)Fx \Leftrightarrow (\forall x)\neg Fx$$

We can show these equivalences with derivations involving metavariables.

| | | | |
|---|---|---|---|
| 1 | $\neg(\forall x)Fx$ | P | $\vdash (\exists x)\neg Fx$ |
| 2 | $\neg(\exists x)\neg Fx$ | A | |
| 3 | $\neg Fa$ | A | |
| 4 | $(\exists x)\neg Fx$ | 3 $\exists$I | |
| 5 | $\neg(\exists x)\neg Fx$ | 2 R | |
| 6 | $Fx$ | 3–5 $\neg$E | |
| 7 | $(\forall x)Fx$ | 6 $\forall$I | |
| 8 | $\neg(\forall x)Fx$ | 1 R | |
| 9 | $(\exists x)\neg Fx$ | 2–8 $\neg$E | |

| | | | |
|---|---|---|---|
| 1 | $(\exists x)\neg Fx$ | P | $\vdash \neg(\forall x)Fx$ |
| 2 | $\neg Fa$ | A | |
| 3 | $(\forall x)Fx$ | A | |
| 4 | $Fa$ | 3 $\forall$E | |
| 5 | $\neg Fa$ | 2 R | |
| 6 | $\neg(\forall x)Fx$ | 3–5 $\neg$I | |
| 7 | $\neg(\forall x)Fx$ | 1, 2–6 $\exists$E | |

This shows the equivalence of $\neg(\forall x)Fx$ and $(\exists x)\neg Fx$. The following two derivations show the equivalence of $\neg(\exists x)Fx$ and $(\forall x)\neg Fx$.

| | | | |
|---|---|---|---|
| 1 | $\neg(\exists x)Fx$ | P | $\vdash (\forall x)\neg Fx$ |
| 2 | $Fa$ | A | |
| 3 | $(\exists x)Fx$ | 2 $\exists$I | |
| 4 | $\neg(\exists x)Fx$ | 1R | |
| 5 | $\neg Fa$ | 2–4 $\neg$I | |
| 6 | $(\forall x)\neg Fx$ | 5 $\forall$I | |

```
 1 │ (∀x)¬Fx                P       ⊢ ¬(∃x)Fx
 2 │┌ (∃x)Fx                A
 3 ││┌ Fa                   A
 4 │││┌ (∃x)Fx              A
 5 ││││ Fa                  3R
 6 ││││ ¬Fa                 1 ∀E
 7 │││ ¬(∃x)Fx              4−6 ¬I
 8 ││ ¬(∃x)Fx              2, 3−7 ∃E
 9 ││ (∃x)Fx                2R
10 │ ¬(∃x)Fx                2−9 ¬I
```

These equivalences license the following Quantifier Negation replacement rule:

QN

─────────────────────────────

¬(∀x)ℙ  ◁▷  (∃x)¬ℙ

¬(∃x)ℙ  ◁▷  (∀x)¬ℙ

This rule can come in handy, especially as it is often easier to deal with a negation that is inside the quantifier than one that is outside. If we take *PD* and add QN, we now have a system employing all the rules: *PDE*.

Here are a few examples employing QN. We will use the full power of *PDE*. Consider this relatively simple one:

```
1 │ (∀x)¬Mx                    P
2 │ (∃y)Jy → (∃x)Mx            P       ⊢ (∀y)¬Jy
3 │ ¬(∃x)Mx                    1 QN
4 │ ¬(∃y)Jy                    2, 3 MT
5 │ (∀y)¬Jy                    4 QN
```

Applying QN to line 1 allows the MT with lines 2 and 3. QN then also allows a quick finish in moving from 3 to 4. This derivation can, of course, be completed without using QN, but it will take 14 lines instead of just five! Here is another:

| 1 | $(\forall x)(Fx \rightarrow (Gx \wedge Hx))$ | P |
|---|---|---|
| 2 | $\neg(\exists z)(Gz \wedge Jz)$ | P |
| 3 | $\neg(\exists y)(Hy \wedge \neg Jy)$ | P    $\vdash (\exists x)\neg Fx$ |
| 4 | $(\forall x)Fx$ | A |
| 5 | $Fa \rightarrow (Ga \wedge Ha)$ | 1 $\forall$E |
| 6 | $Fa$ | 4 $\forall$E |
| 7 | $Ga \wedge Ha$ | 5, 6 $\rightarrow$E |
| 8 | $Ga$ | 7 $\wedge$E |
| 9 | $Ha$ | 7 $\wedge$E |
| 10 | $(\forall z)\neg(Gz \wedge Jz)$ | 2 QN |
| 11 | $(\forall z)(\neg Gz \vee \neg Jz)$ | 10 DeM |
| 12 | $\neg Ga \vee \neg Ja$ | 11 $\forall$E |
| 13 | $\neg Ja$ | 8, 12 DS |
| 14 | $Ha \wedge \neg Ja$ | 9, 13 $\wedge$I |
| 15 | $(\exists y)(Hy \wedge \neg Jy)$ | 14 $\exists$I |
| 16 | $\neg(\exists y)(Hy \wedge \neg Jy)$ | 3R |
| 17 | $\neg(\forall x)Fx$ | 4–16 $\neg$I |
| 18 | $(\exists x)\neg Fx$ | 17 QN |

This is a longer derivation, but it could be worse. The sequence of moves in lines 10–13 employs **SDE** and **PDE** to save a great deal of trouble in working with line 2. Moreover, the main strategy of using $\neg$I to produce '$\neg(\forall x)Fx$', is made possible by the availability of QN to transition from line 17 to 18. So, while we never *need* to use the extended rules, it can often be very useful to do so.

## 7.4.2 Exercises

Note that none of the following derivations *require* the use of **PDE**—in some cases you will likely use only **PD** rules—but many of them will be made significantly easier and more efficient if you use **PDE**. Further note that all of the definitions for **PD** carry over to **PDE** as you would expect.

A. For each of the following, construct a derivation in **PDE** to prove the derivability claim.

1. $(\forall x)Fx \vdash \neg(\forall x)\neg Fx$

2. $\neg(\exists x)Fx \vdash \neg(\forall x)Fx$

3. $\neg(\exists x)(Fx \wedge Gx) \vdash (\forall x)(Fx \rightarrow \neg Gx)$

4. $(\exists y)\neg Ey \vdash \neg(\forall x)(Nx \wedge Ex)$

5. $(\forall x)Fx \wedge P \vdash (\forall x)(Fx \wedge P)$

6. $(\forall x)(Fx \wedge P) \vdash (\forall x)Fx \wedge P$

7. $(\exists x)(Fx \rightarrow P) \vdash (\forall x)Fx \rightarrow P$

8. $(\forall x)Fx \rightarrow P \vdash (\exists x)(Fx \rightarrow P)$

**B.** For each of the following, construct a derivation to show the argument is valid in **PDE**.

1. $(\exists y)Ky \rightarrow (\forall x)Lx$
   $\underline{(\exists x)\neg Lx}$
   $(\forall y)\neg Ky$

2. $\neg(\forall x)Kx$
   $\underline{(\forall y)(Dy \rightarrow Ky)}$
   $\neg(\forall x)Dx$

3. $(\forall x)(Dx \rightarrow Hx)$
   $\underline{\neg(\exists x)Hx}$
   $\neg(\exists x)Dx$

4. $(\forall x)(Tx \rightarrow (\exists y)(\neg Sx \wedge Ryx))$
   $(\exists y)\neg Uyy$
   $\underline{(\forall z)(\neg Sz \rightarrow Uzz)}$
   $(\exists w)\neg Tw$

**C.** For each of the following, construct a derivation to show the wff is a theorem of **PDE**.

1. $\neg((\forall x)Nx \wedge (\forall z)\neg Nz)$

2. $(\forall x)(Fx \rightarrow Gx) \leftrightarrow \neg(\exists x)(Fx \wedge \neg Gx)$

3. $\neg(\exists x)(\forall y)(Rxy \leftrightarrow \neg Ryy)$

4. $(\forall x)((\forall y)(Rxy \leftrightarrow (Fy \wedge \neg Ryy)) \rightarrow \neg Fx)$

**D.** For each of the following, construct derivations to show the pair are equivalent in **PDE**.

1. $(\exists x)Fx \wedge P$          $(\exists x)(Fx \wedge P)$

2. $(\forall x)Fx \vee P$          $(\forall x)(Fx \vee P)$

3. $(\exists x)Fx \vee P$          $(\exists x)(Fx \vee P)$

4. $P \rightarrow (\forall x)Fx$          $(\forall x)(P \rightarrow Fx)$

5. $(\exists x)(P \rightarrow Fx)$          $P \rightarrow (\exists x)Fx$

6. $(\forall x)(Fx \rightarrow P)$          $(\exists x)Fx \rightarrow P$

**E.** For each of the following, construct a derivation to show the set is inconsistent in *PDE*.

    **1.** $\{(\exists x)(Fx \wedge Gx), (\forall x)(\neg Dx \rightarrow \neg Fx), \neg(\exists x)Dx\}$

    **2.** $\{(\exists y)(\forall x)Lxy, \neg(\exists x)Lxx\}$

    **3.** $\{(\forall x)(\forall y)(\forall z)((Fxy \wedge Fyz) \rightarrow Fxz), (\exists x)(\exists y)(Fxy \wedge Fyx), \neg(\exists x)Fxx\}$

# 7.5 **Chapter Glossary**

**Derivable in *PD*, $\Gamma \vdash \mathbb{P}$:**

    A wff $\mathbb{P}$ of *P* is *derivable in PD* from a set $\Gamma$ of wffs of *P* iff there is a derivation in *PD* the primary assumptions of which are members of $\Gamma$ and $\mathbb{P}$ depends on only those assumptions. (257)

**Derivation in *PD*:**

    A *derivation in PD* is a finite sequence of wffs of *P* such that each wff is either an assumption with scope indicated or justified by one of the rules of *PD*. (257)

**Equivalent in *PD*:**

    Two wffs $\mathbb{P}$ and $\mathbb{Q}$ are *equivalent in PD* iff they are interderivable in *PD*; i.e., iff both $\mathbb{P} \vdash \mathbb{Q}$ and $\mathbb{Q} \vdash \mathbb{P}$. (258)

**Inconsistent in *PD*:**

    A set $\Gamma$ of wffs is *inconsistent in PD* iff, for some wff $\mathbb{P}$, both $\Gamma \vdash \mathbb{P}$ and $\Gamma \vdash \neg\mathbb{P}$. (259)

**Substitution Instance:**

    Let $\mathbb{Q}(a/x)$ indicate the wff which is just like $\mathbb{Q}$ except for having the constant a in every position where the variable x appears in $\mathbb{Q}$. Where $\mathbb{P}$ is a closed wff of the form $(\forall x)\mathbb{Q}$ or $(\exists x)\mathbb{Q}$, then $\mathbb{Q}(a/x)$ is a *substitution instance* of $\mathbb{P}$, with a as the *instantiating constant*. (238)

**Theorem of *PD*:**

    A wff $\mathbb{P}$ is a *theorem of PD* iff $\mathbb{P}$ is derivable from the empty set; i.e., iff $\vdash \mathbb{P}$. (258)

**Valid in *PD*:**

    An argument of *P* is *valid in PD* iff the conclusion is derivable from the set consisting of only the premises, otherwise it is invalid in *PD*. (258)

# Part IV
# Advanced Topics

# 8 Basic Set Theory, Paradox, and Infinity

## 8.1 Basics of Sets

This chapter will explore the more accessible levels of naive set theory. This includes not only the basic set theoretic operations, but also some of the interesting and counter-intuitive results of set theory—Russell's and Cantor's Paradoxes, infinite cardinal numbers. The approach is *naive* in the sense that I will not attempt a fully *axiomatic* approach. Though I will speak of axioms and formulate many claims in an augmented version of our language *P*, there will be, in the interest of accessibility to the majority of readers, a certain amount of looseness falling short of full axiomatization.

Initially, the notion of a set is rather intuitive. A set is just a collection or group of things—the coins now in my pocket, a bunch of bananas, a group of logic students. The things *belonging to* or *contained in* the set are its *members* or *elements*—thus, the quarters, dimes, and nickels; the individual bananas; Constance, Ruth, Mary, Bertrand, and Alfred. But the elements of a set need not actually be gathered together in space or time. The students may travel to distant places (each carrying one of the bananas), but the logic students are still a set, as are the bananas. Or perhaps Constance and Mary lived and died long before the others, there is still the set of all five, and we can talk about the set of all bananas, ever. Sets are solely determined by their members regardless of their spatial or temporal locations.

Furthermore, the members of a set need not be concrete objects. We can have sets of abstract objects—sets of numbers, points, lines, sets of sets, sets of sets of sets, and so on—no students, dimes, or bananas need appear. When we deal with sets that contain only other sets, we are doing pure set theory. Indeed, one way of understanding numbers, points, and lines is as certain kinds of sets of sets. So geometry and mathematics as a whole can be founded on set theory and logic.

Intuitively it seems that for any objects we can list, or any condition or description we can state clearly, there is a set containing just those objects in the list, or just those objects that meet the description. And this intuition corresponds to the two basic ways of specifying sets. One is *direct definition*, in which we simply list, between curly braces and in no particular order, all the members of the set. For instance:

$$\Gamma = \{2, 3, 5, 7, 11, 13, 17, 19\}$$
$$\Gamma' = \{11, 3, 17, 5, 2, 19, 7, 13\}$$

Note that a set is determined solely by its members, regardless of their order. So, since order doesn't matter, these two sets are identical—they are the same set. I will typically use uppercase Greek letters for sets that do not already have standard names. So, above, $\Gamma = \Gamma'$.

The second way of specifying a set is using *set builder notation*.[1] We can define, for example, the set of all prime numbers that are less than twenty. That is, the set of all x, such that x is a prime number and x is less than 20. More economically, we use the following notation in which universal quantification over x is implicit, and the vertical bar is read 'such that':

$$\Gamma'' = \{x \mid x \text{ is prime and } x < 20\}$$

Of course, the set defined here is the same as the directly defined set above. That is, $\Gamma = \Gamma' = \Gamma''$. Direct definition is fine for small finite sets, but the latter method is what we will see most of the time, for it allows us to specify sets of even infinite size. For example:

$$\mathbb{N} = \{x \mid x \text{ is a natural number}\}$$
$$\mathbb{P} = \{x \mid x \text{ is prime}\}$$

These are both infinitely large sets,[2] so using set builder notation maximizes clarity. Note, also, that they have standardized symbols '$\mathbb{N}$' and '$\mathbb{P}$' as names.[3]

For recognizable finite sequences and certain commonly recognized infinite sets we can sometimes use the all-powerful ellipsis '...' to specify a set. Here we specify the integers from 1 to 1000, and, again, the natural numbers, and the primes:

$$\Delta = \{1, 2, 3, \ldots, 1000\}$$
$$\mathbb{N} = \{0, 1, 2, 3, \ldots\}.$$
$$\mathbb{P} = \{2, 3, 5, 7, 11, \ldots\}$$

Ellipses are fine in many contexts, but one needs to be sure there is agreement on what they are indicating.

We can make our intuition that any condition determines a set official by giving it a name:

**Axiom Schema of Abstraction:**

For any clearly stated condition, Sx, there exists a set B whose elements are exactly those objects which satisfy Sx:[4]

$$(\exists B)(\forall x)(x \in B \leftrightarrow Sx)$$

---

1. This is also called set *comprehension* or set *abstraction* notation. At this point, we still have in mind the *unrestricted* version. See Sec. 8.3, note 7.
2. Which is larger? See Sec. 8.6.
3. All the standard number sets have symbols, $\mathbb{Z}$: integers, $\mathbb{R}$: reals, $\mathbb{Q}$: rationals, etc.
4. We require that y not be free in S. Moreover, as standard practice, when 'A', 'B', 'C' appear in axioms and definitions, we take them to be implicitly universally quantified unless otherwise indicated. Here, of course, 'B' is existentially quantified.

In set builder notation we can briefly define a set, B:

$$B = \{x \mid Sx\}$$

This is an axiom *schema* because it allows us to produce an unlimited number of claims that such a set exists—one for each condition Sx. Thus, the Axiom Schema of Abstraction is simple, intuitive, and very powerful. But we will soon see that it quickly leads to paradox. Before we get there, though, we need a few more basics.

We use a special epsilon symbol '$\in$' and its negation '$\notin$' to express or deny membership. For example:

| | | | |
|---|---|---|---|
| $2 \in \Gamma$ | $2 \in \mathbb{P}$ | $23 \notin \Gamma$ | $23 \in \mathbb{P}$ |
| $6 \notin \Gamma$ | $6 \notin \mathbb{P}$ | $6 \in \mathbb{N}$ | $\pi \notin \mathbb{N}$ |

The '$\in$' can be read as 'is a member of', 'is an element of', 'is contained in'. Be sure you understand why each of the (non-)membership claims above is true.

Note that the condition used to specify a set might apply to no objects whatsoever. In that case we have specified the *empty set*, symbolized by $\varnothing$ or $\{\}$. Here are some examples:

$$\varnothing = \{\} = \{x \mid x \text{ is the greatest prime number}\}$$
$$\varnothing = \{\} = \{x \mid x \text{ is a unicorn}\}$$
$$\varnothing = \{\} = \{x \mid x \neq x\}$$

Since there is no greatest prime, and there are no unicorns, and no object is distinct from itself, each of these sets is empty. In fact, since set identity is determined by membership, these are all the same set, *the* empty set, $\varnothing$, sometimes called the *null set*.

We can state clearly the idea that set identity is determined by membership as the Axiom of Extensionality:

**Axiom of Extensionality:**

If sets A and B have exactly the same members, then they are identical:

$$(\forall x)((x \in A \leftrightarrow x \in B) \rightarrow A = B)$$

We now have enough of a base to discuss Russell's Paradox, the simplest of the set theoretic paradoxes.

# 8.2 **Russell's Paradox**

Russell's Paradox exposes an inconsistency arising from the intuitive conception of sets enshrined in the Axiom Schema of Abstraction. Abstraction essentially tells us that for any condition Sx, there exists a set. So, consider the following set B, where the condition Sx is non-self-membership '$x \notin x$':

(1)  $B = \{x \mid x \notin x\}$

B is the set of all things that are not members of themselves. This non-self-membership seems to be an unproblematic condition. It is shared by many a set and non-set: I am not a member of myself, the set of all chairs is not a member of itself, the set of all one membered sets is not a member of itself. It even seems that some sets *are* members of themselves: the set of all sets with more than one member, the set of all sets with more than two members... So, at first glance, we seem to be okay.

But consider the following question: is the set B a member of itself? If every condition determines a set, then the condition and the nature of the candidate object should determine whether that object is a member or not. So let's see. By the definition of B, we get

(2)   $(\forall x)(x \in B \leftrightarrow x \notin x)$

This is really just a restatement of (1) above. If we remove the universal quantifier and instantiate x with B, then we get:

(3)   $B \in B \leftrightarrow B \notin B$

which is a contradiction.

> Either a) $B \in B$ or b) $B \notin B$. Case a) and (3) imply that $B \notin B$, so we have $B \in B \wedge B \notin B$. Case b) and (3) imply that $B \in B$, so, again, we have $B \in B \wedge B \notin B$. Either way we have the direct contradiction: $B \in B \wedge B \notin B$.

Given the contradiction, the set B cannot exist. Our intuition that any condition determines a set has led to a contradiction, and our intuition must be wrong. The Axiom Schema of Abstraction has led us astray.

This may seem like just a bit of trickery here, but it goes far deeper. It seemed to just make sense that any condition specified a set. This seemed to be a bit of Reason. But it leads to a contradiction. So we were wrong. Intuitive set theory is inconsistent. This is a big deal because it was once hoped (by Logicists, especially Frege and Russell) that mathematics could be reduced to logic (i.e., that every mathematical statement translates into a statement using only logical notation, and that every truth of mathematics is a truth of logic), and this intuitive set theory was taken to be a part of logic. Among other implications, if we take logic to be analytic, and mathematics reduces to it, then mathematics is apparently analytic—contrary to Kant for one (or, perhaps, logic is synthetic, again contrary to Kant). Moreover, if math reduces to logic, then the concepts and truths of math are as clear and as potentially obvious as those of logic, and its ontology is as clear and respectable as any could be.

Mathematics *does* reduce to logic *plus* set theory—to logic alone if we continue to consider set theory a part of logic. And, if it were not for the paradoxes, it would be sensible to consider set theory a part of logic. Historically, Frege and Russell each had an analogue of set theory stated in a more complex logic, and so they would have been happy to claim that math reduces to logic. However, the most basic and simple set theory, as laid out along Fregean lines, turns out to be inconsistent, as Russell showed via the paradox named after him. Consistent set theories (of which there are various—one of which we'll see—and which in conjunction with logic reduce mathematics) avoid the paradox at the cost of a certain amount of intuitive appeal or obviousness as well as at the cost of a certain sort of generality. One may or may not consider one of these consistent set theories a part of logic. If one does so, one may then claim that math reduces to logic alone. But, because of the lack of obviousness and lack of generality required for a consistent set theory, a reduction of math to such a logic including set theory will not carry the same philosophical weight that Frege and Russell wanted. Rather than cling to the image that math reduces to a simple, general, obvious logic, most philosophers now opt not to include set theory in logic and so would now say that math reduces to logic plus set theory. The reduction is still philosophically important—and not just because of the inconsistency lurking in its original incarnation—but not in exactly the way Frege or Russell had hoped.

A historical note: Frege received Russell's letter in 1902 just as the second volume of his *Grundgesetze der Arithmetik (Basic Laws of Arithmetic)* was going to press. The paradox rocks and destroys the foundations of the Logicist attempt to reduce math to logic. It was too late for Frege to edit the volume, so he added an appendix which opens as follows:

> Hardly anything more unfortunate can befall a scientific writer than to have the foundations of his edifice shaken after the work is finished.
>
> This was the position I was placed in by a letter of Mr. Bertrand Russell, just when the printing of this volume was nearing its completion. It is a matter of my Axiom (V). I have never disguised from myself its lack of the self-evidence that belongs to the other axioms and that must properly be demanded of a logical law... I should gladly have dispensed with this foundation if I had known of any substitute for it. And even now I do not see how arithmetic can be scientifically established; how numbers can be apprehended as logical objects, and brought under review; unless we are permitted—at least conditionally—to pass from a concept to its extension. May I always speak of the extension of a concept—speak of a class? And if not, how are the exceptional cases recognized?... These are the questions raised by Mr. Russell's communication.[5]

But to continue on, how do we avoid the contradiction? There are a number of ways, one is to disallow use of the '$\in$' in stating the defining condition of a set; another (type theory) is to develop a hierarchy of objects and sets such that '$x \in y$' and '$x \notin y$' are well-formed formulas if and only if x is of a lower type in the hierarchy than y. But the former is too extreme, and the latter seems unnatural (though there are interesting theories of this kind) and becomes overly complex. Instead, we can replace the intuitive, but contradictory Axiom Schema of Abstraction with the Axiom Schema of Separation.

## 8.3 The Axiom Schema of Separation

We can avoid the contradiction by putting a minor restriction on the Axiom Schema of Abstraction. Rather than saying that any condition Sx determines a set, we say that, *given*

---

5. Frege, Gottlob. Appendix to *Grundgesetze der Arithmetik*, Vol. II. In *The Frege Reader*, edited by Michael Beaney, 279–80. Oxford: Blackwell, 1997.

*a set* A, any condition Sx determines a *subset* of A.

**Axiom Schema of Separation:**
>    For every set A and every condition Sx there is a set B whose elements are exactly
>    those members of A for which Sx holds:[6]

$$(\exists B)(\forall x)(x \in B \leftrightarrow (x \in A \wedge Sx))$$

Briefly:

$$B = \{x \in A \mid Sx\}$$

The set builder notation is read: B is the set of all x in A such that Sx. This, of course, requires the assumption of some prior set, A, from which we *separate* out the subset determined by Sx. Note, again, that this is an axiom *schema*—i.e., it licenses an unlimited number of existence axioms, one for each S in the language.[7]

How does this help us avoid the contradiction generated by the non-self-member set? Let's see. Now the non-self-member set has to be specified as follows:

(4)   $B = \{x \in A \mid x \notin x\}$

When we ask whether $B \in B$, we get the following:

(5)   $B \in B \leftrightarrow (B \in A \wedge B \notin B)$

What this shows is that for B to be a member of itself, it must be a member of A. If B *can* be a member of A, then we will reproduce the contradiction above, which means it cannot be in A and so cannot be a member of itself. The result is that $B \notin B$. Let's prove it a little more carefully.

>    Suppose $B \in A$. Obviously, either a) $B \in B$ or b) $B \notin B$. In case a), (5) implies
>    $B \notin B$, so we have $B \in B \wedge B \notin B$, a contradiction, so a) cannot be the case.
>    Case b), (5), and our supposition that $B \in A$ imply that $B \in B$, so again we
>    have $B \in B \wedge B \notin B$, a contradiction, as before. But here we can reject the
>    original supposition, $B \in A$. Thus $B \notin A$, and consequently, $B \notin B$.

This can be a little tricky to see. The main point is that, unlike with (3) above, which must be false (making (2) false and the set B impossible), (5) here can be true so long as $B \notin A$ and $B \notin B$. The relevant row of a truth table will help to see this:

$$\underline{B \in B \leftrightarrow (B \in A \wedge B \notin B)}$$

$$\text{F} \quad \boxed{\text{T}} \quad \text{F} \quad \text{F} \quad \text{T}$$

---

6. We again require that y not be free in S.
7. This is also frequently called the Axiom Schema of Specification or the Axiom Schema of Restricted Comprehension.

So the set B can exist under the Axiom Schema of Separation. It is a subset of A, but it is neither a member of A, nor of itself. Note that without the 'B ∈ A' portion there is no truth value assignment on which (5)—i.e., (3)—could be true. This is why we got Russell's Paradox above. But with the condition that there be a set A that we are separating B out of, no contradiction arises. In this formulation B is a respectable set. B is simply a subset of A containing all and only those members of A which are non-self-members. B itself, however, cannot be a member of A. Nor of itself. We showed this by running a version of Russell's Paradox on the assumption that B was in A. But no inconsistency arises, for we simply have shown that B cannot be in A.

Well, what if we just take A to be the set that contains everything—the Universe? We can try that. Take Upsilon, $\Upsilon$, as the set that contains everything. Then run the proof we just finished with $\Upsilon$ in place of A. What we will get is that B ∉ $\Upsilon$. That is, B must be outside of (what we mistakenly took to be) the Universe. Let's then take $\Upsilon'$ to be the set containing everything in $\Upsilon$ plus B. Surely $\Upsilon'$ is the Universe, the set that contains everything... Of course we can run the same argument on $\Upsilon'$ using a set B', and show that B' ∉ $\Upsilon'$. And so on. So nothing contains everything. There is no Universe.

Note: this is related to Cantor's Paradox (1899) that there is no greatest infinite number (roughly speaking). We'll get to this eventually, both imply that there is no set which contains everything. No universe, so to speak.

# 8.4 **Subset, Intersection, Union, Difference**

We mentioned the notion of subset in the explanation of how Separation avoids Russell's Paradox. Here we define it:

**Subset, Inclusion:**
> Set A is a *subset of* or is *included in* B iff every member of A is also a member of B. We use the '⊆' symbol:
>
> $$A \subseteq B \leftrightarrow (\forall x)(x \in A \rightarrow x \in B)$$

Note that every set is a subset of (included in) itself. That is, for every set A, A ⊆ A. That is, subset is a reflexive property. Note that it is also transitive, but not symmetric (see section 6.5). We can distinguish subset/inclusion from proper subset/proper inclusion by requiring that A ≠ B.

**Proper Subset, Proper Inclusion:**
> Set A is a *proper subset of* or is *properly included in* B iff every member of A is also a member of B, but A ≠ B. We use the '⊂' symbol:
>
> $$A \subset B \leftrightarrow (A \subseteq B \land A \neq B)$$

This relation is neither reflexive nor symmetric, but it is, of course, transitive.

It is important not to confuse inclusion with membership. The set of philosophers is a subset of the set of humans, so the set of philosophers is included in the set of humans.

But, since it is a set and not a human, the set of philosophers is *not* a member of the set of humans.

It is always the case that A ⊆ A for every A, but it is not clear how A ∈ A could be the case. Furthermore, set A can be a member of B, A ∈ B, without being a subset, A ⊄ B, and vice versa, A ∉ B and A ⊆ B. The two are not inconsistent, however. As long as A ≠ B, we can have both A ∈ B and A ⊆ B, which implies A ⊂ B. Consider the following sets, and the relations among them:

$$\Gamma = \{a, b, c, \Delta\} \qquad \Delta \in \Gamma \qquad E \notin \Gamma \qquad E \in \Delta$$
$$\Delta = \{b, c, d, E\} \qquad \Delta \not\subseteq \Gamma \qquad E \subset \Gamma \qquad E \subset \Delta$$
$$E = \{b, c\}$$

A further difference is that inclusion is transitive, but membership is not. This will be addressed in the exercises.

Note, finally that ∅ is a subset of every set. That is, for all Γ, ∅ ⊆ Γ. We can prove this easily:

> Suppose, for some Γ, ∅ ⊄ Γ. By the definition of subset, for subset to fail there must be some x ∈ ∅ where x ∉ Γ. But this is impossible, x ∈ ∅ must be false. So, ∅ ⊆ Γ for all Γ.

Is it the case that for all Γ, ∅ ⊂ Γ? Why or why not? This will be treated in the exercises.

Note that '⊆' and '⊂' can be read as 'superset' and 'proper superset'. So, for example, 'A ⊂ B' can be read either as 'A is a proper subset of B' or 'B is a proper superset of A'. We do not, therefore, bother with an additional symbol for superset.

The intersection of two sets is the set containing all and only their common members.

**Intersection:**
> The intersection of a pair of sets A and B, A∩B, the set consisting of all and only the elements in both sets:

$$A \cap B = \{x \mid x \in A \wedge x \in B\}$$

Some simple examples:

$$E = \{b, c\} \qquad\qquad E \cap Z = \{b, c\} = E$$
$$Z = \{a, b, c, d, e\} \qquad\qquad Z \cap H = \{d, e\}$$
$$H = \{d, e, f\} \qquad\qquad H \cap E = \varnothing$$

The definition of pairwise intersection above is actually a special case of the more general notion of intersection that applies to any set of sets:

**Generalized Intersection:**
> For every set of sets A, there is some set C whose elements are all and only the elements in every B ∈ A:

$$(\exists C)(\forall x)(x \in C \leftrightarrow (\forall B)(B \in A \rightarrow x \in B))$$

Or:

$$\cap A = \{x \mid \text{for every } B \in A, \ x \in B\}$$

That is, $\cap A$ is the set of elements that appear in every set in A. For example,

$$A = \{Z, H, \{d, e, j, k\}\} \qquad\qquad \cap A = \{d, e\}$$
$$I = \{E, Z, H, \{d, e, j, k\}\} \qquad\qquad \cap I = \varnothing$$

Consider that for any A, where $\varnothing \in A$, $\cap A = \varnothing$. We can prove this easily:

> Suppose $\varnothing \in A$. Further, suppose that $\cap A \neq \varnothing$. This implies that there is some $x \in \cap A$, which in turn implies that $x \in B$ for every $B \in A$. In particular, $x \in \varnothing$. But this is not possible. So, for all A, if $\varnothing \in A$, then $\cap A = \varnothing$.

As mentioned, pairwise intersection is just a special case of general intersection:

$$A \cap B = \cap\{A, B\}$$

What is $\cap\varnothing$? It is $\{x \mid \text{for every } B \in \varnothing, \ x \in B\}$. But there are no $B \in \varnothing$, so every B satisfies (vacuously) the set definition, so every x is a member of $\cap\varnothing$. This means that $\cap\varnothing = \Upsilon$, but $\Upsilon$, the universe, doesn't exist. So we have a problem. Really we just need to remind ourselves that $\cap$ (indeed all our definitions) must be defined via the Separation Schema. So it should start $\{x \in K \mid \ldots\}$ for some set K. That is, we relativize all this to an assumed set, K, which acts as a temporary or relative universe. Of course it is not really a universe, because using non-self-membership and Separation we can define a set which is not an element of K. Thus, $\cap\varnothing = K$. We keep this in the background for the sake of simplicity. It is only an issue in certain sorts of cases. Note that this is not an issue with pairwise intersection, since its original definition above, $\{x \mid x \in A \wedge x \in B\}$, is equivalent to $\{x \in A \mid x \in B\}$, which is an instance of the Axiom Schema of Separation.

In addition to the intersection of sets, we are interested in the union of sets. First, we'll look at pairwise union, that is, the set containing every member of either of two sets.

**Union:**
> The union of a pair of sets A and B, $A \cup B$, the set consisting of all and only the elements in either set:

$$A \cup B = \{x \mid x \in A \vee x \in B\}$$

Some simple examples:

$$E = \{b, c\} \qquad\qquad E \cup Z = \{a, b, c, d, e\} = Z$$
$$Z = \{a, b, c, d, e\} \qquad\qquad H \cup E = \{b, c, d, e\}$$
$$H = \{d, e, f\} \qquad\qquad H \cup \varnothing = \{d, e, f\} = H$$

Note that pairwise union is a special case of the general Union Axiom:

**Union Axiom:**

For every set of sets A, there is some set C whose elements are all and only the elements of at least one $B \in A$:

$$(\exists C)(\forall x)(x \in C \leftrightarrow (\exists B)(B \in A \wedge x \in B))$$

Or:

$$\cup A = \{x \mid \text{for some } B \in A, \ x \in B\}$$

$\cup A$ is the set of all members of each set B in A. So,

$$A = \{Z, H, \{d, e, j, k\}\} \qquad\qquad \cup A = \{a, b, c, d, e, j, k\}$$
$$I = \{E, Z, H, \{d, e, j, k\}\} \qquad\qquad \cup I = \{a, b, c, d, e, j, k\}$$

Again, we can see pairwise union as a special case of the Union Axiom.

$$A \cup B = \cup\{A, B\}$$

The exercises contain various proofs concerning intersection and union.

Next, we look at the difference between two sets.

**Difference:**

The difference (or relative complement) between sets A and B is the set containing every member of A that is not in B:

$$A - B = \{x \mid x \in A \wedge x \notin B\}$$

Note that, like numerical difference, the operation is not commutative. $A - B \neq B - A$. Some examples:

$$E = \{b, c\} \qquad\qquad Z - E = \{a, d, e\} \qquad\qquad E - Z = \emptyset$$
$$Z = \{a, b, c, d, e\} \qquad\qquad Z - H = \{a, b, c\} \qquad\qquad H - Z = \{f\}$$
$$H = \{d, e, f\} \qquad\qquad H - E = \{d, e, f\} = H \qquad\qquad H - H = \emptyset$$

It can be quickly proved that for any A, $A - A = \emptyset$. I leave this for the exercises.

We can define a temporarily absolute complement, by assuming a set K as an implicit superset of all the sets of current interest.

**Absolute Complement:**

Given K, a superset of each of the sets of current interest, the temporarily absolute complement of a set A, $\overline{A}$ is the set of all elements of K, not in A:

$$\overline{A} = \{x \in K \mid x \notin A\}$$

We suppress reference to K whenever possible to make discussion of complements more tractable. For instance, we might like to prove that:

$$\overline{A \cap B} = \overline{A} \cup \overline{B}$$

Not having to refer to or explicitly specify K makes things easier, but we can if we need to.

Finally, we can define the symmetric difference, or symmetric complement of two sets.

**Symmetric Difference:**

The symmetric difference (or symmetric complement) of sets A and B is the union of the differences:

$$A \ominus B = (A - B) \cup (B - A)$$

Thus, the symmetric difference is the set containing every element that is either in A but not in B or in B but not in A.

| | |
|---|---|
| $E = \{b, c\}$ | $Z \ominus E = \{a, d, e\}$ |
| $Z = \{a, b, c, d, e\}$ | $Z \ominus H = \{a, b, c, f\}$ |
| $H = \{d, e, f\}$ | $H \ominus E = \{b, c, d, e, f\}$ |

Note also that $A \ominus B = (A \cup B) - (A \cap B)$. Compare the examples in the preceding pages. Proof of this is left for the exercises.

## 8.4.1 Exercises

Where only a formula is listed below, prove the formula.

1. Give an example that shows that membership is not transitive.

2. Is it the case that for all $\Gamma$, $\varnothing \subset \Gamma$? Why or why not?

3. $\Gamma \not\subset \Gamma$

4. $(A \subseteq B \wedge B \subseteq C) \rightarrow A \subseteq C$

5. $(A \subseteq B \wedge B \subseteq A) \rightarrow A = B$

6. $A \cup \varnothing = A$

7. $A \cup B = B \cup A$

8. $A \cup (B \cup C) = (A \cup B) \cup C$

9. $A \cup A = A$

10. $A \cap \varnothing = \varnothing$

**11.** $A \cap B = B \cap A$

**12.** $A \cap (B \cap C) = (A \cap B) \cap C$

**13.** $A \cap A = A$

**14.** $A \subseteq B \leftrightarrow A \cup B = B$

**15.** $A \subseteq B \leftrightarrow A \cap B = A$

**16.** $A \cap (B \cup C) = (A \cap B) \cup (A \cap C)$

**17.** $A \cup (B \cap C) = (A \cup B) \cap (A \cup C)$

**18.** Give a simple example that shows that $A - B \neq B - A$

**19.** $A - A = \varnothing$

**20.** $\overline{\overline{A}} = A$[8]

**21.** $\overline{\varnothing} = K$

**22.** $A \cap \overline{A} = \varnothing$

**23.** $A \cup \overline{A} = K$

**24.** $A \subseteq B \leftrightarrow \overline{B} \subseteq \overline{A}$

**25.** $\overline{A \cap B} = \overline{A} \cup \overline{B}$

**26.** $\overline{A \cup B} = \overline{A} \cap \overline{B}$

**27.** $A - B = A \cap \overline{B}$

**28.** $A \subseteq B \leftrightarrow A - B = \varnothing$

**29.** $A - (A - B) = A \cap B$

**30.** $A \cap (B - C) = (A \cap B) - (A \cap C)$

**31.** $A \ominus \varnothing = A$

**32.** $A \ominus A = \varnothing$

**33.** $A \ominus B = (A \cup B) - (A \cap B)$

**34.** $A \ominus B = B \ominus A$

---

8. Recall that when dealing with the temporarily absolute complement, K is a superset of all the sets of current interest. You may invoke it if needed. See, p. 278.

# 8.5 **Pairs, Ordered Pairs, Power Sets, Relations, and Functions**

## Axiom of Pairing:

For any two elements, x and y, there exists a set, A, consisting of just those elements.

$$(\exists A)(\forall z)(z \in A \leftrightarrow z = x \lor z = y)$$

In set builder notation:

$$\{x, y\} = \{z \mid z = x \lor z = y\}$$

The set $\{x, y\}$ is a *unordered pair*. Obviously $\{x, y\} = \{y, x\}$. Moreover, the *singleton* of x, $\{x\} = \{x, x\}$ is a special case of a pair. We can now define the *ordered* pair.

## Ordered Pair:

For any two elements, x and y, there exists a set, $(x, y)$, such that

$$(x, y) = \{\{x\}, \{x, y\}\}$$

What we want here is to capture the notion of a first and a second, which obviously cannot be done just with $\{x, y\}$ and $\{y, x\}$, since they are the same. So we take the ordered pair $(x, y)$ as the unordered pair of the singleton of the first element and the unordered pair of both elements. Note that, assuming $x \neq y$:

$$(x, y) = \{\{x\}, \{x, y\}\}$$
$$(y, x) = \{\{y\}, \{x, y\}\}$$
$$(x, y) \neq (y, x)$$

In this way we define order by appeal to elements that do not simply assume order. We do require, of course, that ordered pairs be unique. That is it must be the case that:

$$(x, y) = (u, v) \rightarrow (x = u \land y = v)$$

Note first that (a) if $\{x, y\} = \{u, v\}$, then either $x = u$ and $y = v$ or $x = v$ and $y = u$. Suppose $(x, y) = (u, v)$. By the definition of ordered pair, $\{\{x\}, \{x, y\}\} = \{\{u\}, \{u, v\}\}$. But given (a), we have:

$$(\{x\} = \{u\} \land \{x, y\} = \{u, v\}) \lor (\{x\} = \{u, v\} \land \{x, y\} = \{u\})$$

If we suppose the left disjunct holds, then we have $x = u$ and $y = v$. If we suppose the right disjunct holds, then we have $x = u = v = y$. In either case we have $x = u \land y = v$.

So the definition of ordered pair guarantees a uniquely recoverable first and second element.

The set of all subsets of A is called the powerset of A. This is because if A has $n$ elements, then its power set, $\mathscr{P}(A)$, has $2^n$ elements.

**Powerset Axiom:**

For each set A, there is a set C containing all subsets B of A:

$$(\exists C)(\forall B)(B \in C \leftrightarrow B \subseteq A)$$

That is:

$$\mathscr{P}(A) = \{B \mid B \subseteq A\}$$

Some examples:

$$E = \{b,c\} \qquad \mathscr{P}(E) = \{\varnothing, \{b\}, \{c\}, \{b,c\}\}$$
$$H = \{d,e,f\} \qquad \mathscr{P}(H) = \{\varnothing, \{d\}, \{e\}, \{f\}, \{d,e\},$$
$$\{d,f\}, \{e,f\}, \{d,e,f\}\}$$
$$Z = \{a,b,c,d,e\} \qquad \mathscr{P}(Z) = \{\varnothing, \{a\}, \{b\}, \{c\}, \{d\}, \{e\},$$
$$\{a,b\}, \{a,c\}, \{a,d\}, \{a,e\}, \{b,c\},$$
$$\{b,d\}, \{b,e\}, \{c,d\}, \{c,e\}, \{d,e\},$$
$$\{a,b,c\}, \{a,b,d\}, \{a,b,e\}, \{a,c,d\}, \{a,c,e\},$$
$$\{a,d,e\}, \{b,c,d\}, \{b,c,e\}, \{b,d,e\}, \{c,d,e\},$$
$$\{a,b,c,d\}, \{a,b,c,e\}, \{a,b,d,e\},$$
$$\{a,c,d,e\}, \{b,c,d,e\}, \{a,b,c,d,e\}\}$$

Note that, at least for finite sets, the power set is always larger than the original. Obviously $n < 2^n$. An interesting question, to be addressed below, is whether or not this holds for infinitely large sets.

We now define the Cartesian product of two sets.

**Cartesian Product:**

The *Cartesian product* of two sets A and B is the set of all ordered pairs $(x,y)$ with $x \in A$ and $y \in B$:

$$A \times B = \{(x,y) \mid x \in A \land y \in B\}$$

Examples:

$$E = \{b,c\} \qquad E \times H = \{(b,d), (b,e), (b,f), (c,d), (c,e), (c,f)\}$$
$$H = \{d,e,f\} \qquad H \times E = \{(d,b), (d,c), (e,b), (e,c), (f,b), (f,c)\}$$
$$H \times H = \{(d,d), (d,e), (d,f), (e,d), (e,e), (e,f), (f,d), (f,e), (f,f)\}$$

Note that $A \times B = B \times A \leftrightarrow A = B$.

We saw the notion of 2-place, or binary, relations in Chapter 6. There, we discussed 2-place predicate letters interpreted via a set of ordered pairs as extension. Here we come at it from the other direction.

**Relation:**

A (2-place) *relation* R is a set of ordered pairs:

$$\text{R is a relation} \leftrightarrow (\forall z)(z \in R \rightarrow (\exists x)(\exists y)(z = (x,y)))$$

So, a set R is a relation iff every member of R is some ordered pair. This means that there are many many relations, as we'll discuss below, and we will not always have an easy description or natural language predicate for the relation in question. But that is not a problem. In terms of the previous chapter we can state the following obvious consequence:

$$(x,y) \in R \leftrightarrow Rxy$$

Here I am using 'R' both to refer to a 2-place relation (on the left) and as a 2-place predicate letter (on the right) interpreted as expressing that very relation. This should not create any confusion and is similar to the practice adopted in Section 6.5.

**Domain, Range, Field:**

For each relation R, we can distinguish the set of elements in the first position, called the *domain* of the relation, the set of elements in the second position, called the *range* of the relation, and the union of these two sets, called the *field* of the relation:

$$\mathscr{D}(R) = \{x \mid (\exists y)Rxy\}$$
$$\mathscr{R}(R) = \{y \mid (\exists x)Rxy\}$$
$$\mathscr{F}(R) = \mathscr{D}(R) \cup \mathscr{R}(R)$$

Each relation, R is the subset of the Cartesian product of its domain and range.

$$R \subseteq \mathscr{D}(R) \times \mathscr{R}(R)$$

Moreover, the domain and range are identical iff they are identical to the field.

$$\mathscr{D}(R) = \mathscr{R}(R) \leftrightarrow (\mathscr{D}(R) = \mathscr{F}(R) \land \mathscr{R}(R) = \mathscr{F}(R))$$

These last two points should be fairly obvious, and can easily be proven. Note that in terms of the semantics of our language **P**, the **UD** of an interpretation provided a superset of the field of all relations. And it should be clear that we can define the various properties of binary relations similarly to Section 6.5.

Given a field F, such that every x, y, z are elements of F we define the following properties of R in F:

**Reflexivity:**

$$(\forall x)((x,x) \in R)$$

**Symmetry:**

$$(\forall x)(\forall x)((x,y) \in R \rightarrow (y,x) \in R)$$

**Transitivity:**

$$(\forall x)(\forall y)(\forall z)(((x,y) \in R \wedge (y,z) \in R) \rightarrow (x,z) \in R)$$

I refrain from defining the other relations discussed in the previous chapter, for, as the reader can see, it is mainly an exercise in variant notation.

We define a few other properties of relations that will be important to the ensuing discussion.

**One-Many:**

A relation R is *one-many* when, if an object x bears R to an object z, then no other object y bears R to z. The relation biological father to son is one-many.

$$(\forall x)(\forall y)(\forall z)((Rxz \wedge Ryz) \rightarrow x = y)$$

**Many-One:**

A relation R is *many-one* when, if an object z bears R to x, then z bears R to no other object y. The relation son to biological father is many-one.

$$(\forall x)(\forall y)(\forall z)((Rzx \wedge Rzy) \rightarrow x = y)$$

**One-One:**

A relation R is *one-one* when it is both one-many and many-one.

$$(\forall x)(\forall y)(\forall z)((Rxz \wedge Ryz) \rightarrow x = y) \wedge (\forall x)(\forall y)(\forall z)((Rzx \wedge Rzy) \rightarrow x = y)$$
$$(\forall x)(\forall y)(\forall z)(((Rxz \wedge Ryz) \vee (Rzx \wedge Rzy)) \rightarrow x = y)$$

**Many-Many:**

A relation R is *many-many* when it is neither one-many nor many-one.

$$\neg(\forall x)(\forall y)(\forall z)((Rxz \wedge Ryz) \rightarrow x = y) \wedge \neg(\forall x)(\forall y)(\forall z)((Rzx \wedge Rzy) \rightarrow x = y)$$
$$(\exists x)(\exists y)(\exists z)((Rxz \wedge Ryz) \wedge x \neq y) \wedge (\exists x)(\exists y)(\exists z)((Rzx \wedge Rzy) \wedge x \neq y)$$

When a relation is many-one, we call it a function. Intuitively a function takes inputs and for each input yields a unique output, but two different inputs may yield the same output. Think of the mathematical function of squaring. If we take the domain to be positive and negative integers, then each input yields a unique output (its square), but both $n$ and $-n$ yield $n^2$. Doubling is also a function. Hence, it is many-one, but it is also one-one—each number has a unique double and no two different numbers have the same double. (Note: squaring is one-one as well, assuming we restrict the domain to the non-negative integers.) This input-output metaphor is misleading however. On the set-theoretic understanding, a function is not a process converting inputs to outputs, it is simply a special type of relation (many-one), and relations are simply sets of ordered pairs, and ordered pairs, as we have seen, are certain sets of sets. So it begins to emerge how we can define certain logical and mathematical notions via set theory.

When we have a function, F, instead of '$(x,y) \in F$', or '$Fxy$', we write:

$$f(x) = y$$

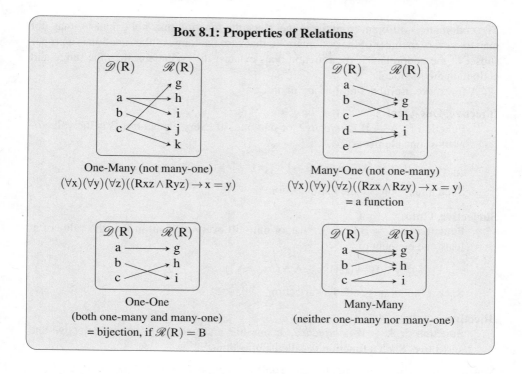

**Box 8.1: Properties of Relations**

One-Many (not many-one)
$(\forall x)(\forall y)(\forall z)((Rxz \wedge Ryz) \to x = y)$

Many-One (not one-many)
$(\forall x)(\forall y)(\forall z)((Rzx \wedge Rzy) \to x = y)$
= a function

One-One
(both one-many and many-one)
= bijection, if $\mathscr{R}(R) = B$

Many-Many
(neither one-many nor many-one)

## Function:

A relation F is a *function* iff it is many-one; that is, iff:

$$(\forall x)(\forall y)(\forall z)((Fzx \wedge Fzy) \to x = y)$$

In this case we write:

$$f(x) = y$$

Here, x is the argument and y is the image (or value) of x under f.

Often, when speaking of a function, we convey it as a mapping from one set to another. The set of inputs, or arguments, is called the domain, as with any relation. The set to which those arguments are mapped, the set of possible values, or *images* of the arguments, is called the *codomain*.

## Codomain:

The *codomain* of a function is the set of possible images. The codomain is a superset of the range, where the range is the set of actual images. We can indicate the domain, A, and codomain, B, of a function, f, as follows:

$$f : A \to B$$

Here, f is a function from A to B.

The codomain is different from the range, since the function may not exhaust—may not map to—every member of the codomain. In that case the range of the function is a proper subset of the codomain. If the function does exhaust the codomain, then the range and codomain are identical.

A few more useful definitions are in order.

### Injective, One-One:

Function f: A → B is *injective*, or one-one, iff every element of B is the value of at most one element of A,

$$(\forall x)(\forall y)(((x \in A \land y \in A) \land f(x) = f(y)) \to x = y)$$

Such a function is called an injection.

### Surjective, Onto:

Function f: A → B is *surjective*, or onto, iff every element in B is the value of at least one element of A,

$$(\forall y)(y \in B \to (\exists x)(x \in A \land f(x) = y))$$

Such a function is called a surjection.

### Bijective:

Function f: A → B is *bijective*, or one-one and onto, iff it is both injective and surjective. Such a function is called a bijection.

Note that, to be a function, a relation need only be many-one. This does not imply anything about whether is it injective for surjective, it may fail to be one or both of these. Boxes 8.1 and 8.2 contain some useful illustrations.

## 8.6  **Infinite Sets and Cantor's Proof**

The concept of infinity has been puzzling to philosophers and mathematicians for thousands of years. Moreover, it has been relied upon to establish positive conclusions (e.g., in proofs of the existence of God) and negative conclusions (e.g., in proofs of the unreality of space, time, multiplicity, motion) often with very little understanding of what was being talked about. It was not until the work of Georg Cantor in the late nineteenth century that we had a clear and coherent notion of infinity. We will introduce and use the notion of *cardinality* of a set to discuss some of the puzzling aspects of infinite sets. Most, if not all, of these are puzzling simply because we continue to attribute certain properties of finite numbers to the infinite. But this is something logic and set theory can allow us to overcome.

### Cardinality:

The number of elements in a set (considered without regard to any order relations on the elements) is called the cardinality of a set. We write the cardinality of a set A as:

$$|A|$$

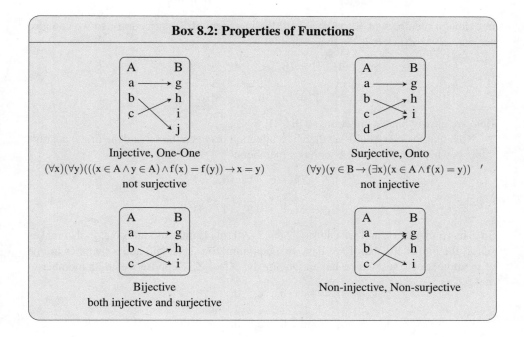

Sets A and B have equal cardinality, or are *equinumerous*,

> $|A| = |B|$ iff there is a bijection from A to B

The cardinality of A is less or equal to that of B,

> $|A| \leq |B|$ iff there is an injection from A to B

The cardinality of A is less than that of B,

> $|A| < |B|$ iff there is an injection but no bijection from A to B

This is rather intuitive in the case of finite sets. Take these sets:

> $E = \{b, c\}$
> $Z = \{a, b, c, d, e\}$
> $H = \{d, e, f\}$

We can easily see that $|E| \leq |Z|$ since there is an injection from E to Z:

$$E = \{b, \quad c\}$$
$$\downarrow \quad \downarrow$$
$$Z = \{a, \quad b, \quad c, \quad d, \quad e\}$$

Every element of Z is the value of at most one element of E, so we have an injection, thus $|E| \leq |Z|$. But it is also clear that there can be no bijection between the two, for there are not enough members of E to exhaust all the members of Z (surjection), so we can also see that $|E| < |Z|$. Here is another example:

$$E \cup H = \{b, \quad c, \quad d, \quad e, \quad f\}$$
$$\downarrow \quad \downarrow \quad \downarrow \quad \downarrow \quad \downarrow$$
$$Z = \{a, \quad b, \quad c, \quad d, \quad e\}$$

Here we clearly have a bijection, so $|E \cup H| = |Z|$.

Note that for every finite set there is a bijection between the set and the first $n$ positive integers which shows that the set has $n$ members:

$$Z = \{a, \quad b, \quad c, \quad d, \quad e\}$$
$$\downarrow \quad \downarrow \quad \downarrow \quad \downarrow \quad \downarrow$$
$$Z = \{1, \quad 2, \quad 3, \quad 4, \quad 5\}$$

So Z has 5 members, and we can say $|Z| = 5$. All this is nice and intuitive. It also makes sense, for finite sets, that if we have two equinumerous sets, and add a member to one of them, that new set will be larger. Obviously, $|Z| = |Z|$. Suppose we add a member to make $Z'$:

$$Z = \{a, \quad b, \quad c, \quad d, \quad e\}$$
$$\downarrow \quad \downarrow \quad \downarrow \quad \downarrow \quad \downarrow$$
$$Z' = \{a, \quad b, \quad c, \quad d, \quad e, \quad f\}$$

So clearly $|Z| < |Z'|$.

Now consider infinite sets. Obviously the set of positive integers $\mathbb{Z}_{>0}$ has the same cardinality as itself:

$$\mathbb{Z}_{>0} = \{1, \quad 2, \quad 3, \quad 4, \quad 5, \ldots\}$$
$$\downarrow \quad \downarrow \quad \downarrow \quad \downarrow \quad \downarrow \cdots$$
$$\mathbb{Z}_{>0} = \{1, \quad 2, \quad 3, \quad 4, \quad 5, \ldots\}$$

We can describe the bijection illustrated above as $f(x) = x$. So, $|\mathbb{Z}_{>0}| = |\mathbb{Z}_{>0}|$. But suppose we add 0 to the positive integers to get the natural numbers:

$$\mathbb{N} = \{0, \quad 1, \quad 2, \quad 3, \quad 4, \ldots\}$$
$$\downarrow \quad \downarrow \quad \downarrow \quad \downarrow \quad \downarrow \cdots$$
$$\mathbb{Z}_{>0} = \{1, \quad 2, \quad 3, \quad 4, \quad 5, \ldots\}$$

Even though we have added a new element in 0, it is clear that $f(x) = x + 1$ is a bijection $f: \mathbb{N} \to \mathbb{Z}_{>0}$. So, $|\mathbb{N}| = |\mathbb{Z}_{>0}|$. This is not a general proof, but we can see that the cardinality of an infinite set is not increased by adding a new element.

Nor is the cardinality of an infinite set decreased by removing half the elements. Let $\mathbb{E}$ be the set of even positive integers:

$$\mathbb{Z}_{>0} = \{1, \quad 2, \quad 3, \quad 4, \quad 5, \ldots\}$$
$$\downarrow \quad \downarrow \quad \downarrow \quad \downarrow \quad \downarrow \cdots$$
$$\mathbb{E} = \{2, \quad 4, \quad 6, \quad 8, \quad 10, \ldots\}$$

The bijection here is $f(x) = 2x$. So, $|\mathbb{E}| = |\mathbb{Z}_{>0}| = |\mathbb{N}|$. A few other examples are to be found in the exercises.

What we are seeing here is that infinite sets, very unlike finite sets, can be put into one-one correspondence with some of their proper subsets (not all though, in particular not with their finite subsets).[9] Since every set is a subset of itself, every set can be put into one-one correspondence with at least one of its subsets (namely, itself). But if we are talking about proper subsets, no finite set can be put into one-one correspondence with one of its *proper* subsets. Note above that Z $\subset$ Z'. Is there any way we could get them in one-one correspondence? Is there any bijection between any finite set and a proper subset? In contrast, the illustrations I have been giving with the infinite sets all show that the set $\mathbb{N}$ has the same cardinality as some of its proper subsets. How many proper subsets of $\mathbb{N}$ are of the same size as $\mathbb{N}$? If all these sets are the same size as $\mathbb{N}$, is $\mathbb{N}$ the "biggest" set possible or are there "bigger" infinities? The answers to these questions are related, but we need to do a little work to get at them.

Cantor gave a name to the cardinal number of $\mathbb{N}$: $\aleph_0$.[10] That is, $|\mathbb{N}| = \aleph_0$. So the final question of the previous paragraph can be rephrased as:

Is there any set A, such that $|A| > \aleph_0$?

First observe that, because the number of elements in the power set of a set of $n$ elements is $2^n$, the power set of a finite set will always be larger than the original. After all, $2^n > n$ for any finite $n$. However, as we've seen with addition and multiplication, we cannot trust that this will hold for infinite sets. So we will need a proof.

Indeed, we will see *two* proofs that $\mathscr{P}(\mathbb{N})$ is of greater cardinality than $\mathbb{N}$ itself. We will prove this by showing that there is an injective function from $\mathbb{N} \to \mathscr{P}(\mathbb{N})$, but that there is no bijection. This will prove that $|\mathbb{N}| < |\mathscr{P}(\mathbb{N})|$ and hence $|\mathscr{P}(\mathbb{N})| > \aleph_0$. The two proofs are essentially the same, but the second is more visual.

> **First Proof:** Cantor gives us a general proof to show that for any set A, $\mathscr{P}(A)$ is of greater cardinality. We prove, for all sets A:
>
> (1) $|A| < |\mathscr{P}(A)|$
>
> Note that h$(x) = \{x\}$ is an injection h: A $\to \mathscr{P}(A)$. So h maps each member of A to its singleton in $\mathscr{P}(A)$. It is clearly one-one but does not necessarily exhaust the members of $\mathscr{P}(A)$. This establishes that,
>
> (2) $|A| \leq |\mathscr{P}(A)|$
>
> We now need to show that there cannot be a bijection f: A $\to \mathscr{P}(A)$. Though f might be an injection, it must fail to be a surjection—it must fail to exhaust all the members of $\mathscr{P}(A)$. We will show this by *reductio ad absurdum*.
>
> Suppose:
>
> (3) $|A| = |\mathscr{P}(A)|$
>
> Then there is a bijection f: A $\to \mathscr{P}(A)$. Next define:

---

9. This is called being Dedekind-infinite, after the German mathematician Richard Dedekind.
10. Pronounced aleph-null, or aleph-zero. '$\aleph$' is the first letter of the Hebrew alphabet.

(4) $B = \{x \in A \mid x \notin f(x)\}$

This is the set of all members of A which are not members of their image in $\mathscr{P}(A)$. Since B is a subset of A, it is a member of $\mathscr{P}(A)$. And since f is a bijection, there must be some member of A whose image is B. That is, there must be an $x \in A$ such that,

(5) $f(x) = B$

Now we ask, is that x an element of B or not an element of B? That is, is this x a member of its image under f? By (4) and (5) we have:

$$x \in f(x) \leftrightarrow (x \in A \wedge x \notin f(x))$$
and, since every $x \in A$, we have,
$$x \in f(x) \leftrightarrow x \notin f(x)$$

But this is a familiar contradiction. It means that there cannot be a bijection from A to $\mathscr{P}(A)$. So our assumption (3) must be false. Thus, we have:

(6) $|A| \neq |\mathscr{P}(A)|$

Obviously, since A was any arbitrary set, (6) and (2) imply (1). Thus, for all sets A,

(1) $|A| < |\mathscr{P}(A)|$

So Cantor's proof establishes that for every set there is at least one of greater cardinality, namely, its powerset. So, in particular:

$$|\mathbb{N}| < |\mathscr{P}(\mathbb{N})|$$
$$\text{and, thus}$$
$$\aleph_0 < |\mathscr{P}(\mathbb{N})|$$

So $\aleph_0$ is not the greatest cardinal number. There are sets larger than $\mathbb{N}$. Moreover,

$$|\mathbb{N}| < |\mathscr{P}(\mathbb{N})| < |\mathscr{P}(\mathscr{P}(\mathbb{N}))| < |\mathscr{P}(\mathscr{P}(\mathscr{P}(\mathbb{N})))| < \cdots$$

There is a series of infinite sets, each of greater cardinality than the previous. There is no greatest infinite set, so no greatest infinite number.

Note two things: First, the structure of this proof is nearly identical to Russell's proof of his Paradox. Cantor's proof was ten years earlier than Russell's, and Russell came up with his proof while trying to find a flaw in Cantor's. Here, Cantor is not showing that intuitive set theory is inconsistent, but he does show, as we did in applying Russell's argument to the Axiom Schema of Separation, that there is no universal set—no set $\Upsilon$ containing everything. Indeed, there is an ever-increasing series of infinite sets. An interesting and unexpected result!

Here is another version of the proof, with a little more visual appeal.

**Box 8.3: A bijection** $f: \mathbb{N} \to \mathscr{P}(\mathbb{N})$?

| x | f(x) contains 0 | 1 | 2 | 3 | 4 | 5 | ... | |
|---|---|---|---|---|---|---|---|---|
| 0 | y | y | y | y | y | y | ... | the natural numbers |
| 1 | n | n | y | n | y | n | ... | the even numbers |
| 2 | n | y | n | y | n | y | ... | the odd numbers |
| 3 | n | y | y | y | y | y | ... | the positive integers |
| 4 | y | n | y | n | n | y | ... | a "random" set |
| 5 | y | n | n | y | n | y | ... | the diagonal set? |
| ⋮ | ⋮ | ⋮ | ⋮ | ⋮ | ⋮ | ⋮ | ⋱ | |

**Second Proof:** The first proof applied to all sets. This version is geared just to $\mathbb{N}$, but is visually more intuitive for some people. As the proofs are essentially the same, this one can be made to generalize. We prove:

(1') $|\mathbb{N}| < |\mathscr{P}(\mathbb{N})|$

Again, first note that $h(x) = \{x\}$ is an injection $h: \mathbb{N} \to \mathscr{P}(\mathbb{N})$. This establishes that,

(2') $|\mathbb{N}| \leq |\mathscr{P}(\mathbb{N})|$

We now need to show that there cannot be a bijection $f: \mathbb{N} \to \mathscr{P}(\mathbb{N})$. Though f might be an injection, it must fail to be a surjection—it must fail to exhaust all the members of $\mathscr{P}(\mathbb{N})$. We will show this by *reductio ad absurdum*.

Suppose:

(3') $|\mathbb{N}| = |\mathscr{P}(\mathbb{N})|$

Then there must be a bijection $f: \mathbb{N} \to \mathscr{P}(\mathbb{N})$. We can represent any supposed bijection via a table which lists the natural numbers across the top and down the left hand side. The first number in each row represents an argument, x, of f, while the rest of that row describes (via a sequence of 'y's and 'n's) f(x), the subset of $\mathbb{N}$ that is the image under f of 'x'.

Take, for example, the function depicted in Box 8.3. There, 0 is assigned the set of natural numbers, 1 is assigned the evens, 2 the odds, and so on. For any such function we can define a set, called the diagonal set, as follows,

$$D = \{x : x \in f(x)\}$$

This is the set of all natural numbers x which are members of their image under f. In the above example $D = \{0, 3, 5, ...\}$. This set is called the diagonal set because the sequence of 'y's and 'n's which would describe it matches the diagonal sequence of 'y's and 'n's in the table (upper-left to lower right) describing the function f. In the above example the sequence describing D is y,n,n,y,n,y,... Clearly D is a subset of the natural numbers, and so $D \in \mathscr{P}(\mathbb{N})$. Hence D may or may not appear in the table. In the table above it looks as if D might be f(5). They both start y,n,n,y,n,y, but, since we have not fully specified f(5), we cannot be sure.[11] It doesn't really matter, because defining D is just an example. We are really interested in the *antidiagonal* set, defined as follows:

(4') $\overline{D} = \{x \in \mathbb{N} \mid x \notin f(x)\}$

This is the set of all natural numbers x which are not members of their image under f. The sequence of 'y's and 'n's which describes it will be the inverse of the sequence which describes D; i.e., wherever 'y' appears in the sequence describing D, an 'n' will appear in the sequence describing $\overline{D}$, and vice versa. In the example above, the sequence describing $\overline{D}$ is n,y,y,n,y,n,...

We complete the proof by showing that, for any f we choose, $\overline{D}$ cannot appear in the table describing f. That is, for any $f : \mathbb{N} \to \mathscr{P}(\mathbb{N})$ there is no natural number x such that $f(x) = \overline{D}$. This will mean that any such function must fail to be surjective, and thus fail to be bijective. That is, it must fail to exhaust the members of $\mathscr{P}(\mathbb{N})$.

Recall that, because of supposition (3'), f is bijective. So there must be an x such that,

(5') $f(x) = \overline{D}$

But is that x an element of $\overline{D}$ or not an element of $\overline{D}$? That is, is this x a member of its image under f? Again, by (4') and (5') we have:

$x \in \overline{D} \leftrightarrow x \notin \overline{D}$

This is our familiar contradiction. It means that for any function f, $\overline{D}$ cannot be an image of any x. So there can be no surjection $f : \mathbb{N} \to \mathscr{P}(\mathbb{N})$, and thus, no bijection.

Another way of looking at this may be more intuitive. Consider the sequence of 'y's and 'n's which describes $\overline{D}$. If it is to appear in the table describing f, it must appear in the $m^{\text{th}}$ row, for some natural number m. Now, consider reading across the $m^{\text{th}}$ row to see the sequence describing $\overline{D}$. When we reach the $m^{\text{th}}$ column (so, the cell of the table corresponding to position m,m) there would have to be both a 'y' and an 'n' at that position, which is not possible. Hence the sequence describing $\overline{D}$ cannot appear in the table describing f. Again, we have shown that f does not exhaust the members of $\mathscr{P}(\mathbb{N})$, it

---

11. Our assumption (3') requires that D appear in the table, but recall that we assumed (3') for *reductio*, so D might not appear.

---

| | | | Box 8.4: $\overline{D}$ cannot appear in $f: \mathbb{N} \to \mathscr{P}(\mathbb{N})$ | | | | | |
|---|---|---|---|---|---|---|---|---|

| | | | $f(x)$ contains | | | | | |
|---|---|---|---|---|---|---|---|---|
| x | 0 | 1 | 2 | 3 | 4 | 5 | $\cdots$ | |
| 0 | y | y | y | y | y | y | $\cdots$ | the natural numbers |
| 1 | n | n | y | n | y | n | $\cdots$ | the even numbers |
| 2 | n | y | n | y | n | y | $\cdots$ | the odd numbers |
| 3 | n | y | y | y | y | y | $\cdots$ | the positive integers |
| 4 | y | n | y | n | n | y | $\cdots$ | a "random" set |
| 5 | n | y | y | n | y | ?! | $\cdots$ | the antidiagonal set? |
| $\vdots$ | $\vdots$ | $\vdots$ | $\vdots$ | $\vdots$ | $\vdots$ | $\vdots$ | $\ddots$ | |

is not surjective, and thus not bijective. See Box 8.4 where we attempt to describe $\overline{D}$ in row 5. At position $5, 5$ in the table, we must place a 'y' if and only if we place a 'n'. No such table can contain $\overline{D}$.

We have proved there can be no bijection $f: \mathbb{N} \to \mathscr{P}(\mathbb{N})$. Thus, our assumption (3′) must be false. Thus,

(6′) $|\mathscr{P}(\mathbb{N})| \neq |\mathbb{N}|$

So (6′) and (2′) imply (1′),

(1′) $|\mathbb{N}| < |\mathscr{P}(\mathbb{N})|$

And, obviously,

$$\aleph_0 < |\mathscr{P}(\mathbb{N})|$$

This version of the proof makes clear why Cantor's and Russell's (and others') results are often said to be achieved by diagonalization. The non-self-member condition (Russell) and the non-member-of-own-image condition (Cantor) are impossible to meet along the diagonal. As with the previous proof, and with the Axiom Schema of Separation, we get the result that there is no universal set—no set $\Upsilon$ containing everything. Again, there is an ever-increasing series of infinite sets.

## 8.6.1 Exercises

1. Let $\mathbb{O}$ be the set of odd positive integers. Describe a bijection $f: \mathbb{O} \to \mathbb{Z}_{>0}$ that shows $|\mathbb{O}| = |\mathbb{Z}_{>0}|$

2. Describe a bijection $f: \mathbb{O} \to \mathbb{N}$ that shows $|\mathbb{O}| = |\mathbb{N}|$

3. Describe a bijection $f: \mathbb{O} \to \mathbb{E}$ that shows $|\mathbb{O}| = |\mathbb{E}|$

4. Could there be a barber who lives in town and shaves all and only those townspeople who do not shave themselves? Prove your answer.

5. Could there be a barber who shaves all and only those townspeople who do not shave themselves? Prove your answer and relate this to the previous question.

# 8.7 **Chapter Glossary**

**Absolute Complement:**

Given K, a superset of all the sets of current interest, the temporarily absolute complement of a set A, $\overline{A}$ is the set of all elements of K, not in A:

$$\overline{A} = \{x \in K \mid x \notin A\} \quad (278)$$

**Axiom of Extensionality:**

If sets A and B have exactly the same members, then they are identical:

$$(\forall x)((x \in A \leftrightarrow x \in B) \to A = B) \quad (271)$$

**Axiom of Pairing:**

For any two elements, x and y, there exists a set, A, consisting of just those elements.

$$(\exists A)(\forall z)(z \in A \leftrightarrow z = x \lor z = y)$$

In set builder notation:

$$\{x, y\} = \{z \mid z = x \lor z = y\} \quad (281)$$

**Axiom Schema of Abstraction:**

For any clearly stated condition, Sx, there exists a set B whose elements are exactly those objects which satisfy Sx:

$$(\exists B)(\forall x)(x \in B \leftrightarrow Sx)$$

In set builder notation we can briefly define a set, B:

$$B = \{x \mid Sx\} \quad (270)$$

**Axiom Schema of Separation:**

For every set A and every condition Sx there is a set B whose elements are exactly those members of A for which Sx holds (y must not be free in S):

$$(\exists B)(\forall x)(x \in B \leftrightarrow (x \in A \land Sx))$$

Briefly:

$$B = \{x \in A \mid Sx\} \quad (274)$$

**Bijective:**

Function f: A $\rightarrow$ B is *bijective*, or one-one and onto, iff it is both injective and surjective. Such a function is called a bijection. (286)

**Cardinality:**

The number of elements in a set (considered without regard to any order relations on the elements) is called the cardinality of a set. We write the cardinality of a set A as:

$$|A| \quad (286)$$

**Cartesian Product:**

The *Cartesian product* of two sets A and B is the set of all ordered pairs $(x, y)$ with $x \in A$ and $y \in B$:

$$A \times B = \{(x, y) \mid x \in A \wedge y \in B\} \quad (282)$$

**Codomain:**

The *codomain* of a function is the set of possible images. The codomain is a superset of the range, where the range is the set of actual images. We can indicate the domain, A, and codomain, B, of a function, f, as follows:

$$f: A \rightarrow B$$

Here, f is a function from A to B. (285)

**Difference:**

The difference (or relative complement) between sets A and B is the set containing every member of A that is not in B:

$$A - B = \{x \mid x \in A \wedge x \notin B\} \quad (278)$$

**Domain, Range, Field:**

For each (2-place) relation R, we can distinguish the set of elements in the first position, called the *domain* of the relation, the set of elements in the second position, called the *range* of the relation, and the union of these two sets, called the *field* of the relation:

$$\mathscr{D}(R) = \{x \mid (\exists y)Rxy\}$$
$$\mathscr{R}(R) = \{y \mid (\exists x)Rxy\}$$
$$\mathscr{F}(R) = \mathscr{D}(R) \cup \mathscr{R}(R) \quad (283)$$

**Field:**

See Domain, Range, Field. (283)

**Function:**
> A relation F is a *function* iff it is many-one; that is, iff:

$$(\forall x)(\forall y)(\forall z)((Fzx \wedge Fzy) \rightarrow x = y)$$

> In this case we write:

$$f(x) = y$$

> Here, x is the argument and y is the image of x under f. (285)

**Generalized Intersection:**
> For every set of sets A, there is some set C whose elements are all and only the elements in every $B \in A$:

$$(\exists C)(\forall x)(x \in C \leftrightarrow (\forall B)(B \in A \rightarrow x \in B))$$

> Or:

$$\bigcap A = \{x \mid \text{for every } B \in A,\ x \in B\} \quad (276)$$

**Injective, One-One:**
> Function $f \colon A \rightarrow B$ is *injective*, or one-one, iff every element of B is the value of at most one element of A,

$$(\forall x)(\forall y)(((x \in A \wedge y \in A) \wedge f(x) = f(y)) \rightarrow x = y)$$

> Such a function is called an injection. (286)

**Intersection:**
> The intersection of a pair of sets A and B, $A \cap B$, the set consisting of all and only the elements in both sets:

$$A \cap B = \{x \mid x \in A \wedge x \in B\} \quad (276)$$

**Many-Many:**
> A relation R is *many-many* when it is neither one-many nor many-one.

$$\neg(\forall x)(\forall y)(\forall z)((Rxz \wedge Ryz) \rightarrow x = y) \wedge \neg(\forall x)(\forall y)(\forall z)((Rzx \wedge Rzy) \rightarrow x = y)$$
$$(\exists x)(\exists y)(\exists z)((Rxz \wedge Ryz) \wedge x \neq y) \wedge (\exists x)(\exists y)(\exists z)((Rzx \wedge Rzy) \wedge x \neq y)$$
$$(284)$$

**Many-One:**
> A relation R is *many-one* when, if an object z bears R to x, then z bears R to no other object y. The relation son to biological father is many-one.

$$(\forall x)(\forall y)(\forall z)((Rzx \wedge Rzy) \rightarrow x = y) \quad (284)$$

**One-Many:**

A relation R is *one-many* when, if an object x bears R to an object z, then no other object y bears R to z. The relation biological father to son is one-many.

$$(\forall x)(\forall y)(\forall z)((Rxz \land Ryz) \to x = y) \quad (284)$$

**One-One:**

A relation R is *one-one* when it is both one-many and many-one.

$$(\forall x)(\forall y)(\forall z)((Rxz \land Ryz) \to x = y) \land (\forall x)(\forall y)(\forall z)((Rzx \land Rzy) \to x = y)$$
$$(\forall x)(\forall y)(\forall z)(((Rxz \land Ryz) \lor (Rzx \land Rzy)) \to x = y) \quad (284)$$

**Onto:**

See Surjective, Onto. (286)

**Ordered Pair:**

For any two elements, x and y, there exists a set, $(x, y)$, such that

$$(x, y) = \{\{x\}, \{x, y\}\} \quad (281)$$

**Powerset Axiom:**

For each set A, there is a set C containing all subsets B of A:

$$(\exists C)(\forall B)(B \in C \leftrightarrow B \subseteq A)$$

That is:

$$\mathscr{P}(A) = \{B \mid B \subseteq A\} \quad (282)$$

**Proper Subset, Proper Inclusion:**

Set A is a *proper subset of* or is *properly included in* B iff every member of A is also a member of B, but $A \neq B$. We use the '$\subset$' symbol:

$$A \subset B \leftrightarrow (A \subseteq B \land A \neq B) \quad (275)$$

**Range:**

See Domain, Range, Field. (283)

**Reflexivity:**

$$(\forall x)((x, x) \in R) \quad (283)$$

**Relation:**

A (2-place) *relation* R is a set of ordered pairs:

$$\text{R is a relation} \leftrightarrow (\forall z)(z \in R \to (\exists x)(\exists y)(z = (x, y))) \quad (283)$$

**Subset, Inclusion:**

Set A is a *subset of* or is *included in* B iff every member of A is also a member of B. We use the '$\subseteq$' symbol:

$$A \subseteq B \leftrightarrow (\forall x)(x \in A \rightarrow x \in B) \qquad (275)$$

**Surjective, Onto:**

Function $f \colon A \rightarrow B$ is *surjective*, or onto, iff every element in B is the value of at least one element of A,

$$(\forall y)(y \in B \rightarrow (\exists x)(x \in A \wedge f(x) = y))$$

Such a function is called a surjection. (286)

**Symmetric Difference:**

The symmetric difference (or symmetric complement) of sets A and B is the union of the differences:

$$A \ominus B = (A - B) \,\dot{\cup}\, (B - A) \qquad (279)$$

**Symmetry:**

$$(\forall x)(\forall x)((x, y) \in R \rightarrow (y, x) \in R) \qquad (283)$$

**Transitivity:**

$$(\forall x)(\forall y)(\forall z)(((x, y) \in R \wedge (y, z) \in R) \rightarrow (x, z) \in R) \qquad (284)$$

**Union:**

The union of a pair of sets A and B, $A \cup B$, the set consisting of all and only the elements in either set:

$$A \cup B = \{x \mid x \in A \vee x \in B\} \qquad (277)$$

**Union Axiom:**

For every set of sets A, there is some set C whose elements are all and only the elements of at least one $B \in A$:

$$(\exists C)(\forall x)(x \in C \leftrightarrow (\exists B)(B \in A \wedge x \in B))$$

Or:

$$\textstyle\bigcup A = \{x \mid \text{for some } B \in A,\ x \in B\} \qquad (278)$$

# 9 Modal Logic

## 9.1 Necessity, Possibility, and Impossibility

### 9.1.1 Modalities

In this chapter we add modal operators to our language $S$ in order to examine sentential modal logic. Chapters 2–4 covered the syntax, semantics, and deductive system for truth-functional logic. The account there, especially regarding the semantics of $S$, centered on interpreting each simple statement as having an unqualified truth value, T or F (and not both). Here we will focus, not just on truth values, but on the *modality* of the truth value of a statement—roughly, the *way in which* the statement is true or false. For example,

(1) *It is necessary that* two plus two equals four
(2) *It is possible that* there are still undiscovered bugs in the code
(3) *It is impossible that* the spacecraft travels faster than the speed of light
(4) *It is necessary that* George W. Bush was the 43$^{rd}$ President of the United States of America

Here the emphasized modal phrase changes a simple subsentence—about math, programming, physics, or history—into a compound claim that is not simply about math, programming, physics, or history. Rather, the compound claim is about the way in which the subsentence *must* be true, *can* be true, or *cannot* be true. That is, the *alethic* modalities necessity, possibility, and impossibility, qualify the way in which a statement is true or false. The word 'alethic' comes from the Greek 'aletheia', meaning truth. Note that these modifiers are non-truth-functional. (1) may well be true, but not simply due to the truth of its subsentence. (4) also has a true subsentence, but (4) itself is not true—various details of history might have gone differently. Since modals are non-truth-functional, the semantics laid out in Chapter 3 will have to be significantly modified.

It is worth briefly noting that the alethic modalities are not the only ones:

(5) *It is obligatory that* people help those in need
(6) *It is permitted that* employees wear jeans on Friday
(7) *It will be the case that* humans colonize Mars
(8) *It has never been the case that* humans coexisted with dinosaurs
(9) *It is believed by some that* humans coexisted with dinosaurs
(10) *It is known that* humans did not coexist with dinosaurs

In each of these cases, the emphasized phrase qualifies its subsentence, changing each from a simple claim into various non-truth-functional modal claims. (5) and (6) express *deontic* modality, making normative claims. (7) and (8) express temporal modality. And (9) and (10) expresses *doxastic* or *epistemic* modality, relating to reasoning, belief, and knowledge. Alethic modality forms the basis for the logical study of these other modalities, and it is our main focus in this chapter. Unless otherwise noted, 'modal' will be used to refer to alethic modalities.

Our aim will be to characterize concepts of alethic modality as clearly as possible. Let's begin with some informal definitions:

**Alethic Modality:**
> The *alethic modalities*—necessity, possibility, contingency, impossibility—qualify the truth value of a statement.

**Necessity:**
> A statement, $\mathbb{P}$, is *necessary*, *necessarily true*, or is a *necessity* iff it could not be false.

**Impossibility:**
> A statement, $\mathbb{P}$, is *impossible*, or is an *impossibility* iff it could not be true.

**Possibility:**
> A statement, $\mathbb{P}$, is *possible*, *possibly true*, or is a *possibility* iff it is not impossible.

**Contingency:**
> A statement, $\mathbb{P}$, is *contingent*, or is a *contingency* iff it is neither necessary nor impossible, that is, if it is possibly true and possibly false.

**Non-Necessity:**
> A statement, $\mathbb{P}$, is *non-necessary*, or is *possibly false* iff it is not necessary, that is, if its negation is possibly true.

Note that these definitions are not simply a question of a statement being true or false. Rather, it has to do with what must, can, or cannot happen regarding the truth value of the statement, $\mathbb{P}$. So we can distinguish the *actual* truth value of $\mathbb{P}$, from its necessity, contingency, or impossibility. In particular, every necessary statement is both possibly true and actually true while every impossible statement is possibly false and actually false. An actually false statement may well be possible. Contingent statements are either actually true or actually false, but could be the other. Box 9.1 illustrates the various interrelationships.

## 9.1.2 **Logical, Metaphysical, Physical**

What does it mean for a certain statement to be necessarily true, or possibly false? How can we know whether some candidate statement is impossible or necessary? These are deep philosophical questions, the various answers to which are much debated, and we will barely scratch the surface of them in this text.

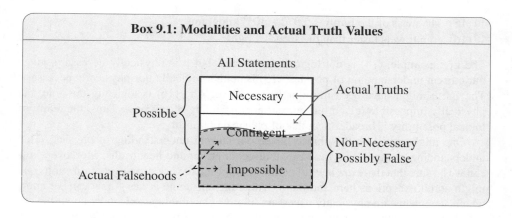

Box 9.1: Modalities and Actual Truth Values

What we are able to do before moving on to the formal work, is give a feel for the types of statements many philosophers have taken to be necessary or possible, and for the different kinds of necessity and possibility philosophers have distinguished. It is interesting to note that, while there is much philosophical dispute surrounding certain aspects of modality, there is a set of logical systems whose structures are clearly defined and well-studied. Since this is a text in logic, we will touch briefly on some philosophy and then get on with the formal work.

We can distinguish different sorts of necessity, possibility, and impossibility: logical, metaphysical, and physical. Logical possibility includes any statement that is consistent with the laws of logic. At minimum, this means the statement is not contradictory. Logical necessities are exactly those statements required by the laws of logic—whose negations are contradictory. Consider these simple examples:

(11)  Either Ann goes to the fair or Ann does not go to the fair
(12)  Bob goes to the fair
(13)  Both Ann goes to the fair and Ann does not go to the fair

Here, (11) is a truth-functionally true statement. Its negation (equivalent to (13)), is contradictory, so (11) is logically necessary. The subsentence of (1), above, is also a truth of logic, since its denial would lead to a contradiction. Thus, the necessity claimed in (1) is true.[1] (11) is also logically possible, of course. (12) is logically possible, but not logically necessary—neither it nor its negation are contradictory. Thus (12) is logically contingent. Finally, (13) is impossible.

We can distinguish logical possibility and necessity from physical possibility and necessity by saying that physical possibilities are statements that are consistent with the laws of nature, while physical necessities are implied by the laws of nature, and physical impossibilities contradict the laws of nature. While anything physically possible must be logically possible, not all logical possibilities are physical possibilities. For instance:

(14)  The speed of light in a vacuum is 299,792,458 meters per second

---

1. I leave aside the possibility of distinguishing mathematical necessity from logical necessity.

(15)  The mass of the Earth is $5.972 \times 10^{24}$ kilograms

(16)  Ann travels faster than the speed of light

The first example, (14) is not logically necessary, but it is physically necessary, given our current understanding of physics. (15) is neither logically nor physically necessary. The Earth might have been more or less massive. And (16) is logically possible, but physically impossible (given our current understanding of physics). Thus, the realm of logical possibility is broader than that of physical possibility.

One might dwell on the phrase 'given our current understanding of physics'. That understanding changes over time as we discover errors and new truths. Moreover, how can we be sure that there *are* laws of nature, as opposed to very strong generalizations that might admit exceptions here and there? These are epistemic issues—how can we know which statements are physically necessary, if any? One might also wonder about the parameters within which we gauge physical necessity, even given our current understanding. What if the universe is deterministic? Then, given the initial conditions of the universe and the laws of nature, the mass of the earth *is* physically necessary. But then would not *every* physical fact be physically necessary, and every falsehood physically impossible? Or can we vary initial conditions while keeping the laws constant? Or perhaps vary certain of the laws, but not others as long as we maintain a consistent set? These are metaphysical and conceptual issues regarding how to understand physical necessity, and I will not discuss various attempts to answer them here.

If one believes in laws that are more general than physical laws, but less general than logical laws, one might want to further distinguish metaphysical possibility as a range of possibility between logical and physical, admitting more possibilities than physical, but fewer than logical. For example, while (16) is physically impossible, it seems entirely metaphysically possible. That is, if we discount the laws of nature, Ann could travel faster than the speed of light. On the other hand, there might be some metaphysical impossibilities, which are logically possible. Some philosophers think this is an example:

(17)  There are events without causes

The idea here is that (17) is not self-contradictory—not contrary to the laws of logic, and so logically possible, but that it is contrary to the laws of metaphysics, and therefore metaphysically impossible. (17) contains no internal *logical* contradiction, so it is logically possible. But, some—determinists—might argue that it contradicts the very concept of an event, so it must be impossible. Since the concepts of event and cause are arguably prior to and more general than any physical concept or law, the impossibility of (17) is metaphysical. Thus, we have the relationship depicted in Box 9.2. Of course, it is also arguable that the concepts of event and cause are part of physics proper, in which case the impossibility of (17), if indeed it is impossible, would be physical impossibility. Thus, exactly how we divide up possibilities is, again, a philosophical issue.

## 9.1.3  **Possible Worlds**

What we mean by 'possible' and 'necessary' is further, if not fully, clarified by the notion of *possible worlds*. Historically associated with the philosopher and mathematician

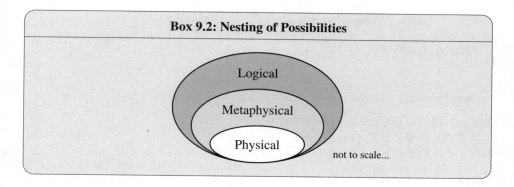

**Box 9.2: Nesting of Possibilities**

Logical

Metaphysical

Physical

not to scale...

Gottfried Wilhelm Leibniz (1646–1716) and systematically developed by the philosopher and logician Saul Kripke (1940– ), each possible world is one of the many ways our world, *the actual world*, could have been. The idea is not of some other planet in this universe. Rather it is the idea of other whole universes—or at least whole descriptions of other universes—where things are at least slightly, and potentially vastly, different from this universe. Thus, if something is necessary, it is so in *all possible worlds*, and if something is possible, it is so in *at least one possible world*. More precisely, if a statement, $\mathbb{P}$, is necessarily true, then it is true in (or at) all possible worlds. And if a statement, $\mathbb{P}$, is possibly true, then it is true in at least one possible world. Note that our world, the actual world, is one among the possible worlds. So, if $\mathbb{P}$ is necessarily true, then it is true in the actual world, and if it is true in the actual world, then it is possible. Further, depending on our focus, we can consider all logically possible worlds, all metaphysically possible worlds, or all physically possible worlds.

This notion of possible worlds, however, does not answer any of the various issues with modality raised in the preceding discussion. This is largely because the idea of possible worlds on its own does not tell us what worlds are logically, metaphysically, or physically possible. Nor does it tell us what they are like—what statements are true at those worlds. Indeed, it seems to introduce additional problems. What is the metaphysical status of these worlds? Are they real, but not actual? Are they imagined by us, stipulated by us, or somehow independent of us? Much has been written on these subjects, but we will leave it to the side here.

We can be comfortable leaving these philosophical issues to the side, because, while the idea of possible worlds might not resolve them, the formal version of possible worlds to be developed in the rest of the chapter *does* help *clarify* them by showing that a rigorous logical structure can ground talk of modalities.

## 9.2 **The Language** $S^\square$

To take account of modality we adjust the syntax of $S$ by adding two sentential operators. We will use the box, '$\square$', to express necessity, and the diamond, '$\lozenge$', to express possibility. These are appended to the left of the statement they modify, exactly like the syntax of the hook, '¬'. Thus the following translations:

(1) Necessarily $\mathbb{P}$          (2) Possibly $\mathbb{P}$

(1') $\Box\mathbb{P}$                (2') $\Diamond\mathbb{P}$

The basic semantic interpretation for the box and diamond are as follows:

**Informal Modal Interpretation:**
> $\Box\mathbb{P}$ is T iff $\mathbb{P}$ is T in every world.
> $\Diamond\mathbb{P}$ is T iff $\mathbb{P}$ is T in at least one world.

In the next section, we will formalize the semantics, but the above is sufficient for present purposes.

Note that the five modal notions—necessity, possibility, contingency, possible false-hood, and impossibility—can all be defined in terms of necessity:

$$\mathbb{P} \text{ is necessary:} \quad\quad \Box\mathbb{P} \Leftrightarrow \neg\Diamond\neg\mathbb{P}$$
$$\mathbb{P} \text{ is possible:} \quad\quad \Diamond\mathbb{P} \Leftrightarrow \neg\Box\neg\mathbb{P}$$
$$\mathbb{P} \text{ is contingent:} \quad \Diamond\mathbb{P} \wedge \Diamond\neg\mathbb{P} \Leftrightarrow \neg(\Box\mathbb{P} \vee \Box\neg\mathbb{P})$$
$$\mathbb{P} \text{ is possibly F:} \quad\quad \Diamond\neg\mathbb{P} \Leftrightarrow \neg\Box\mathbb{P}$$
$$\mathbb{P} \text{ is impossible:} \quad\quad \neg\Diamond\mathbb{P} \Leftrightarrow \Box\neg\mathbb{P}$$

The first two equivalences show that there is redundancy in our symbols. If we wanted, we could eliminate either the box or the diamond, since each is always replaceable by the other flanked by hooks. As typical, we allow the redundancy because it increases ease of expression.[2]

Consider the first equivalence above. Given the informal modal interpretation, we can justify it as follows:

> Suppose $\Box\mathbb{P}$ is T. So $\mathbb{P}$ is T in every world. Thus, there is no world at which ¬$\mathbb{P}$ is T. So $\Diamond\neg\mathbb{P}$ is F. Thus ¬$\Diamond\neg\mathbb{P}$ is T. Each step is an equivalence, so we have proved the equivalence above.

Note that there is a parallel between the interdefinability of the modal operators and that of the quantifiers. Why? The exercises below will address this and ask you to reason through the remaining equivalences.

## 9.2.1 **The Syntax of** $S^{\Box}$

The language $S$ with necessity, which we shall abbreviate, $S^{\Box}$, consists of the following symbols:

**Statement Letters:**
$$A, B, C, \ldots, Z, A_1, B_1, C_1, \ldots, Z_1, A_2, \ldots$$

---

2. Some logic texts opt for even greater redundancy, employing symbols for contingency, necessary implication, and so on.

**Truth-Functional Connectives:**

$\neg \wedge \vee \rightarrow \leftrightarrow$

**Modality Operators:**

$\square \diamond$

**Punctuation Marks:**

$(\,)$

In order to define 'wff of $S^\square$' we first define:

**Expression of $S^\square$:**

An *expression of $S^\square$* is any finite sequence of the symbols of $S^\square$.

Now we can define 'wff of $S^\square$':

**Well-Formed Formula of $S^\square$:**

Where $\mathbb{P}$ and $\mathbb{Q}$ range over expressions of $S^\square$,

      (1) If $\mathbb{P}$ is a statement letter, then $\mathbb{P}$ is a wff of $S^\square$

      (2) If $\mathbb{P}$ and $\mathbb{Q}$ are wffs of $S^\square$, then

          (a) $\neg\mathbb{P}$ is a wff of $S^\square$

          (b) $\square\mathbb{P}$ is a wff of $S^\square$

          (c) $\diamond\mathbb{P}$ is a wff of $S^\square$

          (d) $(\mathbb{P} \wedge \mathbb{Q})$ is a wff of $S^\square$

          (e) $(\mathbb{P} \vee \mathbb{Q})$ is a wff of $S^\square$

          (f) $(\mathbb{P} \rightarrow \mathbb{Q})$ is a wff of $S^\square$

          (g) $(\mathbb{P} \leftrightarrow \mathbb{Q})$ is a wff of $S^\square$

      (3) Nothing is a wff of $S^\square$ unless it can be shown so by a finite number of applications of clauses (1) and (2)

The scope of the modal operators is identical to that of the hook. Moreover, we adopt the same conventions as in Chapter 2 for dropping the outermost parentheses of wffs and for the use of square brackets.

## 9.2.2 Exercises

  **A.** Using the informal modal interpretation of '$\square$' and '$\diamond$', prove each of the following modal equivalences:

      **1.** $\diamond\mathbb{P} \Leftrightarrow \neg\square\neg\mathbb{P}$

      **2.** $\diamond\neg\mathbb{P} \Leftrightarrow \neg\square\mathbb{P}$

      **3.** $\neg\diamond\mathbb{P} \Leftrightarrow \square\neg\mathbb{P}$

      **4.** $\diamond\mathbb{P} \wedge \diamond\neg\mathbb{P} \Leftrightarrow \neg(\square\mathbb{P} \vee \square\neg\mathbb{P})$

      **5.** Explain the parallel between the interdefinability of $\square$ and $\diamond$ and the interdefinability of $(\forall x)$ and $(\exists x)$.

**B.** Using the key provided, symbolize each of the following as wffs of $S^\square$:

> A:  Ann travels faster than light
> M:  Humans colonize Mars
> E:  Energy is conserved
> F:  The universe is finite

1. It is necessary that energy is conserved
2. It is not necessary that the universe is finite
3. It is not impossible that the universe is finite
4. It is impossible for Ann to travel faster than light
5. It is possible that humans colonize Mars
6. It is possible that humans do not colonize Mars
7. It is contingent that the universe is finite
8. Necessarily, both energy is conserved and the universe is finite
9. It is necessary that energy is conserved and it is necessary that the universe is finite
10. It is necessary that either Ann travels faster than light or humans colonize Mars
11. Either it is necessary that Ann travels faster than light or it is necessary that humans colonize Mars
12. Possibly, both humans colonize Mars and energy is conserved
13. Possibly humans colonize Mars and possibly energy is conserved
14. Possibly Ann travels faster than light or possibly the universe is finite
15. Possibly, either Ann travels faster than light or the universe is finite
16. Necessarily, either humans colonize Mars or they do not
17. Either humans necessarily colonize Mars or necessarily humans do not colonize Mars
18. Necessarily, either it's possible for Ann to travel faster than light or it is not possible
19. It's not necessarily impossible that Ann travels faster than light
20. If necessarily the universe is finite, then necessarily energy is conserved
21. Necessarily, if the universe is finite, then energy is conserved
22. If it's impossible for Ann to travel faster than light, then it's not impossible for humans to colonize Mars
23. It is contingent that the universe is not finite if and only if it is not contingent that energy is conserved
24. Necessarily, if energy is conserved and the universe is finite, then it's impossible for Ann to travel faster than light
25. If it is necessary that both energy is conserved and the universe is finite, then it's impossible for Ann to travel faster than light

# 9.3 **Basic Possible Worlds Semantics for** $S^\square$

Since the evaluation of modal wffs requires not just assessment of truth value assignments, but truth value assignments at some or all possible worlds, we need a more complex semantics.

**Interpretation of $S^\square$:**

An *interpretation M of $S^\square$* consists of:

(1) a non-empty set, **W**, of worlds, w

(2) a function, **E**, which for each $w \in \mathbf{W}$, assigns T or F to each statement letter of $S^\square$

For instance:

**M1. W**: $\{w_1, w_2, w_3, \ldots\}$

| | | | |
|---|---|---|---|
| $\mathbf{E}(\text{'A'}, w_1) = \mathsf{T}$ | $\mathbf{E}(\text{'A'}, w_2) = \mathsf{F}$ | $\mathbf{E}(\text{'A'}, w_3) = \mathsf{F}$ | $\cdots$ |
| $\mathbf{E}(\text{'B'}, w_1) = \mathsf{F}$ | $\mathbf{E}(\text{'B'}, w_2) = \mathsf{T}$ | $\mathbf{E}(\text{'B'}, w_3) = \mathsf{T}$ | $\cdots$ |
| $\mathbf{E}(\text{'C'}, w_1) = \mathsf{T}$ | $\mathbf{E}(\text{'C'}, w_2) = \mathsf{T}$ | $\mathbf{E}(\text{'C'}, w_3) = \mathsf{T}$ | $\cdots$ |

In this example we see only 3 worlds and three statement letters, but you can see that at each world, each statement letter is assigned a truth value by the assignment function **E**. This, of course, gives us the truth value of each atomic wff at each world, but does not tell us how to compute truth values at worlds for more complex wffs of $S^\square$. For this we define truth at a world:

**Truth at a World in $S^\square$:**

Truth in $S^\square$ is determined by a valuation function **V**, such that, for a given a model **M**:

(1) If $\mathbb{P}$ is a statement letter, $\mathbf{V}(\mathbb{P}, w) = \mathsf{T}$ iff $\mathbf{E}(\mathbb{P}, w) = \mathsf{T}$

(2) $\mathbf{V}(\neg\mathbb{P}, w) = \mathsf{T}$ iff $\mathbf{E}(\mathbb{P}, w) = \mathsf{F}$

(3) $\mathbf{V}(\mathbb{P} \wedge \mathbb{Q}, w) = \mathsf{T}$ iff $\mathbf{E}(\mathbb{P}, w) = \mathsf{T}$ and $\mathbf{E}(\mathbb{Q}, w) = \mathsf{T}$

(4) $\mathbf{V}(\mathbb{P} \vee \mathbb{Q}, w) = \mathsf{T}$ iff $\mathbf{E}(\mathbb{P}, w) = \mathsf{T}$ or $\mathbf{E}(\mathbb{Q}, w) = \mathsf{T}$

(5) $\mathbf{V}(\mathbb{P} \rightarrow \mathbb{Q}, w) = \mathsf{T}$ iff $\mathbf{E}(\mathbb{P}, w) = \mathsf{F}$ or $\mathbf{E}(\mathbb{Q}, w) = \mathsf{T}$

(6) $\mathbf{V}(\mathbb{P} \leftrightarrow \mathbb{Q}, w) = \mathsf{T}$ iff $\mathbf{E}(\mathbb{P}, w) = \mathbf{E}(\mathbb{Q}, w)$

(7) $\mathbf{V}(\Diamond\mathbb{P}, w) = \mathsf{T}$ iff there is at least one $x \in \mathbf{W}$ such that $\mathbf{E}(\mathbb{P}, x) = \mathsf{T}$

(8) $\mathbf{V}(\square\mathbb{P}, w) = \mathsf{T}$ iff for every $x \in \mathbf{W}$, $\mathbf{E}(\mathbb{P}, x) = \mathsf{T}$

Note, first, that truth at a world is relative to an interpretation **M** which consists of set of worlds **W** and assignment **E**. Thus, clause (1) here simply reiterates that the truth values of statement letters at worlds are assigned directly by the function **E**. Clauses (2)–(6) reiterate, in compressed form and relative to worlds, the familiar truth functions.

Finally, (7) and (8) formalize the interpretation of the box and diamond. Let's look at some examples:

**M2.  W**: $\{w_1, w_2, w_3\}$

| | | |
|---|---|---|
| $\mathbf{E}(J, w_1) = \mathsf{T}$ | $\mathbf{E}(J, w_2) = \mathsf{F}$ | $\mathbf{E}(J, w_3) = \mathsf{F}$ |
| $\mathbf{E}(K, w_1) = \mathsf{T}$ | $\mathbf{E}(K, w_2) = \mathsf{T}$ | $\mathbf{E}(K, w_3) = \mathsf{T}$ |
| $\mathbf{E}(L, w_1) = \mathsf{F}$ | $\mathbf{E}(L, w_2) = \mathsf{F}$ | $\mathbf{E}(L, w_3) = \mathsf{T}$ |

Here **W** contains only 3 worlds, and we restrict our attention to just 3 statement letters. Moreover, I have dispensed with the quotation marks around the statement letters, as doing so will cause no confusion. Even so, we can be still more elegant in presenting interpretation **M2**. Observe:

**M2.**

| •$w_1$ | •$w_2$ | •$w_3$ |
|---|---|---|
| J | ¬J | ¬J |
| K | K | K |
| ¬L | ¬L | L |

Here we have each world in **W** represented by a labeled dot. Under each world we list either the statement letter or its negation, indicating whether **E** assigns T or F, respectively, to that statement letter. This graphical method makes it easy to see how **E** varies across worlds and is the basis of the tree approach to be introduced below.

Given interpretation **M2**, determining the truth values of other truth-functional combinations is not difficult. Consider the truth values of, for example,

(1) $J \land K$
(2) $J \lor L$

At $w_1$ (1) is T, but at $w_2$ and $w_3$ it is F, while (2) is T at $w_1$ and $w_3$, but F at $w_2$. At this stage, the truth values of modal wffs are not much more difficult to calculate.

(3) $\Box J$
(4) $\Box K$
(5) $\Diamond \neg J$
(6) $\neg \Diamond L$
(7) $\Box(\neg J \lor \neg L)$

Let us focus on truth values at $w_1$ for the time being. Because 'J' is F at $w_2$ and $w_3$, it is not T at every $w \in \mathbf{W}$, so '$\Box J$' is F at $w_1$. In contrast, 'K' is T at every $w \in \mathbf{W}$, so (4), '$\Box K$', is T at $w_1$. (5) is clearly T at $w_1$, because '$\neg J$' is T in at least one world—indeed in both $w_2$ and $w_3$. (6) is F at $w_1$ because '$\Diamond L$' is T at $w_1$, since 'L' is T in at least one world: $w_3$. Finally, (7) is true at $w_1$ since '$\neg J \lor \neg L$' is T at every world.

Thus the truth values of some example modal wffs *at $w_1$*. What about the truth values of (3)–(7) at $w_2$ and $w_3$? It turns out that they are the same as at $w_1$. With the relatively simple modal semantics we are working with right now, if $\mathbb{P}$ is T at any world, then $\Diamond \mathbb{P}$

is T at every world, and if $\mathbb{P}$ is T at every world, then $\square\mathbb{P}$ is also T at every world. This might make intuitive sense: even if $\mathbb{P}$ isn't T at the actual world, if it is T *somewhere*, then it is possible, not just here, but *everywhere*. And, of course if $\mathbb{P}$ is T at every world, then, in addition to being necessary here, it is necessary *everywhere*. To clarify a bit, we have the following equivalences:

(8)  $\square\mathbb{P} \Leftrightarrow \square\square\mathbb{P}$

(9)  $\Diamond\mathbb{P} \Leftrightarrow \square\Diamond\mathbb{P}$

(10)  $\square\mathbb{P} \Leftrightarrow \Diamond\square\mathbb{P}$

(11)  $\Diamond\mathbb{P} \Leftrightarrow \Diamond\Diamond\mathbb{P}$

These equivalences hold regardless of what world we consider in an interpretation. Here is a more careful argument justifying (8):

> Suppose $\square\mathbb{P}$ is T at some $w \in \mathbf{W}$, this means that $\mathbb{P}$ is T at every $x \in \mathbf{W}$. Thus, no matter what $x \in \mathbf{W}$ we consider, $\square\mathbb{P}$ is T, which means $\square\square\mathbb{P}$ is T at w. Now suppose $\square\square\mathbb{P}$ is T at some $w \in \mathbf{W}$, thus $\square\mathbb{P}$ is T at every $x \in \mathbf{W}$, in particular w. So the equivalence in (8) holds for any world w.

## 9.3.1  Semantic Properties of Wffs and Sets of Wffs

We can now define logical properties of modal wffs and sets of wffs.

**Modally True Wff:**
> A wff $\mathbb{P}$ of $S^\square$ is *modally true* iff on every interpretation $\mathbb{P}$ is true in every world.

**Modally False Wff:**
> A wff $\mathbb{P}$ of $S^\square$ is *modally false* iff on every interpretation $\mathbb{P}$ is false in every world.

**Modally Contingent Wff:**
> A wff $\mathbb{P}$ of $S^\square$ is *modally contingent* iff $\mathbb{P}$ is neither modally true nor modally false; i.e., iff on some interpretation it is false in some world and on some interpretation it is true in some world.

**Modally Satisfiable Wff:**
> A wff $\mathbb{P}$ of $S^\square$ is *modally satisfiable* iff $\mathbb{P}$ is not modally false; i.e., on some interpretation it is true in some world.

**Modally Falsifiable Wff:**
> A wff $\mathbb{P}$ of $S^\square$ is *modally falsifiable* iff $\mathbb{P}$ is not modally true; i.e., on some interpretation it is false in some world.

**Modally Equivalent Pair of Wffs:**
> Wffs $\mathbb{P}$ and $\mathbb{Q}$ of $S^\square$ are *modally equivalent* iff there is no interpretation containing a world in which they differ in truth value.

### Modally Contradictory Pair of Wffs:

Wffs $\mathbb{P}$ and $\mathbb{Q}$ of $S^{\square}$ are *modally contradictory* iff there is no interpretation containing a world in which they have the same truth value.

### Modally Consistent Set of Wffs:

A set $\Gamma$ of wffs of $S^{\square}$ is *modally consistent* iff on at least one interpretation there is a world in which all members of $\Gamma$ are true. $\Gamma$ is *modally inconsistent* iff it is not modally consistent.

### Modally Entails:

A set $\Gamma$ of wffs of $S^{\square}$ *modally entails* a wff $\mathbb{P}$, $\Gamma \vDash \mathbb{P}$, iff there is no interpretation containing a world in which all the members of $\Gamma$ are true and $\mathbb{P}$ is false.

### Modally Valid Argument:

An argument of $S^{\square}$ is *modally valid* iff there is no interpretation containing a world in which all the premises are true and the conclusion is false. An argument of $S^{\square}$ is *modally invalid* iff it is not modally valid.

These definitions should be largely familiar. The main modification is that they are now relative not simply to truth value assignments, but to truth value assignments varying across worlds and interpretations.

We will see a few examples involving some of these properties, leaving most for the exercises. Moreover in the next section we will extend the tree method developed in Section 3.5, which will allow us to systematically test for the properties.

For example, we can easily prove the following modal entailment claim:

(12)  $\square C \vDash \Diamond C$

> Suppose '$\square C$' is T at some $w \in \mathbf{W}$ on some interpretation $\mathbf{M}$. Then for every $x \in \mathbf{W}$, 'C' is T at x. So for at least one $x \in \mathbf{W}$, 'C' is T at x. So '$\Diamond C$' is T at w on interpretation $\mathbf{M}$. So there is no interpretation containing a world at which '$\square C$' is T and '$\Diamond C$' is F.

Show that the following entailment claim fails:

(13)  $\Diamond B \vDash \square B$

Here, we need only produce an interpretation on which '$\Diamond B$' is T and '$\square B$' is F. It is easy to see how this can happen. We need a model where 'B' is T at some world, but not all. Here is such an interpretation:

**M3.**                       •$w_1$          •$w_2$
                               B              $\neg$B

Note that since 'B' is T at $w_1$, then '$\Diamond B$' is T at $w_1$ as well. But, since 'B' is F at $w_2$, then '$\square B$' is F at $w_1$. So we have an interpretation, **M3**, containing a world in which the left hand side is T, but the target statement is F.

Prove that the following set is inconsistent:

(14) $\{\square(A \rightarrow B), \square\neg B, A\}$

Here, we cannot provide an interpretation, since the set is, in fact, inconsistent. But we can show by reductio that there is no interpretation containing a world at which all members of set (14) are T.

> Suppose the set is consistent. Then there must be an interpretation **M** with a world w at which every member of the set is T. That means that both 'A' and 'A → B' are T at w, which implies that 'B' is T at w. But '$\square\neg$B' is T at w, implying that '¬B' is T at w, contradicting the previous sentence. So there is no interpretation containing a world at which every member of (14) is T, so (14) is inconsistent.

As a final example, construct an interpretation to show the following wff is not modally true.

(15) $(\lozenge D \wedge \lozenge C) \rightarrow \lozenge(C \wedge D)$

We need an interpretation that makes the antecedent T and the consequent F. For the antecedent to be T, we need at least one world where 'D' is T and at least one (possibly different) world where 'C' is T. For the consequent to be F, there must be no world at which both 'C' and 'D' are true. This works:

**M4.**

|  $\bullet w_1$  |  $\bullet w_2$  |
|:---:|:---:|
| D | ¬D |
| ¬C | C |

## 9.3.2  Exercises

**A.** Using the definition of truth at a world in $S^\square$, prove each of the following modal equivalences:

 1. $\square P \Leftrightarrow \square\square P$
 2. $\lozenge P \Leftrightarrow \square\lozenge P$
 3. $\square P \Leftrightarrow \lozenge\square P$
 4. $\lozenge P \Leftrightarrow \lozenge\lozenge P$

**B.** For each of the following, construct an interpretation that shows the modal property holds:

 1. Show modally satisfiable: $\lozenge(A \wedge B) \wedge \lozenge\neg(A \wedge B)$
 2. Show modally falsifiable: $\square(G \vee H) \rightarrow (\square G \vee \square H)$
 3. Show modally consistent: $\{\square(A \rightarrow B), \square(B \rightarrow C), \lozenge A, \neg\square C\}$
 4. Show the entailment fails: $\{\square(H \rightarrow J), \lozenge\neg(J \vee H)\} \nvDash \square\neg J$

### 9.3.3 **Possible Worlds and Trees**

We can extend the tree method of Section 3.5 to accommodate modal statements. Recall that the tree method involves listing a set of wffs and decomposing them into statement letters and their negations. The aim is to generate a truth value assignment that shows the original set consistent—if the tree is open, then the set is consistent. If the tree is closed, then the set is not consistent. Adding modals into the mix increases the level of complexity slightly since we will need to consider truth value assignments at multiple worlds. This will give us trees at each world we consider, resulting in tree structures typically involving multiple worlds. Here is how it works.

First, we allow ourselves all the tree rules for the language $S$. To these we will add decomposition rules for '$\neg\Box$', '$\neg\Diamond$', '$\Box$', and '$\Diamond$'. The negated modal rules are very simple, so we will begin with those. Negated Possibility Decomposition, $\neg\Diamond$D, and Negated Necessity Decomposition, $\neg\Box$D.

One thing you will note here is that our tree is "planted" on a particular world, $w_i$. We will see more of what this entails with the rules below, for now just keep in mind that every diagram will have a tree at some world, and frequently trees at multiple worlds. Given this, the rules above are nice and simple. $\neg\Diamond$D tells us that if we have a negated possibility, we rewrite it as a necessary negation, and check off the original. Similarly, $\neg\Box$D, tells us to rewrite a negated necessity as a possible negation, checking off the original.

Note that both of the above rules leave us with a modal as the main operator of the result. So, clearly, we need rules for dealing with the modal operators. Here are Possibility Decomposition, $\Diamond$D, and Necessity Decomposition, $\Box$D:

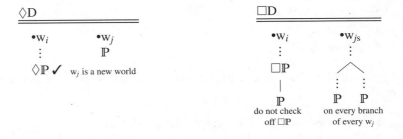

To decompose a wff with the diamond as the main operator, $\Diamond\mathbb{P}$, we introduce a new world, $w_j$, and write the subformula, $\mathbb{P}$, without the diamond, at the base of that world.

We check off the original wff under $w_i$. We know that if $\Diamond \mathbb{P}$ is T at $w_i$, then $\mathbb{P}$ must be T at some world x. Since there is no guarantee that x is one of the worlds already in the structure, we must introduce a new one. In the case of a wff with the box as the main operator, $\square \mathbb{P}$, we do not need to add any worlds, but we do need to write the subformula, $\mathbb{P}$, on every branch of every existing world, including $w_i$ itself. We cannot check off $\square \mathbb{P}$ because if we add worlds to the diagram (by applying $\Diamond$D) we must add $\mathbb{P}$ to those as well.

As a first example, take set (14) from above. There we argued that set was inconsistent, so we should get a closed tree. Recall that we begin by listing all members of the set at the base of the tree:

$$\bullet w_1$$
$$\square(A \rightarrow B)$$
$$\square \neg B$$
$$A$$

With two applications of $\square$D, we arrive at this:

$$\bullet w_1$$
$$\square(A \rightarrow B)$$
$$\square \neg B$$
$$A$$
$$|$$
$$\neg B$$
$$|$$
$$A \rightarrow B$$

Finally applying $\rightarrow$D, we have:

$$\bullet w_1$$
$$\square(A \rightarrow B)$$
$$\square \neg B$$
$$A$$
$$|$$
$$\neg B$$
$$|$$
$$A \rightarrow B \checkmark$$

$$\neg A \qquad B$$
$$\times \qquad \times$$

Since there is no need for $\Diamond$D here, we have just one world in our diagram. We can easily see that each branch has a contradiction, the tree structure is closed, so there is no way to make every member of set (14) true. (14) is inconsistent.

Here is another example:

(16) $\{\Diamond K, \Diamond \neg K, \neg \Diamond (J \vee H)\}$

We begin as usual:

$$\begin{array}{c} \bullet w_1 \\ \Diamond K \\ \Diamond \neg K \\ \neg \Diamond (J \vee H) \end{array}$$

Since decomposing $\Diamond$ statements creates new worlds, it is usually a good policy to take care of any diamonds first. We apply $\Diamond D$ once to each of the first two wffs to achieve:

$$\begin{array}{ccc} \bullet w_1 & \bullet w_2 & \bullet w_3 \\ \Diamond K \checkmark & K & \neg K \\ \Diamond \neg K \checkmark & & \\ \neg \Diamond (J \vee H) & & \end{array}$$

We still have to deal with the third wff at $w_1$, so we apply $\neg \Diamond D$ to get:

$$\begin{array}{ccc} \bullet w_1 & \bullet w_2 & \bullet w_3 \\ \Diamond K \checkmark & K & \neg K \\ \Diamond \neg K \checkmark & & \\ \neg \Diamond (J \vee H) \checkmark & & \\ | & & \\ \Box \neg (J \vee H) & & \end{array}$$

Now we have apply $\Box D$, remembering to apply it to all existing worlds:

$$\begin{array}{ccc} \bullet w_1 & \bullet w_2 & \bullet w_3 \\ \Diamond K \checkmark & K & \neg K \\ \Diamond \neg K \checkmark & \neg (J \vee H) & \neg (J \vee H) \\ \neg \Diamond (J \vee H) \checkmark & & \\ | & & \\ \Box \neg (J \vee H) & & \\ | & & \\ \neg (J \vee H) & & \end{array}$$

Finally, we apply $\neg \vee D$ at each world:

$$\bullet w_1 \qquad\qquad \bullet w_2 \qquad\qquad \bullet w_3$$
$$\Diamond K \checkmark \qquad\qquad K \qquad\qquad \neg K$$
$$\Diamond \neg K \checkmark \qquad \neg(J \vee H) \checkmark \qquad \neg(J \vee H) \checkmark$$
$$\neg \Diamond (J \vee H) \checkmark \qquad\quad | \qquad\qquad\quad |$$
$$| \qquad\qquad \neg J \qquad\qquad \neg J$$
$$\square \neg (J \vee H) \qquad \neg H \qquad\qquad \neg H$$
$$|$$
$$\neg(J \vee H) \checkmark$$
$$|$$
$$\neg J$$
$$\neg H$$

Clearly, we have an open branch and so an open tree at each world. Thus the whole tree structure is open, and set (16) is consistent. We can read an interpretation right off the tree structure by looking at the statement letters and negations on the open branches:

**M5.**

$$\bullet w_1 \qquad \bullet w_2 \qquad \bullet w_3$$
$$\neg J \qquad\quad K \qquad\quad \neg K$$
$$\neg H \qquad\quad \neg J \qquad\quad \neg J$$
$$\qquad\quad \neg H \qquad\quad \neg H$$

Note that the tree did not commit us to a truth value for 'K' at $w_1$. This means that it could either be T or F at $w_1$. We can simply choose. Or, since whatever we choose will mean that either $w_1 = w_2$ or $w_1 = w_3$, we can choose the latter, and compress the model to:

**M5′.**

$$\bullet w_1 \qquad \bullet w_2$$
$$\neg K \qquad\quad K$$
$$\neg J \qquad\quad \neg J$$
$$\neg H \qquad\quad \neg H$$

Consider the following entailment claim:

(17)  $\{\square(E \to G), \square \neg G\} \vDash \neg \Diamond E$

Recall that to test entailment we create a tree with the members of the set and the negation of the target. If that set is consistent (tree structure is open), then it is possible for the members of the set to be T with the target F, showing failure of entailment. If that set is inconsistent (tree structure is closed), then it is not possible for the members of the set to be T with the target F, showing the entailment holds. We begin:

$$\bullet w_1$$
$$\square(E \to G)$$
$$\square \neg G$$
$$\neg \neg \Diamond E$$

We can do a quick ¬¬D to get '◇E' at the bottom of $w_1$, and then apply ◇D to generate $w_2$:

```
        •w₁              •w₂
     □(E → G)             E
       □¬G
      ¬¬◇E ✓
        |
       ◇E ✓
```

Two applications of □D give us the following:

```
        •w₁              •w₂
     □(E → G)             E
       □¬G               ¬G
      ¬¬◇E ✓            E → G
        |
       ◇E ✓
        |
       ¬G
        |
      E → G
```

We now apply →D and find that the tree on $w_2$ is closed:

```
        •w₁                 •w₂
     □(E → G)                E
       □¬G                  ¬G
      ¬¬◇E ✓              E → G ✓
        |                   ╱╲
       ◇E ✓              ¬E    G
        |                 ✗    ✗
       ¬G
        |
     E → G ✓
        ╱╲
     ¬E    G
        ✗
```

The tree on $w_1$ is still open, but '◇E' is T at $w_1$, requiring $w_2$, where 'E' is T. But this forces a contradiction on all branches of the $w_2$ tree. Since there is at least one world with a closed tree, the whole tree structure is closed, and the set is inconsistent, showing that the entailment holds.

One complication we have to deal with is when a modal wff appears on some but not all open branches of a tree. These branches always occur as the result of decomposing some sort of disjunction.[3] We need a way to develop the modal side of the disjunct across worlds in the structure, while distinguishing it as one way among others for a tree structure to develop. We do this by decomposing that branch in a box extending across the structure:

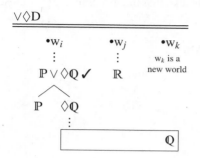

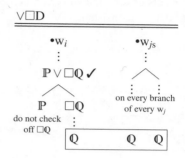

With $\vee\Diamond$D, we are indicating that in developing the branch from $\Diamond$Q, we need a new world, $w_k$ at which Q is T, but that world depends on the branch at $w_i$. With $\vee\square$D, we do not introduce new worlds, but we append Q to all branches of all existing worlds within the box to show that the results depend on the branch at $w_i$. We do not check off $\square$Q because if new worlds emerge, we will need to extend the box and append Q.

When a wff in a box contradicts a wff on a branch directly above it, we place an ✗ in the box under that wff. We can close a modal disjunct branch when either i) there is a contradiction on the originating branch, or ii) for some world in the structure, there is an ✗ in the box under every open branch of that world's tree. Some examples:

(18) $\{\neg G, \Diamond J, A \rightarrow \square G\}$

(19) $\{A, \Diamond\neg G, A \rightarrow \square G\}$

The tree structure for set (18) demonstrates an application of $\vee\square$D where we immediately obtain a contradiction. This is an instance for case i). Thus, an ✗ goes in the box under 'G' and as a result we place an ✗ directly under '$\square$G' in the disjunction of $w_1$. This may seem a bit redundant, but it makes clear that that branch of that tree at the originating world is closed. Note that the branch containing '$\neg$A' is still open, so set (18) is consistent. In the

---

3. This includes applications of not just $\vee$D, but also $\rightarrow$D, $\neg\wedge$D, $\leftrightarrow$D, and $\neg\leftrightarrow$D. These decomposition rules always cause trees to branch.

structure for set (19) we do not get a contradiction directly on the '□G' branch itself. But we do get a completely closed tree in the box under $w_2$, so we place an ✗ both under G in $w_2$ and directly under '□G' in the disjunction of $w_1$. The fact that a closed tree occurs in $w_2$ warrants closing the whole branch at $w_1$. Note also that the left branch of $w_1$ closes, so set (19) is inconsistent.

(20)  {A, ◊(¬G ∨ J), A → □G}                **M6.**

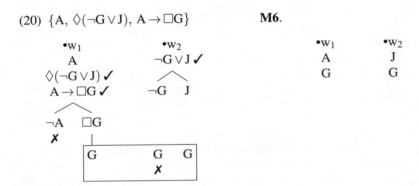

Set (20) gives us an example of both an open and a closed branch in the modal disjunct box. Even though the left branch of $w_2$ is closed in the box, this does not close the '□G' branch of $w_1$, because the right branch of $w_2$ is open. So the right branch of $w_1$ is open. Thus, the structure is open and set (20) is consistent. In fact the model is just as displayed in **M6**. Note that since the truth values of 'J' at $w_1$ and of 'A' at $w_2$ are not determined by the tree, we can choose.

Here are two examples involving ∨◊D:

(21)  {¬A, □¬B, ◊B ∨ A}                (22)  {¬A, ¬B, ◊B ∨ A}

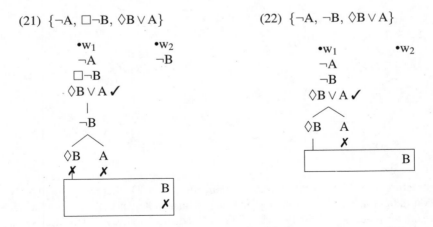

Note that in building the structure for set (21) we start with just $w_1$. We decompose '□¬B' by adding '¬B' to the tree at $w_1$. When we decompose the disjunction '◊B ∨ A', the right branch closes and we wind up with a modal disjunct on the left branch. This requires us to introduce $w_2$ and place 'B' in the box below $w_2$. Since we still have '□¬B' at $w_1$, we need to decompose it by adding '¬B' to the tree at $w_2$. And this results in a contradiction, so the left branch of $w_1$ is closed. Thus, the tree at $w_1$ is closed. As a result, the whole structure is closed. Note that with (22), which differs slightly from (21), we are not forced into putting '¬B' at $w_2$, so the left branch at $w_1$ does not close, and we have an open tree. Set (22) is consistent.

Consider the following entailment claim:

(23) $\square\lozenge A \vDash A$

If we set up a tree to test this and carry out $\square D$ and $\lozenge D$ at $w_1$, it will look like this:

$$
\begin{array}{cc}
{}^\bullet w_1 & {}^\bullet w_2 \\
\square\lozenge A & A \\
\neg A & \\
\lozenge A\ \checkmark &
\end{array}
$$

But, of course, this tree structure is not complete. Now that $w_2$ is in the picture, we have to apply $\square D$ to '$\square\lozenge A$', producing '$\lozenge A$' at $w_2$. But then we have to decompose that, producing $w_3$, and requiring us to place not only 'A' there, but also '$\lozenge A$', starting the cycle again...

$$
\begin{array}{cccccc}
{}^\bullet w_1 & {}^\bullet w_2 & {}^\bullet w_3 & \cdots & {}^\bullet w_n & \cdots \\
\square\lozenge A & A & A & \cdots & A & \cdots \\
\neg A & \lozenge A\ \checkmark & \lozenge A\ \checkmark & \cdots & \lozenge A\ \checkmark & \cdots \\
\lozenge A\ \checkmark & & & & &
\end{array}
$$

So this tree structure never stops growing! Is the set at $w_1$ consistent (showing the entailment fails)? Well, yes, but the tree really doesn't show that, because the tree never completes. This is because of the pattern $\square\lozenge$ sets up a cascade of worlds if we follow our rules mechanically. Looking at this tree, even if incomplete, we can see how to give an interpretation to show that entailment (23) fails:

**M7.**

$$
\begin{array}{cc}
{}^\bullet w_1 & {}^\bullet w_2 \\
\neg A & A
\end{array}
$$

**M7** makes clear that all we need are the first two worlds to show that $\square\lozenge A \nvDash A$, contrary to (23). Since 'A' is T at $w_2$, '$\lozenge A$' is T at both worlds. Hence, '$\square\lozenge A$' is T at both worlds, but in particular at $w_1$, where '¬A' is T, that is, where the target 'A' is F. So we have an interpretation with a world where the members of the left side are all true, but the target is false. Thus, the entailment fails.

Will the $\square\lozenge$ pattern always create an ever-growing tree? Consider this slightly different case:

(24) $\square\lozenge A \vDash \lozenge A$

Here there are two different tree structures. The one on the left, which just has one world, closes if we do not require full decomposition into statement letters and their negations. If we do want to be strict about that, a few more steps causes $w_2$ to close, closing the whole structure. In either case we can tell that the entailment claim (24) holds, as expected. Note that unlike the tree for (23), the presence of '$\neg\Diamond A$' prevents this tree from generating a cascade of worlds. Many other complications are to be found in modal trees. These are left for the exercises.

Here are all the rules in one place (also see Appendix C):

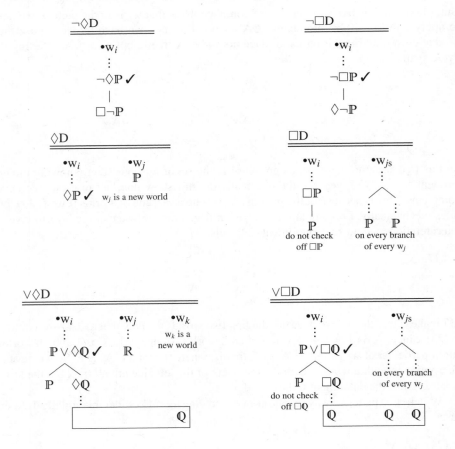

Here are some definitions of terms we have been using:

**Modal Tree Structure:**
> A *modal tree structure* for a set of wffs is a structure that lists all the wffs, and all the well-formed components that must be true in attempting to show the set modally consistent, typically including trees at various worlds.

**Closed Branch:**
> A branch on a modal tree is a *closed branch* iff the branch contains both some atomic wff $\mathbb{P}$ and its negation $\neg\mathbb{P}$.

**Open Branch:**
> A branch on a tree is an *open branch* iff it is not closed.

**Closed Tree:**
> A modal tree is a *closed tree* iff every branch is closed.

**Open Tree:**
> A modal tree is an *open tree* iff at least one branch is open.

**Closed Tree Structure:**
> A modal tree structure is a *closed structure* iff there is some world at which the tree is closed.

**Open Tree Structure:**
> A modal tree structure is an *open structure* iff there is an open tree at every world.

## Tests with Truth Trees

Below is a list of the various tests we can perform using modal tree structures. Recall that we are always testing some set of wffs for consistency. Thus to test for other properties, we must understand them in terms of the consistency or inconsistency of the appropriate set.

**Modal Tree Test for Modally Consistent Set of Wffs:**
> A set $\Gamma$ is modally consistent iff it has an open tree structure.

**Modal Tree Test for Modally Inconsistent Set of Wffs:**
> A set $\Gamma$ is modally inconsistent iff it has a closed tree structure.

**Modal Tree Test for Modally True Wff:**
> A wff $\mathbb{P}$ is modally true iff $\{\neg\mathbb{P}\}$ has a closed tree structure.

**Modal Tree Test for Modally False Wff:**
> A wff $\mathbb{P}$ is modally false iff $\{\mathbb{P}\}$ has a closed tree structure.

**Modal Tree Test for Modally Contingent Wff:**
> A wff $\mathbb{P}$ is modally contingent iff $\{\mathbb{P}\}$ has an open tree structure and $\{\neg\mathbb{P}\}$ has an open tree structure.

**Modal Tree Test for Modally Satisfiable Wff:**
    A wff $\mathbb{P}$ is modally satisfiable iff $\{\mathbb{P}\}$ has an open tree structure.

**Modal Tree Test for Modally Falsifiable Wff:**
    A wff $\mathbb{P}$ is modally falsifiable iff $\{\neg\mathbb{P}\}$ has an open tree structure.

**Modal Tree Test for Modally Equivalent Pair of Wffs:**
    Wffs $\mathbb{P}$ and $\mathbb{Q}$ are modally equivalent iff $\{\neg(\mathbb{P}\leftrightarrow\mathbb{Q})\}$ has a closed tree structure.

**Modal Tree Test for Modally Contradictory Pair of Wffs:**
    Wffs $\mathbb{P}$ and $\mathbb{Q}$ are modally contradictory iff $\{\mathbb{P}\leftrightarrow\mathbb{Q}\}$ has a closed tree structure.

**Modal Tree Test for Modally Entails:**
    A set of wffs, $\Gamma$, modally entails a target wff, $\mathbb{P}$, iff $\Gamma\cup\{\neg\mathbb{P}\}$ has a closed tree structure.

**Modal Tree Test for Modally Valid Argument:**
    An argument is modally valid iff the set consisting of all and only the premises and the negation of the conclusion has a closed tree structure.

## 9.3.4  Exercises

**A.** For each of the following construct modal tree structure(s) to determine which of the five properties of individual wffs apply to the wff.

    **1.** $A \to \Box A$
    **2.** $A \to \Diamond A$
    **3.** $\Diamond(A \land B) \land (\neg\Diamond A \lor \neg\Diamond B)$

**B.** For each of the following construct a modal tree structure to determine whether the members of the pair are modally equivalent, modally contradictory, or neither.

    **1.** $A$                 $\Box A$
    **2.** $\Box(C \land D)$       $\Box C \land \Box D$
    **3.** $\Box(G \land H)$       $\Diamond(\neg G \lor \neg H)$

**C.** For each of the following construct a modal tree structure to determine whether the set is consistent.

    **1.** $\{\Box(A \to B), \Box(A \land C), \Diamond\neg B\}$
    **2.** $\{\Diamond(D \land E), \Diamond\neg E, \neg\Box D, \Box(D \lor E)\}$

**D.** For each of the following construct a modal tree structure to determine whether the entailment holds.

    **1.** $A \vDash \Box\Diamond A$
    **2.** $B \to \Diamond B \vDash \Diamond B \to \Box B$
    **3.** $\Diamond D \land \Diamond E \vDash \Diamond(D \land E)$

    **4.** $\Diamond(D \wedge E) \vDash \Diamond D \wedge \Diamond E$
    **5.** $\square A \vDash \square(B \to A)$
    **6.** $A \to \Diamond A \vDash \Diamond A \to \square A$
    **7.** $\Diamond\Diamond H \vDash \Diamond H$
    **8.** $\{\square(A \to B), \square B \to \square C, \neg\square C\} \vDash \Diamond \neg A$

## 9.4 **Natural Deduction in** $S^\square$

As with $S$ and $P$ we can perform natural deductions or derivations for the language $S^\square$. We simply need to add rules dealing with the modal operators. But which rules? In fact, there are a variety of systems of differing strengths, which have been proposed for different purposes. This section will review some of the better known systems and explore how they interrelate. The basis of all the systems we will look at is system $K$.

### 9.4.1 **System** $K$

We will take system $K$ to include all the rules of $SDE$. Plus the following:

Dual
___

$\Diamond \mathbb{P} \quad \lhd\rhd \quad \neg\square\neg\mathbb{P}$
$\square \mathbb{P} \quad \lhd\rhd \quad \neg\Diamond\neg\mathbb{P}$

Dual simply expresses the dual interdefinability of the modal operators, and it functions as do all the equivalence rules found in $SDE$. Next we have:

N
___

$i \mid \mathbb{P}$             $\vdash$

$\rhd \mid \square\mathbb{P}$        $i$ N

Rule N (for Necessitation) tells us that if a wff $\mathbb{P}$ is a theorem then we may derive $\square\mathbb{P}$ from it, citing the N rule. So every theorem of $SD$ is a necessity in $K$. Here is a simple example:

| | | |
|---|---|---|
| 1 | $\boxed{G}$ | A |
| 2 | $G$ | 1 R |
| 3 | $G \to G$ | $1\text{–}2 \to$I |
| 4 | $\square(G \to G)$ | 3 N |

A quick $\to$I shows that '$G \to G$' is a theorem. As a result we may apply N to arrive at line 4. Note that for efficiency, we will frequently allow ourselves to skip the process of deriving the theorem. If the wff is easily recognizable as a theorem, or has previously been derived elsewhere, we can simply introduce it at any line of the derivation with the single turnstile in the justification column:

$$
\begin{array}{l|ll}
1 & G \to G & \vdash \\
2 & \Box(G \to G) & 1\ N
\end{array}
$$

This derivation becomes rather trivial, but this convention can save us significant time and space in longer derivations. We can, of course, always check to see if what we are citing as a theorem actually qualifies, either in the current derivation or in an independent derivation.

Finally, the K Rule:

K
_____

$$
\begin{array}{l|ll}
i & \Box(\mathbb{P} \to \mathbb{Q}) & \\
\rhd & \Box\mathbb{P} \to \Box\mathbb{Q} & i\ K
\end{array}
$$

K tells us that if a material conditional is necessary, then the consequent is necessary if the antecedent is. Necessity distributes across the material conditional. So we could extend the derivation above:

$$
\begin{array}{l|ll}
1 & G \to G & \vdash \\
2 & \Box(G \to G) & 1\ N \\
3 & \Box G \to \Box G & 2\ K
\end{array}
$$

Again, not terribly interesting, but it demonstrates how the rule works.

System **K** is a very weak system, but we can add some derived rules. Like the rules of **SDE** relative to **SD**, these derived rules do not add any actual power, but they do add convenience. Here are the derived rules of **K**:

K◇
_____

$$
\begin{array}{l|ll}
i & \Box(\mathbb{P} \to \mathbb{Q}) & \\
\rhd & \Diamond\mathbb{P} \to \Diamond\mathbb{Q} & i\ K\Diamond
\end{array}
$$

Dist□◇
_____

$$
\Box(\mathbb{P} \wedge \mathbb{Q}) \quad \lhd\rhd \quad \Box\mathbb{P} \wedge \Box\mathbb{Q}
$$
$$
\Diamond(\mathbb{P} \vee \mathbb{Q}) \quad \lhd\rhd \quad \Diamond\mathbb{P} \vee \Diamond\mathbb{Q}
$$

ME
_____

$$
\neg\Diamond\mathbb{P} \quad \lhd\rhd \quad \Box\neg\mathbb{P}
$$
$$
\neg\Box\mathbb{P} \quad \lhd\rhd \quad \Diamond\neg\mathbb{P}
$$

We can easily show how the Modal Exchange rules, ME, can be derived from Dual. I will produce just the top one. The other appears in the exercises.

| 1 | $\neg\lozenge\mathbb{P}$ | P |
|---|---|---|
| 2 | $\neg\square\neg\mathbb{P}$ | A |
| 3 | $\lozenge\mathbb{P}$ | 2 Dual |
| 4 | $\neg\lozenge\mathbb{P}$ | 1 R |
| 5 | $\square\neg\mathbb{P}$ | 2–4 ¬I |

| 1 | $\square\neg\mathbb{P}$ | P |
|---|---|---|
| 2 | $\neg\neg\square\neg\mathbb{P}$ | 1 DN |
| 3 | $\neg\lozenge\mathbb{P}$ | 2 Dual |

Next we prove the $K\lozenge$ rule:

| 1 | $\square(\mathbb{P}\to\mathbb{Q})$ | P |
|---|---|---|
| 2 | $\square(\neg\mathbb{Q}\to\neg\mathbb{P})$ | 1 Trans |
| 3 | $\square\neg\mathbb{Q}\to\square\neg\mathbb{P}$ | 2 K |
| 4 | $\neg\lozenge\mathbb{Q}\to\square\neg\mathbb{P}$ | 3 ME |
| 5 | $\neg\lozenge\mathbb{Q}\to\neg\lozenge\mathbb{P}$ | 4 ME |
| 6 | $\lozenge\mathbb{P}\to\lozenge\mathbb{Q}$ | 5 Trans |

As a final example, we prove one direction of the Dist$\square\lozenge$ rule. The others are left as exercises.

| 1 | $\square(\mathbb{P}\wedge\mathbb{Q})$ | P |
|---|---|---|
| 2 | $(\mathbb{P}\wedge\mathbb{Q})\to\mathbb{P}$ | $\vdash$ |
| 3 | $(\mathbb{P}\wedge\mathbb{Q})\to\mathbb{Q}$ | $\vdash$ |
| 4 | $\square((\mathbb{P}\wedge\mathbb{Q})\to\mathbb{P})$ | 2 N |
| 5 | $\square((\mathbb{P}\wedge\mathbb{Q})\to\mathbb{Q})$ | 3 N |
| 6 | $\square(\mathbb{P}\wedge\mathbb{Q})\to\square\mathbb{P}$ | 4 K |
| 7 | $\square(\mathbb{P}\wedge\mathbb{Q})\to\square\mathbb{Q}$ | 5 K |
| 8 | $\square\mathbb{P}$ | 1, 6 →E |
| 9 | $\square\mathbb{Q}$ | 1, 7 →E |
| 10 | $\square\mathbb{P}\wedge\square\mathbb{Q}$ | 8, 9 ∧I |

It is worth reflecting, briefly, on the shortcomings of **K**. First, note the following failures of derivability in **K**:

(1) $\square\mathbb{P}\nvdash\mathbb{P}$

(2) $\mathbb{P}\nvdash\lozenge\mathbb{P}$

(3) $\square\mathbb{P}\nvdash\lozenge\mathbb{P}$

(4) $\neg\mathbb{P}\nvdash\neg\square\mathbb{P}$

As (1) points out, in **K** we cannot derive $\mathbb{P}$ from $\square\mathbb{P}$. Intuitively, this means that necessity does not imply truth. Nor, as per (2), can we derive the possibility of $\mathbb{P}$ from $\mathbb{P}$ itself. Truth does not imply possibility. (3) points out that, even if we assume or can derive the necessity of $\mathbb{P}$, we cannot then derive the possibility of $\mathbb{P}$. Finally, (4) indicates that the negation of $\mathbb{P}$ does not imply the non-necessity of $\mathbb{P}$. All of these would be intuitive results for alethic modality, so system **K** is too weak to capture our intuitive understanding of the alethic, or other, modalities. It is, however, the ground for a series of further systems we will look at now.

## 9.4.2 **System** $D$

$K + \mathrm{D}$

Primitive Rule:

$$
\begin{array}{c|l}
\multicolumn{2}{l}{\mathrm{D}} \\
\hline\hline
i & \Box \mathbb{P} \\
\\
\rhd & \Diamond \mathbb{P} \qquad\qquad i\,\mathrm{D}
\end{array}
$$

System $D$ addresses (3) above by adding it as a derivation rule. Note, however, that the failure of derivability in (1) still holds, so $D$ is still too weak for alethic modality. This can be seen as a feature, however. $D$ can function as a basic system for *deontic* modality if we interpret '$\Box$' as 'it is obligatory that' and '$\Diamond$' as 'it is permissible that'. We then get from the D rule that $\vdash \Box\mathbb{P} \rightarrow \Diamond\mathbb{P}$, or: if $\mathbb{P}$ is obligatory, then $\mathbb{P}$ is permissible. Moreover, the failure in (1) is as it should be for deontic modality: that $\mathbb{P}$ is obligatory should not imply that $\mathbb{P}$ is the case. We often fail to do what is obligatory!

## 9.4.3 **System** $T$

$K + \mathrm{T}$

Primitive Rule:

$$
\begin{array}{c|l}
\multicolumn{2}{l}{\mathrm{T}} \\
\hline\hline
i & \Box \mathbb{P} \\
\\
\rhd & \mathbb{P} \qquad\qquad i\,\mathrm{T}
\end{array}
$$

Derived Rules: D, and …

$$
\begin{array}{c|l}
\multicolumn{2}{l}{\mathrm{T}\Diamond} \\
\hline\hline
i & \mathbb{P} \\
\\
\rhd & \Diamond \mathbb{P} \qquad\qquad i\,\mathrm{T}\Diamond
\end{array}
$$

System $T$ adds a rule to take care of (1), thereby adding one of the most intuitive alethic principles: if $\mathbb{P}$ is necessary, then $\mathbb{P}$ is the case. (2) is taken care of by T$\Diamond$, which is a derived rule. Note that $T$ does not contain the D rule as a primitive rule, but D can be derived with the help of T$\Diamond$, thereby taking care of (3). Finally (4) can be shown to be derivable in $T$. I leave these for the exercises.

Since the failures (1)–(4) above are now all derivable in $T$, $T$ is a minimal system for alethic (**Truth**-related) modalities. The remaining systems deal with the interactions among the modalities.

## 9.4.4  System *B*

*T* + B

Primitive Rule:                                     Derived Rule:

B                                                    B◇□
───────────────                                     ───────────────
*i* | ℙ                                             *i* | ◇□ℙ

▷ | □◇ℙ          *i* B                              ▷ | ℙ          *i* B◇□

System ***B*** represents one branch of development of modal logics. The primitive rule B tells us that if ℙ is the case, then it is necessarily possible that ℙ is the case. This is intuitive. Given the semantics discussed above, if ℙ is T at this world, then ◇ℙ is T at every world, thus □◇ℙ is T (at this world). But it is interesting to reflect that B implies B◇□, which can be read: if it is possibly necessary that ℙ, then ℙ is the case. This is much less intuitive for some. Suppose ◇□ℙ is T. This means that there is at least one world, w, at which □ℙ is T, so ℙ is T at every world, including this one. But it seems that we could claim that ℙ is possibly necessary, without it being actually true at this world. One example, in the realm of physical necessity and possibility, might be a physical law, 𝕃, which, while F at this and any world with the actual physical laws, would be physically necessary in any world in which it (and an alternate set of physical laws) held. Thus we would have ◇□𝕃 ⊭ 𝕃, and we would not want B◇□ as a derivation rule. This suggests that for physical necessity we might need to distinguish relevant sets of worlds.[4]

## 9.4.5  System *S4*

*T* + S4

Primitive Rule:                                     Derived Rule:

S4                                                   S4◇
───────────────                                     ───────────────
*i* | □ℙ                                            *i* | ◇◇ℙ

▷ | □□ℙ          *i* S4                             ▷ | ◇ℙ          *i* S4◇

***S4*** represents the other branch of modal logics and is independent of ***B*** (see the diagram below). ***S4*** tells us that every necessity is necessarily necessary. Given the S4 rule and the T rule, strings of boxes of any length are equivalent to just one box. The same

─────────────────────────

4. In fact, we can do this by modifying our semantics via an accessibility relation—a topic not covered in the current volume.

applies for the diamond, given S4◇ and T◇. That is:

$$\Box\Box\ldots\Box\mathbb{P} \Leftrightarrow \Box\mathbb{P}$$
$$\Diamond\Diamond\ldots\Diamond\mathbb{P} \Leftrightarrow \Diamond\mathbb{P}$$

Note that in *S4*, while every necessity is necessary, possibilities are only possibly possible, not necessarily possible.

### 9.4.6 **System** *S5*

*T* + S5 rule

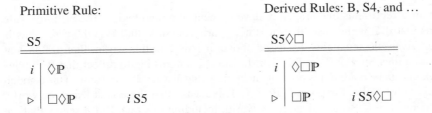

Primitive Rule:                    Derived Rules: B, S4, and …

With the S5 rule, system *S5* adds the principle that possibilities are necessarily possible, something not contained in *S4*. Even though we do not add the S4 or the B rule as primitive rules here, they can be derived here, as can S5◇□. These are left for the exercises.

*S5* represents the strongest version of alethic modality. Indeed, it matches up with both the semantics and the tree methods explored earlier in the chapter. Further, given T, T◇, S4, S4◇, S5, and S5◇□, all of which are part of *S5*, we are able to reduce any string of modal operators, even mixed combinations of the box and diamond, to the rightmost operator. That is, where ⬦ represents either a box or a diamond:

$$\boxtimes\boxtimes\ldots\Box\mathbb{P} \Leftrightarrow \Box\mathbb{P}$$
$$\boxtimes\boxtimes\ldots\Diamond\mathbb{P} \Leftrightarrow \Diamond\mathbb{P}$$

### 9.4.7 **Relations Between Modal Systems**

S → S′ indicates that S′ is a proper extension of S. That is, everything derivable in S is derivable in S′, but some things derivable in S′ are not derivable in S. Thus, S′ is stronger than S.

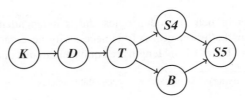

## 9.4.8 Exercises

**A.** Complete the following derivations using system **K**.

    **1.** $\neg\Diamond P \vdash \square\neg P$    Do not use ME; we are proving it here.

    **2.** $\square\neg P \vdash \neg\Diamond P$    Do not use ME; we are proving it here.

    **3.** $\neg\square P \vdash \Diamond\neg P$    Do not use ME; we are proving it here.

    **4.** $\Diamond\neg P \vdash \neg\square P$    Do not use ME; we are proving it here.

    **5.** $\square(P \to Q) \vdash \Diamond P \to \Diamond Q$    You may use ME, but do not use K$\Diamond$; we are proving it here.

    **6.** $\square(P \wedge Q) \vdash \square P \wedge \square Q$    Do not use Dist$\square\Diamond$; we are proving it here.

    **7.** $\square P \wedge \square Q \vdash \square(P \wedge Q)$    Do not use Dist$\square\Diamond$; we are proving it here.

    **8.** $\Diamond(P \vee Q) \vdash \Diamond P \vee \Diamond Q$    Do not use the $\Diamond$ Dist$\square\Diamond$; we are proving it here. You *may* use the $\square$ version of Dist$\square\Diamond$.

    **9.** $\Diamond P \vee \Diamond Q \vdash \Diamond(P \vee Q)$    Do not use Dist$\square\Diamond$; we are proving it here.

**B.** Complete the following derivations using system **T**.

    **1.** $P \vdash \Diamond P$    Do not use the T$\Diamond$ rule; we are proving it here.

    **2.** $\square P \vdash \Diamond P$    You may use T$\Diamond$; but do not use the D rule, we are proving it here.

    **3.** $\neg P \vdash \neg\square P$    Here and below you may use all of system **T**.

    **4.** $P \vdash \Diamond\Diamond\Diamond P$

**C.** Complete the following derivations using system **B**.

    **1.** $\Diamond\square P \vdash P$    Do not use the B$\Diamond\square$ rule; we are proving it here.

**D.** Complete the following derivations using system **S4**.

    **1.** $\Diamond\Diamond P \vdash \Diamond P$    Do not use the S4$\Diamond$ rule; we are proving it here.

    **2.** $\square\square\square\square P \vdash \square P$    Here and below you may use all rules of **S4**.

    **3.** $\square P \vdash \square\square\square\square P$

    **4.** $\Diamond\Diamond\Diamond\Diamond P \vdash \Diamond P$

    **5.** $\Diamond P \vdash \Diamond\Diamond\Diamond\Diamond P$

**E.** Complete the following derivations using system **S5**.

    **1.** $P \vdash \square\Diamond P$    Do not use the B rule; we are proving it here.

    **2.** $\Diamond\square P \vdash \square P$    Do not use the S5$\Diamond\square$ rule; we are proving it here.

    **3.** $\square P \vdash \square\square P$    Do not use the S4 rule; we are proving it here. You may use B and S5$\Diamond\square$.

    **4.** $\square\Diamond\square P \vdash \square P$    Here and below you may use all rules of **S5**.

    **5.** $\Diamond\square\Diamond P \vdash \Diamond P$

## 9.5 **Chapter Glossary**

**Alethic Modality:**
>The *alethic modalities*—necessity, possibility, contingency, impossibility—qualify the truth value of a statement. (300)

**Closed Branch:**
>A branch on a modal tree is a *closed branch* iff the branch contains both some atomic wff $\mathbb{P}$ and its negation $\neg\mathbb{P}$. (321)

**Closed Tree:**
>A modal tree is a *closed tree* iff every branch is closed. (321)

**Closed Tree Structure:**
>A modal tree structure is a *closed structure* iff there is some world at which the tree is closed. (321)

**Contingency:**
>A statement, $\mathbb{P}$, is *contingent*, or is a *contingency* iff it is neither necessary nor impossible, that is, if it is possibly true and possibly false. (300)

**Expression of $S^{\square}$:**
>An *expression of $S^{\square}$* is any finite sequence of the symbols of $S^{\square}$. (305)

**Impossibility:**
>A statement, $\mathbb{P}$, is *impossible*, or is an *impossibility* iff it could not be true. (300)

**Informal Modal Interpretation:**
>$\square\mathbb{P}$ is T iff $\mathbb{P}$ is T in every world.
>$\Diamond\mathbb{P}$ is T iff $\mathbb{P}$ is T in at least one world. (304)

**Interpretation of $S^{\square}$:**
>An *interpretation $M$ of $S^{\square}$* consists of:

>>(1) a non-empty set, $\mathbf{W}$, of worlds, w
>>(2) a function, $\mathbf{E}$, which for each w $\in \mathbf{W}$, assigns T or F to each statement letter of $S^{\square}$ (307)

**Modal Tree Structure:**
>A *modal tree structure* for a set of wffs is a structure that lists all the wffs, and all the well-formed components that must be true in attempting to show the set modally consistent, typically including trees at various worlds. (321)

**Modality Operators:**
>$\square\,\Diamond$    (305)

**Modally Consistent Set of Wffs:**
>A set $\Gamma$ of wffs of $S^{\square}$ is *modally consistent* iff on at least one interpretation there

is a world in which all members of $\Gamma$ are true. $\Gamma$ is *modally inconsistent* iff it is not modally consistent. (310)

**Modal Tree Test:** A set $\Gamma$ is modally consistent iff it has an open tree structure. (321)

**Modal Tree Test:** A set $\Gamma$ is modally inconsistent iff it has a closed tree structure. (321)

## Modally Contingent Wff:

A wff $\mathbb{P}$ of $S^\square$ is *modally contingent* iff $\mathbb{P}$ is neither modally true nor modally false; i.e., iff on some interpretation it is false in some world and on some interpretation it is true in some world. (309)

**Modal Tree Test:** A wff $\mathbb{P}$ is modally contingent iff $\{\mathbb{P}\}$ has an open tree structure and $\{\neg\mathbb{P}\}$ has an open tree structure. (321)

## Modally Contradictory Pair of Wffs:

Wffs $\mathbb{P}$ and $\mathbb{Q}$ of $S^\square$ are *modally contradictory* iff there is no interpretation containing a world in which they have the same truth value. (310)

**Modal Tree Test:** Wffs $\mathbb{P}$ and $\mathbb{Q}$ are modally contradictory iff $\{\mathbb{P} \leftrightarrow \mathbb{Q}\}$ has a closed tree structure. (322)

## Modally Entails:

A set $\Gamma$ of wffs of $S^\square$ *modally entails* a wff $\mathbb{P}$, $\Gamma \vDash \mathbb{P}$, iff there is no interpretation containing a world in which all the members of $\Gamma$ are true and $\mathbb{P}$ is false. (310)

**Modal Tree Test:** A set of wffs, $\Gamma$, modally entails a target wff, $\mathbb{P}$, iff $\Gamma \cup \{\neg\mathbb{P}\}$ has a closed tree structure. (322)

## Modally Equivalent Pair of Wffs:

Wffs $\mathbb{P}$ and $\mathbb{Q}$ of $S^\square$ are *modally equivalent* iff there is no interpretation containing a world in which they differ in truth value. (309)

**Modal Tree Test:** Wffs $\mathbb{P}$ and $\mathbb{Q}$ are modally equivalent iff $\{\neg(\mathbb{P} \leftrightarrow \mathbb{Q})\}$ has a closed tree structure. (322)

## Modally False Wff:

A wff $\mathbb{P}$ of $S^\square$ is *modally false* iff on every interpretation $\mathbb{P}$ is false in every world. (309)

**Modal Tree Test:** A wff $\mathbb{P}$ is modally false iff $\{\mathbb{P}\}$ has a closed tree structure. (321)

## Modally Falsifiable Wff:

A wff $\mathbb{P}$ of $S^\square$ is *modally falsifiable* iff $\mathbb{P}$ is not modally true; i.e., on some interpretation it is false in some world. (309)

**Modal Tree Test:** A wff $\mathbb{P}$ is modally falsifiable iff $\{\neg\mathbb{P}\}$ has an open tree structure. (322)

## Modally Inconsistent Set of Wffs:

See Modally Consistent Set of Wffs. (310)

**Modally Satisfiable Wff:**

A wff $\mathbb{P}$ of $S^\square$ is *modally satisfiable* iff $\mathbb{P}$ is not modally false; i.e., on some interpretation it is true in some world. (309)

**Modal Tree Test:** A wff $\mathbb{P}$ is modally satisfiable iff $\{\mathbb{P}\}$ has an open tree structure. (322)

**Modally True Wff:**

A wff $\mathbb{P}$ of $S^\square$ is *modally true* iff on every interpretation $\mathbb{P}$ is true in every world. (309)

**Modal Tree Test:** A wff $\mathbb{P}$ is modally true iff $\{\neg\mathbb{P}\}$ has a closed tree structure. (321)

**Modally Valid Argument:**

An argument of $S^\square$ is *modally valid* iff there is no interpretation containing a world in which all the premises are true and the conclusion is false. An argument of $S^\square$ is *modally invalid* iff it is not modally valid. (310)

**Modal Tree Test:** An argument is modally valid iff the set consisting of all and only the premises and the negation of the conclusion has a closed tree structure. (322)

**Necessity:**

A statement, $\mathbb{P}$, is *necessary*, *necessarily true*, or is a *necessity* iff it could not be false. (300)

**Non-Necessity:**

A statement, $\mathbb{P}$, is *non-necessary*, or is *possibly false* iff it is not necessary, that is, if its negation is possibly true. (300)

**Open Branch:**

A branch on a tree is an *open branch* iff it is not closed. (321)

**Open Tree:**

A modal tree is an *open tree* iff at least one branch is open. (321)

**Open Tree Structure:**

A modal tree structure is an *open structure* iff there is an open tree at every world. (321)

**Possibility:**

A statement, $\mathbb{P}$, is *possible*, *possibly true*, or is a *possibility* iff it is not impossible. (300)

**Truth at a World in $S^\square$:**

Truth in $S^\square$ is determined by a valuation function **V**, relative to an interpretation **M**. See text for full definition. (307)

**Well-Formed Formula, Wff, of $S^{\square}$:**

A *well-formed formula* or *wff* of $S^{\square}$ is a grammatical sentence of the language $S^{\square}$. See full definition in text. (305)

# Part V
# Appendices

# A Answers to Exercises

## A.1 Answers to 1.6

1. A statement is a declarative sentence—not a question, command, or exclamation. We assume it is either true or false, but not both.

2. An argument is a set of statements, one of which—the conclusion—is supposed to be supported by the others—the premises.

3. The truth value of a statement is just its truth or falsehood. For our purposes there are two truth values: True and False. Every statement has either the one value or the other, but not both.

4. An argument is deductively valid if and only if it is not possible for all the premises to be true and the conclusion false.

5. An argument is sound if and only if it is valid and all the premises are true.

6. Knowing that an argument is valid does not tell us anything about the actual truth values of the component statements; but it does tell us that if the premises are all true, then the conclusion is true as well.

7. If an argument is sound, then it must be valid as well.

8. If an argument is sound then all the component statements are true. If it is sound then it is valid and has true premises, but this also means that the conclusion is true.

9. Knowing only that all the component statements are true tells us nothing about the validity or soundness of the argument (except that if it is valid, then it is also sound).

10. An argument is inductively strong to the degree to which the premises provide evidence to make the truth of the conclusion plausible or probable.

11. Deductive validity requires that it is not possible for all the premises to be true and the conclusion false. The validity or invalidity of a deductive argument is a matter of form. In inductive arguments there is no attempt at guaranteed truth preservation, but the assumed truth of the premises is supposed to make the conclusion plausible or probable. The strength of an inductive argument is not a matter of form; much background knowledge is relevant to the assessment of strength.

**12.** Abductive reasoning is a kind of induction in which the conclusion is supposed to explain the premises. Abduction is also known as inference to the best explanation.

**13.–16.** Check with a classmate or your instructor.

## A.2  **Answers to 2.3.3**

**1.** Not an expression: neither 'b' nor '$\mathbb{F}$' are part of the language $S$.

**2.** Not a wff: the hook is not a binary connective.

**3.**

$$A \quad B$$
$$A \wedge B$$
$$\neg(A \wedge B) \quad B$$
$$\neg(A \wedge B) \rightarrow B$$

**4.**

$$A \quad B$$
$$A \wedge B \quad C$$
$$(A \wedge B) \wedge C \quad D$$
$$((A \wedge B) \wedge C) \wedge D$$

**5.**

$$G$$
$$\neg G \quad J$$
$$K \quad \neg G \rightarrow J$$
$$F \quad K \wedge (\neg G \rightarrow J) \quad F \quad K$$
$$F \vee (K \wedge (\neg G \rightarrow J)) \quad\quad F \rightarrow K$$
$$(F \vee (K \wedge (\neg G \rightarrow J))) \leftrightarrow (F \rightarrow K)$$

**6.** Not a wff: a right parenthesis is missing immediately after the 'O'.

## A.3 **Answers to 2.4.2**

**Note:** For every symbolization there are multiple equivalent (and thus correct) answers. Indeed, there are an infinite number of symbolizations equivalent to any given symbolization. In the following answer keys I only list multiple answers if the equivalents are likely to occur to the reader, or have some instructional value. Thus, if your answer does not appear, but you believe it to be equivalent and correct, check with your instructor.

1. $\neg$G
2. W $\wedge$ V
3. F $\vee$ M
4. S $\wedge \neg$K
5. K $\vee \neg$K
6. $\neg$(F $\vee$ M)
   $\neg$F $\wedge \neg$M
7. $\neg$F $\wedge \neg$M
   $\neg$(F $\vee$ M)
8. $\neg$(F $\wedge$ M)
   $\neg$F $\vee \neg$M
9. $\neg$F $\vee \neg$M
   $\neg$(F $\wedge$ M)
10. (S $\wedge$ C) $\wedge \neg$Z
11. (N $\wedge$ K) $\vee$ O
12. N $\wedge$ (K $\vee$ O)
13. K $\vee$ (N $\vee$ S)
    (K $\vee$ N) $\vee$ S
14. (G $\wedge$ H) $\vee \neg$(F $\vee$ M)
    (G $\wedge$ H) $\vee$ ($\neg$F $\wedge \neg$M)
15. (S $\wedge$ C) $\wedge$ (G $\vee \neg$F)

## A.4 **Answers to 2.4.4**

1. N $\rightarrow$ F
2. K $\rightarrow$ G
3. G $\rightarrow$ K
4. G $\leftrightarrow$ K
5. $\neg$Z $\rightarrow$ S
6. F $\rightarrow$ W

7. $L \rightarrow F$
8. $L \leftrightarrow F$
9. $M \vee G$
   $\neg M \rightarrow G$
   $\neg G \rightarrow M$
10. $\neg S \vee \neg M$
    $S \rightarrow \neg M$
    $M \rightarrow \neg S$

## A.5  **Answers to 2.4.6**

**A.**    **1.** $(S \vee Z) \wedge \neg C$
    **2.** $(N \wedge K) \vee Z$
    **3.** $(F \vee M) \wedge \neg (F \wedge M)$
    **4.** $(W \wedge V) \wedge (\neg M \vee \neg F)$
    **5.** $(S \wedge G) \vee (O \wedge H)$
    **6.** $F \leftrightarrow (N \vee S)$
    **7.** $(V \wedge \neg M) \rightarrow \neg T$
    **8.** $\neg F \vee N$
        $F \rightarrow N$
        $\neg N \rightarrow \neg F$
    **9.** $(S \wedge K) \vee M$
        $\neg M \rightarrow (S \wedge K)$
        $\neg (S \wedge K) \rightarrow M$
  **10.** $(L \wedge F) \rightarrow (N \wedge \neg H)$
  **11.** $G \rightarrow (S \wedge K)$
  **12.** $(S \wedge C) \rightarrow (L \wedge F)$
  **13.** $[(K \wedge N) \wedge S] \rightarrow (\neg H \wedge \neg M)$
      $[(K \wedge N) \wedge S] \rightarrow \neg (H \vee M)$
  **14.** $(G \wedge F) \leftrightarrow (S \wedge K)$
      $[(S \wedge K) \rightarrow (G \wedge F)] \wedge [(G \wedge F) \rightarrow (S \wedge K)]$
  **15.** $(O \wedge Z) \rightarrow [\neg N \rightarrow (F \rightarrow \neg H)]$

**B.**    **1.** $D \rightarrow B$
    **2.** $A \rightarrow (L \wedge R)$
    **3.** $A \leftrightarrow (Q \wedge E)$
    **4.** $(B \wedge C) \rightarrow D$
    **5.** $\neg A \wedge \neg C$
        $\neg (A \vee C)$
    **6.** $\neg (A \vee C)$
        $\neg A \wedge \neg C$
    **7.** $\neg A \vee \neg C$

¬(A ∧ C)
8. ¬(A ∧ C)
   ¬A ∨ ¬C
9. X → (R ∧ L)
10. X ∨ L
   ¬X → L
   ¬L → X
11. Y → (¬D ∨ (Q ∧ P))
   Y → (D → (Q ∧ P))
   Y → (¬(Q ∧ P) → ¬D)
12. [(S ∧ M) → (Q ∨ O)] ∧ [(S ∧ M) → Y]
   (S ∧ M) → [(Q ∨ O) ∧ Y]
13. R → (S ∨ T)
14. (A ∧ B) ∧ C
15. (A ∨ B) ∨ C
16. [(A ∧ B) ∨ (B ∧ C)] ∨ (A ∧ C)
17. [(¬A ∧ ¬B) ∨ (¬B ∧ ¬C)] ∨ (¬A ∧ ¬C)
18. ¬[(A ∧ B) ∧ C]
19. [(A ∨ B) ∨ C] ∧ ¬[(A ∧ B) ∧ C]
20. [(A ∨ B) ∨ C] ∧ ([(¬A ∧ ¬B) ∨ (¬B ∧ ¬C)] ∨ (¬A ∧ ¬C))

# A.6   **Answers to 3.2.1**

**1.**

| A | A ∨ ¬A |
|---|--------|
| T |    T   F |
| F |    T   T |

T-Fly Satisfiable, T-Fly True

**2.**

| A | A ∧ ¬A |
|---|--------|
| T |    F   F |
| F |    F   T |

T-Fly Falsifiable, T-Fly False

**3.**

| F G H | F → ¬(G ∨ H) |
|-------|--------------|
| T T T |   F   F    T |
| T T F |   F   F    T |
| T F T |   F   F    T |
| T F F |   T   T    F |
| F T T |   T   F    T |
| F T F |   T   F    T |
| F F T |   T   F    T |
| F F F |   T   T    F |

T-Fly Satisfiable, T-Fly Falsifiable, T-Fly Contingent

**4.**

| J K | ¬(J ∧ K) | ↔ | (J → ¬K) |
|-----|----------|---|----------|
| T T | F    T   | T | F    F   |
| T F | T    F   | T | T    T   |
| F T | T    F   | T | T    F   |
| F F | T    F   | T | T    T   |

T-Fly Satisfiable, T-Fly True

**5.**

| D | (D → (D → D)) | ↔ | D |
|---|---------------|---|---|
| T | T        T    | T | T |
| F | T        T    | T | F |

T-Fly Satisfiable, T-Fly Falsifiable, T-Fly Contingent

## A.7  **Answers to 3.3.1**

**A.**  **1.**

| D E | ¬(D ↔ E) | (D ∧ ¬E) ∨ (¬D ∧ E) |
|-----|----------|---------------------|
| T T | F   T    | F F  | F | F F    |
| T F | T   F    | T T  | T | F F    |
| F T | T   F    | F F  | T | T T    |
| F F | F   T    | F T  | F | T F    |

Equivalent.

**2.**

| I J | I ∧ J | ¬(J ∨ I) |     |
|-----|-------|----------|-----|
| T T | T     | F    T   | ⇐   |
| T F | F     | F    T   | ⇐   |
| F T | F     | F    T   |     |
| F F | F     | T    F   |     |

Neither.

**3.**

| D E F | D → (E → F) | D ∧ (E ∧ ¬F) |
|-------|-------------|--------------|
| T T T | T     T     | F | F   F    |
| T T F | F     F     | T | T   T    |
| T F T | T     T     | F | F   F    |
| T F F | T     T     | F | F   T    |
| F T T | T     T     | F | F   F    |
| F T F | T     F     | F | T   T    |
| F F T | T     T     | F | F   F    |
| F F F | T     T     | F | F   T    |

Contradictory.

**B.   1.**

| A B C | A ∧ B | (A ∧ B) → C | ¬C | ¬C ∧ A | A → B |
|-------|-------|-------------|----|--------|-------|
| T T T | T | T | F | F | T |
| T T F | T | F | T | T | T |
| T F T | F | T | F | F | F |
| T F F | F | T | T | T | F |
| F T T | F | T | F | F | T |
| F T F | F | T | T | F | T |
| F F T | F | T | F | F | T |
| F F F | F | T | T | F | T |

Inconsistent.

**2.**

| K L | ¬(K ∧ L) | K ∧ L | ¬(K ∨ L) | K ∨ L | K → ¬L | ¬L |
|-----|----------|-------|----------|-------|--------|----|
| T T | F | T | F | T | F | F |
| T F | T | F | F | T | T | T |
| F T | T | F | F | T | T | F |
| F F | T | F | T | F | T | T | ⟸

Consistent.

**C.   1.**

| C D E | C ∨ D | ¬E | ¬E → ¬D | ¬D | C → E | E |
|-------|-------|----|---------|----|-------|---|
| T T T | T | F | T | F | T | T |
| T T F | T | T | F | F | F | F |
| T F T | T | F | T | T | T | T |
| T F F | T | T | T | T | F | F |
| F T T | T | F | T | F | T | T |
| F T F | T | T | F | F | T | F |
| F F T | F | F | T | T | T | T |
| F F F | F | T | T | T | T | F |

{C∨D, ¬E→¬D, C→E} ⊨ E

**2.**

| A B | A → B | B | A |
|-----|-------|---|---|
| T T | T | T | T |
| T F | F | F | T |
| F T | T | T | F | ⟸
| F F | T | F | F |

{A→B, B} ⊭ A

**D.    1.**

| B C D | ¬D → C | | B → ¬C | | ¬D → ¬B | |
|-------|--------|---|--------|---|---------|---|
| T T T | F | T | F | F | F | T | F |
| T T F | T | T | F | F | T | F | F |
| T F T | F | T | T | T | F | T | F |
| T F F | T | F | T | T | T | F | F |
| F T T | F | T | T | F | F | T | T |
| F T F | T | T | T | F | T | T | T |
| F F T | F | T | T | T | F | T | T |
| F F F | T | F | T | T | T | T | T |

Valid.

**2.**

| G H I | G ↔ H | | ¬H ∧ I | | ¬G ∧ I | |
|-------|-------|---|--------|---|--------|---|
| T T T | T | | F | F | F | F |
| T T F | T | | F | F | F | F |
| T F T | F | | T | T | F | F |
| T F F | F | | T | F | F | F |
| F T T | F | | F | F | T | T |
| F T F | F | | F | F | T | F |
| F F T | T | | T | T | T | T |
| F F F | T | | T | F | T | F |

Valid.

# A.8 **Answers to 3.4.1**

**A.**    1. All truth-functionally true wffs are satisfiable, so $\mathbb{P}$ is satisfiable.

2. All truth-functionally false wffs are falsifiable, so $\mathbb{P}$ is falsifiable.

3. All truth-functionally contingent wffs are both satisfiable and falsifiable, so $\mathbb{P}$ is satisfiable and falsifiable.

4. Contingency could be defined as both satisfiable and falsifiable.

5. It is truth-functionally true (and therefore also satisfiable), for $\mathbb{P}$ is true on every tva, making $\mathbb{P} \lor \mathbb{Q}$ true on every tva.

6. It is truth-functionally true (and therefore also satisfiable), for $\mathbb{P}$ is false on every tva, making $\mathbb{P} \to \mathbb{Q}$ true on every tva.

7. It is satisfiable, for $\mathbb{P}$ is false on at least one tva, making $\mathbb{P} \to \mathbb{Q}$ true on at least one tva.

8. It is truth-functionally false (and therefore also falsifiable), for on every tva $\mathbb{P}$ is true and $\mathbb{Q}$ false, making $\mathbb{P} \to \mathbb{Q}$ false on every tva.

9. It is falsifiable, for $\mathbb{P}$ is true on every tva and $\mathbb{Q}$ is false on at least one tva, making $\mathbb{P} \to \mathbb{Q}$ false on at least one tva.

10. It is satisfiable, for $\mathbb{P}$ is false on at least one tva, making $\mathbb{P} \to \mathbb{Q}$ true on at least one tva.

**B.**

1. It is truth-functionally true (and therefore satisfiable), for $\mathbb{P}$ and $\mathbb{Q}$ have the same truth value on every tva and $\mathbb{P} \to \mathbb{Q}$ is false only if $\mathbb{P}$ is true and $\mathbb{Q}$ is false, hence $\mathbb{P} \to \mathbb{Q}$ is true on every tva.

2. It is truth-functionally true (and therefore satisfiable), for $\mathbb{P}$ and $\mathbb{Q}$ have the same truth value on every tva, hence $\mathbb{P} \leftrightarrow \mathbb{Q}$ is true on every tva.

3. Since $\mathbb{P} \land \mathbb{Q}$ can be either true or false when the conjuncts have the same truth value, the equivalence of $\mathbb{P}$ and $\mathbb{Q}$ tells us only that $\mathbb{P} \land \mathbb{Q}$ is equivalent to both $\mathbb{P}$ and $\mathbb{Q}$.

4. Since $\mathbb{P} \to \mathbb{Q}$ can be either true or false when the antecedent and consequent have opposite truth values, that $\mathbb{P}$ and $\mathbb{Q}$ are contradictory tells us only that $\mathbb{P} \to \mathbb{Q}$ is equivalent to $\mathbb{Q}$.

5. It is truth-functionally false (and therefore also falsifiable), for $\mathbb{P}$ and $\mathbb{Q}$ have opposite truth values on every tva, hence $\mathbb{P} \leftrightarrow \mathbb{Q}$ is false on every tva.

6. It is truth-functionally false (and therefore also falsifiable), for exactly one of $\mathbb{P}$ and $\mathbb{Q}$ is false on every tva, hence $\mathbb{P} \land \mathbb{Q}$ is false on every tva.

7. It is inconsistent, for $\mathbb{P}$ is false on every tva, hence there is no tva on which all members of $\{\mathbb{P}\}$ are true.

8. It is consistent, for $\mathbb{P}$ is true on at least one tva, hence there is at least one tva on which all members of $\{\mathbb{P}\}$ are true.

9. $\Gamma \vDash \mathbb{P}$, for there is no tva on which all members of $\Gamma$ are true, hence there is no tva on which all members of $\Gamma$ are true and $\mathbb{P}$ false.

10. $\vDash \mathbb{P}$ iff there are no tvas on which all members of $\emptyset$ are true and $\mathbb{P}$ is false; i.e., (since there are no members of $\emptyset$) iff there are no tvas on which $\mathbb{P}$ is false; i.e., iff $\mathbb{P}$ is truth-functionally true.

11. $\vDash \neg\mathbb{P}$ iff there are no tvas on which all members of $\emptyset$ are true and $\neg\mathbb{P}$ is false; i.e., (since there are no members of $\emptyset$) iff there are no tvas on which $\neg\mathbb{P}$ is false; i.e., iff there are no tvas on which $\mathbb{P}$ is true; i.e., iff $\mathbb{P}$ is truth-functionally false.

12. Any such argument is valid, for there is no tva on which $\mathbb{P}$ is false; hence there is no tva on which all the premises are true and the conclusion, $\mathbb{P}$, is false.

13. It is valid, for there is no tva on which all members of $\{\mathbb{P}_1, \ldots, \mathbb{P}_n, \neg\mathbb{Q}\}$ are true; but this is equivalent to there being no tva on which all of $\mathbb{P}_1, \ldots, \mathbb{P}_n$ are true and $\mathbb{Q}$ false.

14. It is a truth-functionally true wff (and therefore also satisfiable), for there is no tva on which all of $\mathbb{P}_1, \ldots, \mathbb{P}_n$ are true and $\mathbb{Q}$ false, but this is equivalent to there being no tva on which the antecedent of the conditional is true and the consequent false; i.e., no tva on which the conditional is false.

15. $\mathbb{Q} \vDash \mathbb{R}$, for $\mathbb{P}$ and $\mathbb{Q}$ have the same truth value on every tva, and there is no tva on which $\mathbb{P}$ is true and $\mathbb{R}$ is false, hence there is no tva on which $\mathbb{Q}$ is true and $\mathbb{R}$ is false.

## A.9 **Answers to 3.5.2**

**A.**    **1.** A ∨ ¬A

'A ∨ ¬A' is T-Fly True and T-Fly Satisfiable. (Note: only the left tree is required; if you do the left tree first, then the right tree is unnecessary.)

**2.** A ∧ ¬A

'A ∧ ¬A' is T-Fly False and T-Fly Falsifiable. (Note: only the left tree is required; if you do the left tree first, then the right tree is unnecessary.)

**3.** F → ¬(G ∨ H)

F → ¬(G ∨ H) ✓
⌒
¬F    ¬(G ∨ H) ✓
|
¬G
¬H

¬(F → ¬(G ∨ H)) ✓
|
F
¬¬(G ∨ H) ✓
|
G ∨ H ✓
⌒
G    H

'F → ¬(G ∨ H)' is T-Fly Satisfiable, T-Fly Falsifiable, and T-Fly Contingent. (Note: both trees are required.)

**4.** ¬(J ∧ K) ↔ (J → ¬K)

¬(¬(J ∧ K) ↔ (J → ¬K)) ✓

¬(J ∧ K) ✓    ¬¬(J ∧ K) ✓
¬(J → ¬K) ✓   J → ¬K ✓
|             |
J             J ∧ K ✓
¬¬K ✓         |
|             J
K             K
⌒             ⌒
¬J    ¬K    ¬J    ¬K
✗     ✗     ✗     ✗

¬(J ∧ K) ↔ (J → ¬K)

¬(J ∧ K) ✓    ¬¬(J ∧ K) ✓
J → ¬K ✓      ¬(J → ¬K) ✓
                |
¬J      ¬K    J ∧ K ✓
⌒       ⌒      |
¬J  ¬K  ¬J  ¬K  J
                K
                |
                J
                ¬¬K ✓
                |
                K

'¬(J ∧ K) ↔ (J → ¬K)' is T-Fly True and T-Fly Satisfiable. (Note: only the left tree is required.)

**5.** $(D \rightarrow (D \rightarrow D)) \leftrightarrow D$

'$(D \rightarrow (D \rightarrow D)) \leftrightarrow D$' is T-Fly Satisfiable, T-Fly Falsifiable, and T-Fly Contingent. (Note: both trees are required.)

**B.    1.** $\neg(D \leftrightarrow E) \qquad (D \wedge \neg E) \vee (\neg D \wedge E)$

$$\neg(D \leftrightarrow E) \leftrightarrow ((D \wedge \neg E) \vee (\neg D \wedge E)) \checkmark$$

```
                    ¬(D↔E) ✓                        ¬¬(D↔E) ✓
            (D∧¬E)∨(¬D∨E) ✓                  ¬((D∧¬E)∨(¬D∧E)) ✓
                                                        |
              D              ¬D                      D↔E ✓
             ¬E              E                          |
                                               ¬(D∧¬E) ✓
      D∧¬E✓ ¬D∧E✓ D∧¬E✓ ¬D∧E✓                ¬(¬D∧E) ✓
        |      |     |     |
        D     ¬D     D    ¬D     D                    ¬D
       ¬E     E     ¬E    E     E                     ¬E
              ✗     ✗
                                  ¬D     ¬¬E ✓    ¬D        ¬¬E ✓
                                  |              |
                                ¬¬D✓ ¬¬D✓ ¬E   ¬¬D✓    ¬¬D✓ ¬E
                                     |              |
                                 D  ¬E  D       E  D  ¬E   D      E
                                 ✗  ✗   |       ✗  ✗      |      ✗
                                        E                 E
                                                          ✗
```

The members of the pair are T-Fly Equivalent. (Note: only the first tree is required.)

**2.** $I \wedge J \qquad \neg(J \vee I)$

```
        (I∧J)↔¬(J∨I) ✓                      ¬((I∧J)↔¬(J∨I)) ✓

    I∧J✓        ¬(I∧J) ✓                  I∧J✓    ¬(I∧J) ✓
  ¬(J∨I) ✓     ¬¬(J∨I) ✓                ¬¬(J∨I) ✓  ¬(J∨I) ✓
     |             |                        |          |
     I           J∨I✓                       I          ¬J
     J                                      J          ¬I
     |          J        I                  |
    ¬J                                     J∨I✓      ¬I    ¬J
    ¬I      ¬I   ¬J    ¬I   ¬J
     ✗      ✗    ✗                        J    I
```

The members of the pair are neither T-Fly Equivalent nor T-Fly Contradictory. (Note: both trees are required.)

**3.** $D \to (E \to F)$ $\qquad$ $D \land (E \land \neg F)$

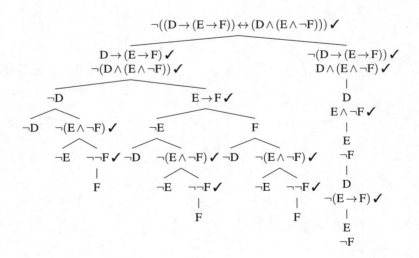

The members of the pair are T-Fly Contradictory. (Note: only the first tree is required.)

**C.** **1.** $\{(A \wedge B) \rightarrow C, \neg C \wedge A, A \rightarrow B\}$

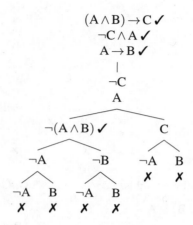

The set is T-Fly Inconsistent.

**2.** $\{\neg(K \wedge L), \neg(K \vee L), K \rightarrow \neg L\}$

$$\neg(K \wedge L) \checkmark$$
$$\neg(K \vee L) \checkmark$$
$$K \rightarrow \neg L \checkmark$$
$$|$$
$$\neg K$$
$$\neg L$$

```
              ┌──────────┴──────────┐
             ¬K                    ¬L
          ┌───┴───┐            ┌───┴───┐
         ¬K      ¬L           ¬K      ¬L
```

The set is T-Fly Consistent.

**D.**    **1.** {C∨D, ¬E→¬D, C→E} ⊨ E

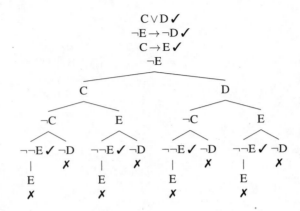

The entailment holds, {C∨D, ¬E→¬D, C→E} ⊨ E.

**2.** {A→B, B} ⊨ A

$$A→B ✓$$
$$B$$
$$¬A$$

¬A    B

The entailment does not hold, {A→B, B} ⊭ A

**E.**    **1.** ¬D→C
B→¬C
——————
¬D→¬B

¬D→C ✓
B→¬C ✓
¬(¬D→¬B) ✓
|
¬D
¬¬B ✓
|
B

¬B        ¬C

¬¬D    C    ¬¬D    C
|      ✗    |      ✗
D           D
✗           ✗

The argument is T-Fly valid.

**2.** G ↔ H
$\dfrac{\neg H \wedge I}{\neg G \wedge I}$

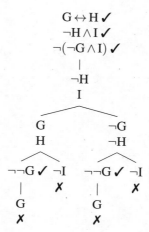

The argument is T-Fly valid.

# A.10 **Answers to 4.1.7**

**A.** **1.**

| | | | |
|---|---|---|---|
| 1 | F | P | |
| 2 | E ∧ G | P | ⊢ F ∧ G |
| 3 | G | 2 ∧E | |
| 4 | F ∧ G | 1, 3 ∧I | |

**2.**

| | | | |
|---|---|---|---|
| 1 | A | P | |
| 2 | A → B | P | ⊢ B ∧ A |
| 3 | B | 1, 2 →E | |
| 4 | B ∧ A | 3, 1 ∧I | |

**3.**

| | | | |
|---|---|---|---|
| 1 | A ∧ E | P | |
| 2 | B ∧ C | P | |
| 3 | (A ∧ B) → D | P | ⊢ E ∧ D |
| 4 | A | 1 ∧E | |
| 5 | B | 2 ∧E | |
| 6 | A ∧ B | 4, 5 ∧I | |
| 7 | D | 3, 6 →E | |
| 8 | E | 1 ∧E | |
| 9 | E ∧ D | 7, 8 ∧I | |

**4.**    1. | $A \wedge B$                                   P
         2 | $(B \wedge A) \rightarrow C$                   P
         3 | $C \rightarrow D$                              P          $\vdash D$
         _____
         4 | A                                              1 $\wedge$E
         5 | B                                              1 $\wedge$E
         6 | $B \wedge A$                                   4, 5 $\wedge$I
         7 | C                                              2, 6 $\rightarrow$E
         8 | D                                              3, 7 $\rightarrow$E

**5.**   1 | $(D \wedge G) \wedge H$                         P
         2 | $(H \wedge D) \rightarrow I$                    P
         3 | $(I \wedge G) \rightarrow J$                    P          $\vdash J$
         _____
         4 | H                                              1 $\wedge$E
         5 | $D \wedge G$                                   1 $\wedge$E
         6 | D                                              5 $\wedge$E
         7 | $H \wedge D$                                   4, 6 $\wedge$I
         8 | I                                              2, 7 $\rightarrow$E
         9 | G                                              5 $\wedge$E
        10 | $I \wedge G$                                   8, 9 $\wedge$I
        11 | J                                              3, 10 $\rightarrow$E

**B.**  **1.**   1 | $A \rightarrow B$                       P
                 2 | $B \rightarrow C$                       P          $\vdash A \rightarrow C$
                 3 |   | A                                   A
                 _____
                 4 |   | B                                   1, 3 $\rightarrow$E
                 5 |   | C                                   2, 4 $\rightarrow$E
                 6 | $A \rightarrow C$                       3–5 $\rightarrow$I

        **2.**   1 | $A \rightarrow B$                       P
                 2 | $\neg B$                                P          $\vdash \neg A$
                 3 |   | A                                   A
                 _____
                 4 |   | B                                   1, 3 $\rightarrow$E
                 5 |   | $\neg B$                            2 R
                 6 | $\neg A$                                3–5 $\neg$I

**3.**

| | | | |
|---|---|---|---|
| 1 | ¬F ∧ G | P | |
| 2 | ¬C → D | P | |
| 3 | D → F | P | ⊢ C |
| 4 | ¬C | A | |
| 5 | D | 2, 4 →E | |
| 6 | F | 3, 5 →E | |
| 7 | ¬F | 1 ∧E | |
| 8 | C | 4−7 ¬E | |

**4.**

| | | | |
|---|---|---|---|
| 1 | G → H | P | |
| 2 | (H ∧ G) → K | P | |
| 3 | K → L | P | ⊢ G → (K ∧ L) |
| 4 | G | A | |
| 5 | H | 1, 4 →E | |
| 6 | H ∧ G | 4, 5 ∧I | |
| 7 | K | 2, 6 →E | |
| 8 | L | 3, 7 →E | |
| 9 | K ∧ L | 7, 8 ∧I | |
| 10 | G → (K ∧ L) | 4−9 →I | |

**5.**

| | | | |
|---|---|---|---|
| 1 | A → B | P | |
| 2 | ¬B | P | ⊢ A → ¬C |
| 3 | A | A | |
| 4 | C | A | |
| 5 | B | 1, 3 →E | |
| 6 | ¬B | 2 R | |
| 7 | ¬C | 4−6 ¬I | |
| 8 | A → ¬C | 3−7 →I | |

**C.** **1.**

| | | | |
|---|---|---|---|
| 1 | A | P | |
| 2 | (A ∨ B) → C | P | ⊢ C |
| 3 | A ∨ B | 1 ∨I | |
| 4 | C | 2, 3 →E | |

**2.**

| | | | |
|---|---|---|---|
| 1 | T ↔ K | P | |
| 2 | J → T | P | |
| 3 | J ∨ K | P | ⊢ T |
| 4 |   J | A | |
| 5 |   T | 2, 4 →E | |
| 6 |   K | A | |
| 7 |   T | 1, 6 ↔E | |
| 8 | T | 3, 4−5, 6−7 ∨E | |

**3.**

| | | | |
|---|---|---|---|
| 1 | A ↔ B | P | |
| 2 | C ↔ B | P | ⊢ A ↔ C |
| 3 |   A | A | |
| 4 |   B | 1, 3 ↔E | |
| 5 |   C | 2, 4 ↔E | |
| 6 |   C | A | |
| 7 |   B | 2, 6 ↔E | |
| 8 |   A | 1, 7 ↔E | |
| 9 | A ↔ C | 3−5, 6−8 ↔I | |

**4.**

| | | | |
|---|---|---|---|
| 1 | (G ∨ H) → I | P | |
| 2 | K → H | P | ⊢ K → (I ∨ (M ∧ N)) |
| 3 |   K | A | |
| 4 |   H | 2, 3 →E | |
| 5 |   G ∨ H | 4 ∨I | |
| 6 |   I | 1, 5 →E | |
| 7 |   I ∨ (M ∧ N) | 6 ∨I | |
| 8 | K → (I ∨ (M ∧ N)) | 3−7 →I | |

**5.**

| | | | |
|---|---|---|---|
| 1 | $A \wedge K$ | P | |
| 2 | $A \rightarrow (I \vee J)$ | P | |
| 3 | $(I \wedge K) \rightarrow L$ | P | |
| 4 | $L \leftrightarrow J$ | P | $\vdash L \wedge A$ |
| 5 | A | $1 \wedge E$ | |
| 6 | K | $1 \wedge E$ | |
| 7 | $I \vee J$ | $2, 5 \rightarrow E$ | |
| 8 | I | A | |
| 9 | $I \wedge K$ | $6, 8 \wedge I$ | |
| 10 | L | $3, 9 \rightarrow E$ | |
| 11 | J | A | |
| 12 | L | $4, 11 \leftrightarrow E$ | |
| 13 | L | $7, 8-10, 11-12 \vee E$ | |
| 14 | $L \wedge A$ | $5, 13 \wedge I$ | |

# A.11 **Answers to 4.3.1**

**A.** **1.**

| | | | |
|---|---|---|---|
| 1 | $F \wedge G$ | P | |
| 2 | $H \leftrightarrow F$ | P | $\vdash G \wedge H$ |
| 3 | F | $1 \wedge E$ | |
| 4 | H | $2, 3 \leftrightarrow E$ | |
| 5 | G | $1 \wedge E$ | |
| 6 | $G \wedge H$ | $4, 5 \wedge I$ | |

**2.**

| | | | |
|---|---|---|---|
| 1 | $A \wedge (B \wedge C)$ | P | |
| 2 | $(A \wedge B) \rightarrow D$ | P | $\vdash C \wedge D$ |
| 3 | A | $1 \wedge E$ | |
| 4 | $B \wedge C$ | $1 \wedge E$ | |
| 5 | B | $4 \wedge E$ | |
| 6 | C | $4 \wedge E$ | |
| 7 | $A \wedge B$ | $3, 5 \wedge I$ | |
| 8 | D | $2, 7 \rightarrow E$ | |
| 9 | $C \wedge D$ | $6, 8 \wedge I$ | |

**3.**

| | | | |
|---|---|---|---|
| 1 | F → K | P | |
| 2 | F → (G ∧ J) | P | |
| 3 | (K ∧ J) → H | P | ⊢ F → (G ∧ H) |
| 4 |   F | A | |
| 5 |   K | 1, 4 →E | |
| 6 |   G ∧ J | 2, 4 →E | |
| 7 |   G | 6 ∧E | |
| 8 |   J | 6 ∧E | |
| 9 |   K ∧ J | 5, 8 ∧I | |
| 10 |   H | 3, 9 →E | |
| 11 |   G ∧ H | 7, 10 ∧I | |
| 12 | F → (G ∧ H) | 4–11 →I | |

**4.**

| | | | |
|---|---|---|---|
| 1 | ¬L → (G ∧ C) | P | |
| 2 | D → (¬L ∧ J) | P | |
| 3 | G → ¬J | P | ⊢ ¬D |
| 4 |   D | A | |
| 5 |   ¬L ∧ J | 2, 4 →E | |
| 6 |   ¬L | 5 ∧E | |
| 7 |   G ∧ C | 1, 6→E | |
| 8 |   G | 7 ∧E | |
| 9 |   ¬J | 3, 8 →E | |
| 10 |   J | 5 ∧E | |
| 11 | ¬D | 4–10 ¬I | |

**B.**  **1.**

| | | | |
|---|---|---|---|
| 1 | I ∧ J | P | |
| 2 | (J ∨ H) → G | P | ⊢ F ∨ G |
| 3 | J | 1 ∧E | |
| 4 | J ∨ H | 3 ∨I | |
| 5 | G | 2, 4 →E | |
| 6 | F ∨ G | 5 ∨I | |

**2.**

| | | | |
|---|---|---|---|
| 1 | ¬H ↔ (F ∨ ¬D) | P | |
| 2 | J ∧ H | P | ⊢ D |
| 3 |   ¬D | A | |
| 4 |   F ∨ ¬D | 3 ∨I | |
| 5 |   ¬H | 1, 4 ↔E | |
| 6 |   H | 2 ∧E | |
| 7 | D | 3–6 ¬E | |

**3.**

| | | |
|---|---|---|
| 1 | $O \wedge (K \rightarrow M)$ | P |
| 2 | $(N \rightarrow M) \wedge (K \vee N)$ | P     $\vdash M \wedge O$ |
| 3 | $K \vee N$ | 2 $\wedge$E |
| 4 | $\quad$ K | A |
| 5 | $\quad$ $K \rightarrow M$ | 1 $\wedge$E |
| 6 | $\quad$ M | 4, 5 $\rightarrow$E |
| 7 | $\quad$ N | A |
| 8 | $\quad$ $N \rightarrow M$ | 2 $\wedge$E |
| 9 | $\quad$ M | 7, 8 $\rightarrow$E |
| 10 | M | 3, 4–6, 7–9 $\vee$E |
| 11 | O | 1 $\wedge$E |
| 12 | $M \wedge O$ | 10, 11 $\wedge$I |

**4.**

| | | |
|---|---|---|
| 1 | $D \vee (K \wedge L)$ | P |
| 2 | $(E \wedge J) \leftrightarrow (F \wedge D)$ | P |
| 3 | $F \wedge (K \rightarrow E)$ | P     $\vdash E \wedge F$ |
| 4 | $\quad$ D | A |
| 5 | $\quad$ F | 3 $\wedge$E |
| 6 | $\quad$ $F \wedge D$ | 4, 5 $\wedge$I |
| 7 | $\quad$ $E \wedge J$ | 2, 6 $\leftrightarrow$E |
| 8 | $\quad$ E | 7 $\wedge$E |
| 9 | $\quad$ $K \wedge L$ | A |
| 10 | $\quad$ K | 9 $\wedge$E |
| 11 | $\quad$ $K \rightarrow E$ | 3 $\wedge$E |
| 12 | $\quad$ E | 10, 11 $\rightarrow$E |
| 13 | E | 1, 4–8, 9–12 $\vee$E |
| 14 | F | 3 $\wedge$E |
| 15 | $E \wedge F$ | 13, 14 $\wedge$I |

**C. 1.**

| | | |
|---|---|---|
| 1 | $C \rightarrow E$ | P |
| 2 | $(G \rightarrow I) \leftrightarrow (A \wedge E)$ | P     $\vdash (A \wedge C) \rightarrow (G \rightarrow I)$ |
| 3 | $\quad$ $A \wedge C$ | A |
| 4 | $\quad$ C | 3 $\wedge$E |
| 5 | $\quad$ E | 1, 4 $\rightarrow$E |
| 6 | $\quad$ A | 3 $\wedge$E |
| 7 | $\quad$ $A \wedge E$ | 5, 6 $\wedge$I |
| 8 | $\quad$ $G \rightarrow I$ | 2, 7 $\leftrightarrow$E |
| 9 | $(A \wedge C) \rightarrow (G \rightarrow I)$ | 3–8 $\rightarrow$I |

**2.**

| | | | | |
|---|---|---|---|---|
| 1 | A → B | | P | |
| 2 | ¬B | | P | ⊢ ¬A |
| 3 | A | | A | |
| 4 | B | | 1, 3 →E | |
| 5 | ¬B | | 2 R | |
| 6 | ¬A | | 3−5 ¬I | |

**3.**

| | | | |
|---|---|---|---|
| 1 | A → (D ∨ F) | P | |
| 2 | G ↔ ¬D | P | |
| 3 | F → ¬G | P | ⊢ A → ¬G |
| 4 | A | A | |
| 5 | D ∨ F | 1, 4 →E | |
| 6 | F | A | |
| 7 | ¬G | 3, 6 →E | |
| 8 | D | A | |
| 9 | G | A | |
| 10 | ¬D | 2, 9 ↔E | |
| 11 | D | 8 R | |
| 12 | ¬G | 9−11 ¬I | |
| 13 | ¬G | 5, 6−7, 8−12 ∨E | |
| 14 | A → ¬G | 4−13 →I | |

**4.**

| | | | |
|---|---|---|---|
| 1 | (A ∧ D) → E | P | |
| 2 | E → (A → D) | P | ⊢ A → (D ↔ E) |
| 3 | A | A | |
| 4 | D | A | |
| 5 | A ∧ D | 3, 4 ∧I | |
| 6 | E | 1, 5 →E | |
| 7 | E | A | |
| 8 | A → D | 2, 7 →E | |
| 9 | D | 3, 8 →E | |
| 10 | D ↔ E | 4−6, 7−9 ↔I | |
| 11 | A → (D ↔ E) | 3−10 →I | |

**D.  1.**

|   |   |   | $\vdash \neg(A \wedge \neg A)$ |
|---|---|---|---|
| 1 | | $A \wedge \neg A$ | A |
| 2 | | A | $1 \wedge E$ |
| 3 | | $\neg A$ | $1 \wedge E$ |
| 4 | $\neg(A \wedge \neg A)$ | | $1-3 \neg I$ |

**2.**

|   |   |   | $\vdash A \rightarrow (B \rightarrow A)$ |
|---|---|---|---|
| 1 | | A | A |
| 2 | | | B | A |
| 3 | | | A | 1 R |
| 4 | | $B \rightarrow A$ | $2-3 \rightarrow I$ |
| 5 | $A \rightarrow (B \rightarrow A)$ | | $1-4 \rightarrow I$ |

**3.**

|   |   | $\vdash (B \rightarrow C) \rightarrow ((A \rightarrow B) \rightarrow (A \rightarrow C))$ |
|---|---|---|
| 1 | $B \rightarrow C$ | A |
| 2 | $A \rightarrow B$ | A |
| 3 | A | A |
| 4 | B | $2, 3 \rightarrow E$ |
| 5 | C | $1, 4 \rightarrow E$ |
| 6 | $A \rightarrow C$ | $3-5 \rightarrow I$ |
| 7 | $(A \rightarrow B) \rightarrow (A \rightarrow C)$ | $2-6 \rightarrow I$ |
| 8 | $(B \rightarrow C) \rightarrow ((A \rightarrow B) \rightarrow (A \rightarrow C))$ | $1-7 \rightarrow I$ |

**4.**

|   |   | $\vdash D \vee \neg D$ |
|---|---|---|
| 1 | $\neg(D \vee \neg D)$ | A |
| 2 | D | A |
| 3 | $D \vee \neg D$ | $2 \vee I$ |
| 4 | $\neg(D \vee \neg D)$ | 1 R |
| 5 | $\neg D$ | $2-4 \neg I$ |
| 6 | $D \vee \neg D$ | $5 \vee I$ |
| 7 | $\neg(D \vee \neg D)$ | 1 R |
| 8 | $D \vee \neg D$ | $1-7 \neg E$ |

**E.    1.**

| | | | |
|---|---|---|---|
| 1 | A → B | P | ⊢ ¬B → ¬A |
| 2 | ¬B | A | |
| 3 | A | A | |
| 4 | B | 1, 3 →E | |
| 5 | ¬B | 2 R | |
| 6 | ¬A | 3–5 ¬I | |
| 7 | ¬B → ¬A | 2–6 →I | |

| | | | |
|---|---|---|---|
| 1 | ¬B → ¬A | P | ⊢ A → B |
| 2 | A | A | |
| 3 | ¬B | A | |
| 4 | ¬A | 1, 3 →E | |
| 5 | A | 2 R | |
| 6 | B | 3–5 ¬E | |
| 7 | A → B | 2–6 →I | |

**2.**

| | | | |
|---|---|---|---|
| 1 | A | P | ⊢ ¬¬A |
| 2 | ¬A | A | |
| 3 | A | 1 R | |
| 4 | ¬A | 2 R | |
| 5 | ¬¬A | 2–4 ¬I | |

| | | | |
|---|---|---|---|
| 1 | ¬¬A | P | ⊢ A |
| 2 | ¬A | A | |
| 3 | ¬A | 2 R | |
| 4 | ¬¬A | 1 R | |
| 5 | A | 2–4 ¬E | |

**3.**

| | | | |
|---|---|---|---|
| 1 | A → (B → C) | P | ⊢ (A ∧ B) → C |
| 2 | A ∧ B | A | |
| 3 | A | 2 ∧E | |
| 4 | B | 2 ∧E | |
| 5 | B → C | 1, 3 →E | |
| 6 | C | 4, 5 →E | |
| 7 | (A ∧ B) → C | 2–6 →I | |

| 1 | $(A \wedge B) \rightarrow C$ | P | $\vdash A \rightarrow (B \rightarrow C)$ |
|---|---|---|---|
| 2 | A | A | |
| 3 | B | A | |
| 4 | $A \wedge B$ | 2, 3 $\wedge$I | |
| 5 | C | 1, 4 $\rightarrow$E | |
| 6 | $B \rightarrow C$ | 3–5 $\rightarrow$I | |
| 7 | $A \rightarrow (B \rightarrow C)$ | 2–6 $\rightarrow$I | |

**4.**

| 1 | $\neg A \vee \neg B$ | P | $\vdash \neg(A \wedge B)$ |
|---|---|---|---|
| 2 | $\neg A$ | A | |
| 3 | $A \wedge B$ | A | |
| 4 | A | 3 $\wedge$E | |
| 5 | $\neg A$ | 2 R | |
| 6 | $\neg(A \wedge B)$ | 3–5 $\neg$I | |
| 7 | $\neg B$ | A | |
| 8 | $A \wedge B$ | A | |
| 9 | B | 8 $\wedge$E | |
| 10 | $\neg B$ | 7 R | |
| 11 | $\neg(A \wedge B)$ | 8–10 $\neg$I | |
| 12 | $\neg(A \wedge B)$ | 1, 2–6, 7–11 $\vee$E | |

| 1 | $\neg(A \wedge B)$ | P | $\vdash \neg A \vee \neg B$ |
|---|---|---|---|
| 2 | $\neg(\neg A \vee \neg B)$ | A | |
| 3 | A | A | |
| 4 | B | A | |
| 5 | $A \wedge B$ | 3, 4 $\wedge$I | |
| 6 | $\neg(A \wedge B)$ | 1 R | |
| 7 | $\neg B$ | 4–6 $\neg$I | |
| 8 | $\neg A \vee \neg B$ | 7 $\vee$I | |
| 9 | $\neg(\neg A \vee \neg B)$ | 2 R | |
| 10 | $\neg A$ | 3–9 $\neg$I | |
| 11 | $\neg A \vee \neg B$ | 10 $\vee$I | |
| 12 | $\neg(\neg A \vee \neg B)$ | 2 R | |
| 13 | $\neg A \vee \neg B$ | 2–12 $\neg$E | |

**F.**  **1.**

| | | |
|---|---|---|
| 1 | C ∧ A | P |
| 2 | B ↔ A | P |
| 3 | B → ¬C | P          ⊢ ℙ, ¬ℙ |
| 4 | A | 1 ∧E |
| 5 | B | 2, 4 ↔E |
| 6 | ¬C | 3, 5 →E |
| 7 | C | 1 ∧E |

**2.**

| | | |
|---|---|---|
| 1 | (¬C ∨ ¬D) → ¬E | P |
| 2 | ¬D ∧ E | P          ⊢ ℙ, ¬ℙ |
| 3 | ¬D | 2 ∧E |
| 4 | ¬C ∨ ¬D | 3 ∨I |
| 5 | ¬E | 1, 4 →E |
| 6 | E | 2 ∧E |

**3.**

| | | |
|---|---|---|
| 1 | S ∨ T | P |
| 2 | ¬T | P |
| 3 | ¬S | P          ⊢ ℙ, ¬ℙ |
| 4 | T | A |
| 5 | T | 4 R |
| 6 | S | A |
| 7 | ¬T | A |
| 8 | ¬S | 3 R |
| 9 | S | 6 R |
| 10 | T | 7–9 ¬E |
| 11 | T | 1, 4–5, 6–10 ∨E |
| 12 | ¬T | 2 R |

**4.**

| | | |
|---|---|---|
| 1 | A ↔ ¬A | P          ⊢ ℙ, ¬ℙ |
| 2 | A | A |
| 3 | ¬A | 1, 2 ↔E |
| 4 | A | 2 R |
| 5 | ¬A | 2–4 ¬I |
| 6 | A | 1, 5 ↔E |

# A.12  **Answers to 4.4.2**

**1.**  
1 | $C \vee \neg B$     P  
2 | $(\neg A \vee B) \wedge \neg C$     P     $\vdash \neg A$  
3 | $\neg C$     2 $\wedge$E  
4 | $\neg B$     1, 3 DS  
5 | $\neg A \vee B$     2 $\wedge$E  
6 | $\neg A$     4, 5 DS  

**2.**  
1 | $J \rightarrow L$     P  
2 | $K \rightarrow J$     P  
3 | $L \rightarrow M$     P     $\vdash \neg M \rightarrow \neg K$  
4 | $K \rightarrow L$     1, 2 HS  
5 | $K \rightarrow M$     3, 4 HS  
6 | $\neg M$     A  
7 | $\neg K$     5, 6 MT  
8 | $\neg M \rightarrow \neg K$     6–7 $\rightarrow$I  

**3.**  
1 | $(G \vee H) \wedge \neg J$     P  
2 | $(\neg I \rightarrow J) \wedge (H \rightarrow \neg I)$     P     $\vdash G$  
3 | $G \vee H$     1 $\wedge$E  
4 | $\neg J$     1 $\wedge$E  
5 | $\neg I \rightarrow J$     2 $\wedge$E  
6 | $\neg \neg I$     4, 5 MT  
7 | $H \rightarrow \neg I$     2 $\wedge$E  
8 | $\neg H$     6, 7 MT  
9 | $G$     3, 8 DS

**4.**

| | | | |
|---|---|---|---|
| 1 | K→C | P | |
| 2 | H→(S∨T) | P | |
| 3 | S→K | P | ⊢H→(C∨T) |
| 4 | H | A | |
| 5 | S∨T | 2, 4 →E | |
| 6 | T | A | |
| 7 | C∨T | 6 ∨I | |
| 8 | S | A | |
| 9 | S→C | 1, 3 HS | |
| 10 | C | 8, 9 →E | |
| 11 | C∨T | 10 ∨I | |
| 12 | C∨T | 5, 6−7, 8−11 ∨E | |
| 13 | H→(C∨T) | 4−12 →I | |

## A.13 **Answers to 4.4.4**

**A.**

**1.**

| | | | |
|---|---|---|---|
| 1 | H∨(G∧F) | P | ⊢ (F∧G)∨H |
| 2 | H∨(F∧G) | 1 Com | |
| 3 | (F∧G)∨H | 2 Com | |

**2.**

| | | | |
|---|---|---|---|
| 1 | A→B | P | ⊢ B∨¬A |
| 2 | ¬A∨B | 1 Impl | |
| 3 | B∨¬A | 2 Com | |

**3.**

| | | | |
|---|---|---|---|
| 1 | ¬A→B | P | |
| 2 | ¬B | P | ⊢ A |
| 3 | ¬¬A | 1, 2 MT | |
| 4 | A | 3 DN | |

**4.**

| | | | |
|---|---|---|---|
| 1 | ¬(K∧J) | P | |
| 2 | J | P | ⊢ ¬K |
| 3 | ¬K∨¬J | 1 DeM | |
| 4 | ¬¬J | 2 DN | |
| 5 | ¬K | 3, 4 DS | |

**5.** 

| 1 | $(A \lor C) \lor B$ | P | |
|---|---|---|---|
| 2 | $\neg A$ | P | |
| 3 | $\neg C$ | P | $\vdash B$ |
| 4 | $A \lor (C \lor B)$ | 1 Assoc | |
| 5 | $C \lor B$ | 2, 4 DS | |
| 6 | B | 3, 5 DS | |

**6.**

| 1 | $(Z \land Z) \lor Z$ | P | $\vdash Z$ |
|---|---|---|---|
| 2 | $Z \lor Z$ | 1 Idem | |
| 3 | Z | 2 Idem | |

**7.**

| 1 | $I \rightarrow (J \land K)$ | P | |
|---|---|---|---|
| 2 | $J \rightarrow (K \rightarrow L)$ | P | $\vdash I \rightarrow L$ |
| 3 | $(J \land K) \rightarrow L$ | 2 Exp | |
| 4 | $I \rightarrow L$ | 1, 3 HS | |

**8.**

| 1 | $F \leftrightarrow G$ | P | |
|---|---|---|---|
| 2 | $\neg(F \land G)$ | P | $\vdash \neg F \land \neg G$ |
| 3 | $(F \land G) \lor (\neg F \land \neg G)$ | 1 Equiv | |
| 4 | $\neg F \land \neg G$ | 2, 3 DS | |

**9.**

| 1 | $F \rightarrow G$ | P | |
|---|---|---|---|
| 2 | $\neg F \rightarrow \neg H$ | P | $\vdash \neg G \rightarrow \neg H$ |
| 3 | $\neg G \rightarrow \neg F$ | 1 Trans | |
| 4 | $\neg G \rightarrow \neg H$ | 2, 3 HS | |

**10.**

| 1 | $(A \lor B) \land (A \lor C)$ | P | |
|---|---|---|---|
| 2 | $\neg(B \land C)$ | P | $\vdash A$ |
| 3 | $A \lor (B \land C)$ | 1 Dist | |
| 4 | A | 2, 3 DS | |

**B.** **1.**

| 1 | $(A \land B) \rightarrow C$ | P | |
|---|---|---|---|
| 2 | $C \rightarrow D$ | P | |
| 3 | $\neg(D \lor E)$ | P | $\vdash \neg A \lor \neg B$ |
| 4 | $\neg D \land \neg E$ | 3 DeM | |
| 5 | $\neg D$ | 4 $\land$E | |
| 6 | $\neg C$ | 2, 5 MT | |
| 7 | $\neg(A \land B)$ | 1, 6 MT | |
| 8 | $\neg A \lor \neg B$ | 7 DeM | |

**2.**

| | | | |
|---|---|---|---|
| 1 | ¬C | P | ⊢ ¬(B ∧ C) |
| 2 | ¬B ∨ ¬C | 1 ∨I | |
| 3 | ¬(B ∧ C) | 2 DeM | |

**3.**

| | | | |
|---|---|---|---|
| 1 | (A ∨ B) → (C ∧ D) | P | |
| 2 | C → ¬D | P | ⊢ ¬B |
| 3 | ¬C ∨ ¬D | 2 Impl | |
| 4 | ¬(C ∧ D) | 3 DeM | |
| 5 | ¬(A ∨ B) | 1, 4 MT | |
| 6 | ¬A ∧ ¬B | 5 DeM | |
| 7 | ¬B | 6 ∧E | |

**4.**

| | | | |
|---|---|---|---|
| 1 | F → G | P | |
| 2 | H ↔ G | P | |
| 3 | I ∨ ¬H | P | ⊢ ¬F ∨ I |
| 4 | (H → G) ∧ (G → H) | 2 Equiv | |
| 5 | G → H | 4 ∧E | |
| 6 | F → H | 1, 5 HS | |
| 7 | ¬H ∨ I | 3 Com | |
| 8 | H → I | 7 Impl | |
| 9 | F → I | 6, 8 HS | |
| 10 | ¬F ∨ I | 9 Impl | |

**C.**   **1.**

| | | | |
|---|---|---|---|
| 1 | D → (E → F) | P | |
| 2 | ¬F | P | ⊢ ¬E ∨ ¬D |
| 3 | (D ∧ E) → F | 1 Exp | |
| 4 | ¬(D ∧ E) | 2, 3 MT | |
| 5 | ¬D ∨ ¬E | 4 DeM | |
| 6 | ¬E ∨ ¬D | 5 Com | |

**2.**

| | | | |
|---|---|---|---|
| 1 | ¬(D ∨ ¬A) | P | |
| 2 | (¬A ∨ ¬B) ∨ (¬A ∨ ¬C) | P | ⊢ ¬(B ∧ C) |
| 3 | ¬D ∧ ¬¬A | 1 DeM | |
| 4 | ¬¬A | 3 ∧E | |
| 5 | ¬A ∨ (¬B ∨ (¬A ∨ ¬C)) | 2 Assoc | |
| 6 | ¬B ∨ (¬A ∨ ¬C) | 4, 5 DS | |
| 7 | ¬B ∨ (¬C ∨ ¬A) | 6 Com | |
| 8 | (¬B ∨ ¬C) ∨ ¬A | 7 Assoc | |
| 9 | ¬B ∨ ¬C | 4, 8 DS | |
| 10 | ¬(B ∧ C) | 9 DeM | |

**3.**

| | | |
|---|---|---|
| 1 | $\neg J \wedge \neg K$ | P |
| 2 | $(L \vee K) \vee (J \wedge M)$ | P    $\vdash L$ |
| 3 | $\neg J$ | 1 $\wedge$E |
| 4 | $\neg J \vee \neg M$ | 3 $\vee$I |
| 5 | $\neg (J \wedge M)$ | 4 DeM |
| 6 | $L \vee K$ | 2, 5 DS |
| 7 | $\neg K$ | 1 $\wedge$E |
| 8 | L | 6, 7 DS |

**4.**

| | | |
|---|---|---|
| 1 | $C \vee D$ | P |
| 2 | $(\neg E \wedge \neg H) \vee F$ | P |
| 3 | $\neg D \vee E$ | P |
| 4 | $\neg F \vee G$ | P    $\vdash C \vee G$ |
| 5 | $\neg \neg C \vee D$ | 1 DN |
| 6 | $\neg C \rightarrow D$ | 5 Impl |
| 7 | $\neg (E \vee H) \vee F$ | 2 DeM |
| 8 | $(E \vee H) \rightarrow F$ | 7 Impl |
| 9 | $(\neg D \vee E) \vee H$ | 3 $\vee$I |
| 10 | $\neg D \vee (E \vee H)$ | 9 Assoc |
| 11 | $D \rightarrow (E \vee H)$ | 10 Impl |
| 12 | $F \rightarrow G$ | 4 Impl |
| 13 | $\neg C \rightarrow (E \vee H)$ | 6, 11 HS |
| 14 | $\neg C \rightarrow F$ | 8, 13 HS |
| 15 | $\neg C \rightarrow G$ | 12, 14 HS |
| 16 | $\neg \neg C \vee G$ | 15 Impl |
| 17 | $C \vee G$ | 16 DN |

**D.**    **1.**

                                  $\vdash A \vee \neg A$

| | | |
|---|---|---|
| 1 | $\neg (A \vee \neg A)$ | A |
| 2 | $\neg A \wedge \neg \neg A$ | 1 DeM |
| 3 | $\neg A$ | 2 $\wedge$E |
| 4 | $\neg \neg A$ | 2 $\wedge$E |
| 5 | $A \vee \neg A$ | 1–4 $\neg$E |

**2.**                                                              ⊢ (B → C) ↔ ¬(B ∧ ¬C)

| 1 | B → C | A |
|---|---|---|
| 2 | ¬B ∨ C | 1 Impl |
| 3 | ¬B ∨ ¬¬C | 2 DN |
| 4 | ¬(B ∧ ¬C) | 3 DeM |
| 5 | ¬(B ∧ ¬C) | A |
| 6 | ¬B ∨ ¬¬C | 5 DeM |
| 7 | ¬B ∨ C | 6 DN |
| 8 | B → C | 7 Impl |
| 9 | (B → C) ↔ ¬(B ∧ ¬C) | 1–4, 5–8 ↔I |

**E.   1.**

| 1 | H ∧ ¬G | P | ⊢ ¬(¬H ∨ G) |
|---|---|---|---|
| 2 | ¬¬H ∧ ¬G | 1 DN | |
| 3 | ¬(¬H ∨ G) | 2 DeM | |

| 1 | ¬(¬H ∨ G) | P | ⊢ H ∧ ¬G |
|---|---|---|---|
| 2 | ¬¬H ∧ ¬G | 1 DeM | |
| 3 | H ∧ ¬G | 2 DN | |

**2.**

| 1 | ¬(D ↔ E) | P | ⊢ (¬D ∧ E) ∨ (D ∧ ¬E) |
|---|---|---|---|
| 2 | ¬((D → E) ∧ (E → D)) | 1 Equiv | |
| 3 | ¬((¬D ∨ E) ∧ (E → D)) | 2 Impl | |
| 4 | ¬((¬D ∨ E) ∧ (¬E ∨ D)) | 3 Impl | |
| 5 | ¬(¬D ∨ E) ∨ ¬(¬E ∨ D) | 4 DeM | |
| 6 | (¬¬D ∧ ¬E) ∨ ¬(¬E ∨ D) | 5 DeM | |
| 7 | (¬¬D ∧ ¬E) ∨ (¬¬E ∧ ¬D) | 6 DeM | |
| 8 | (D ∧ ¬E) ∨ (¬¬E ∧ ¬D) | 7 DN | |
| 9 | (D ∧ ¬E) ∨ (E ∧ ¬D) | 8 DN | |
| 10 | (E ∧ ¬D) ∨ (D ∧ ¬E) | 9 Com | |
| 11 | (¬D ∧ E) ∨ (D ∧ ¬E) | 10 Com | |

| 1 | $(\neg D \wedge E) \vee (D \wedge \neg E)$ | P | $\vdash \neg(D \leftrightarrow E)$ |
|---|---|---|---|
| 2 | $\neg\neg((\neg D \wedge E) \vee (D \wedge \neg E))$ | 1 DN | |
| 3 | $\neg(\neg(\neg D \wedge E) \wedge \neg(D \wedge \neg E))$ | 2 DeM | |
| 4 | $\neg((\neg\neg D \vee \neg E) \wedge \neg(D \wedge \neg E))$ | 3 DeM | |
| 5 | $\neg((\neg\neg D \vee \neg E) \wedge (\neg D \vee \neg\neg E))$ | 4 DeM | |
| 6 | $\neg((D \vee \neg E) \wedge (\neg D \vee \neg\neg E))$ | 5 DN | |
| 7 | $\neg((D \vee \neg E) \wedge (\neg D \vee E))$ | 6 DN | |
| 8 | $\neg((\neg E \vee D) \wedge (\neg D \vee E))$ | 7 Com | |
| 9 | $\neg((E \rightarrow D) \wedge (\neg D \vee E))$ | 8 Impl | |
| 10 | $\neg((E \rightarrow D) \wedge (D \rightarrow E))$ | 9 Impl | |
| 11 | $\neg((D \rightarrow E) \wedge (E \rightarrow D))$ | 10 Com | |
| 12 | $\neg(D \leftrightarrow E)$ | 11 Equiv | |

**F.  1.**

| 1 | $(L \vee \neg M) \wedge (L \vee N)$ | P | |
|---|---|---|---|
| 2 | $J \leftrightarrow (N \wedge \neg M)$ | P | |
| 3 | $\neg L \wedge \neg J$ | P | $\vdash \mathbb{P}, \neg\mathbb{P}$ |
| 4 | $L \vee (\neg M \wedge N)$ | 1 Dist | |
| 5 | $\neg L$ | 3 $\wedge$E | |
| 6 | $\neg M \wedge N$ | 4, 5 DS | |
| 7 | $N \wedge \neg M$ | 6 Com | |
| 8 | $J$ | 2, 7 $\leftrightarrow$E | |
| 9 | $\neg J$ | 3 $\wedge$E | |

**2.**

| 1 | $(\neg A \vee \neg B) \leftrightarrow (\neg C \vee \neg D)$ | P | |
|---|---|---|---|
| 2 | $(C \wedge D) \wedge (\neg B \vee \neg C)$ | P | $\vdash \mathbb{P}, \neg\mathbb{P}$ |
| 3 | $C \wedge D$ | 2 $\wedge$E | |
| 4 | $C$ | 3 $\wedge$E | |
| 5 | $\neg\neg C$ | 4 DN | |
| 6 | $\neg B \vee \neg C$ | 2 $\wedge$E | |
| 7 | $\neg B$ | 5, 6 DS | |
| 8 | $\neg A \vee \neg B$ | 7 $\vee$I | |
| 9 | $\neg C \vee \neg D$ | 1, 8 $\leftrightarrow$E | |
| 10 | $\neg D$ | 5, 9 DS | |
| 11 | $D$ | 3 $\wedge$E | |

# A.14  **Answers to 5.2.3**

**1.** Not a wff. The variable in the quantifier does not appear in the remainder of the expression, which violates clause (3) of the definition of wff of *P* (155).

**2.**

$$Fx\underline{z}$$
$$|$$
$$(\forall z)Fx\underline{z} \quad \text{Open, 'x' free.}$$

**3.**

$$Fxz$$
$$|$$
$$(\forall z)Fxz$$
$$|$$
$$(\forall x)(\forall z)Fxz \quad \text{Closed.}$$

**4.**

$$Aw$$
$$|$$
$$(\forall w)Aw \quad Kjw$$
$$\smile$$
$$(\forall w)Aw \land Kj\underline{w} \quad \text{Open, 'w' free.}$$

**5.**

$$Aw \quad Kjw$$
$$\smile$$
$$Aw \land Kjw$$
$$|$$
$$(\forall w)(Aw \land Kjw) \quad \text{Closed.}$$

**6.**

$$Aw \quad Kx\underline{w}$$
$$\smile$$
$$Aw \land Kxw$$
$$|$$
$$(\forall w)(Aw \land K\underline{x}w) \quad \text{Open, 'x' free.}$$

**7.**

$$Kxw$$
$$|$$
$$Aw \quad (\exists x)Kxw$$
$$\smile$$
$$Aw \land (\exists x)Kxw$$
$$|$$
$$(\forall w)(Aw \land (\exists x)Kxw) \quad \text{Closed.}$$

**8.**

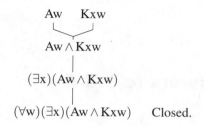

$$(\forall w)(\exists x)(Aw \wedge Kxw) \quad \text{Closed.}$$

## A.15  **Answers to 5.3.2**

1. $Gf \wedge Wf$

2. $Bfs \rightarrow Mfs$

3. $(\forall x)Wx$

4. $(\exists x)Wx$

5. $(\exists x)\neg Wx$
   $\neg(\forall x)Wx$

6. $(\forall x)\neg Wx$
   $\neg(\exists x)Wx$

7. $\neg(\forall y)Oy$
   $(\exists y)\neg Oy$

8. $(\forall y)\neg Oy$
   $\neg(\exists y)Oy$

9. $(\forall z)(Mz \vee Wz)$

10. $(\forall z)Mz \vee (\forall z)Wz$

11. $(\exists x)Wx \rightarrow (\exists x)Tx$

12. $(\exists x)(Ox \wedge \neg Tx)$

13. $Mft \rightarrow \neg(\exists z)Mzf$

14. $\neg Lbs \leftrightarrow \neg(\exists x)Lxs$

15. $Lsb \rightarrow (\forall w)Lwb$

16. $\neg Ltt \rightarrow (\exists x)\neg Lxx$

17. $(\exists x)Lxf \wedge \neg(\forall y)Lyf$

**18.** $(\forall z)Lzz$

**19.** $(\forall z)Lzz \rightarrow (\exists x)Lxf$

**20.** $\neg(\exists x)\neg Tx$

# A.16 **Answers to 5.3.4**

**1.** $(\forall x)Ax$

**2.** $(\forall x)Lxk$

**3.** $(\forall z)(Az \rightarrow Dz)$

**4.** $(\forall z)(Az \rightarrow \neg Dz)$
   $\neg(\exists z)(Az \wedge Dz)$

**5.** $(\exists z)(Az \wedge Dz)$

**6.** $(\exists z)(Az \wedge \neg Dz)$
   $\neg(\forall z)(Az \rightarrow Dz)$

**7.** $\neg(\forall z)(Az \rightarrow Dz)$

**8.** $(\forall w)(Cw \rightarrow \neg Dw)$

**9.** $(\exists x)(Cx \wedge \neg Dx)$

**10.** $(\forall y)(Wyk \rightarrow Lyk)$

**11.** $(\forall y)[(Wyk \vee \neg Lyf) \rightarrow Lyk]$

**12.** $\neg(\forall x)(Lxk \rightarrow Wxk)$
   $(\exists x)(Lxk \wedge \neg Wxk)$

**13.** $\neg(\exists z)(Wzf \wedge Lzf)$
   $(\forall z)(Wzf \rightarrow \neg Lzf)$

**14.** $(\exists x)[Ax \wedge (Lxf \wedge Lxk)]$

**15.** $(\exists x)[(Ax \wedge Dx) \wedge (Lxf \wedge \neg Lxk)]$

**16.** $(\exists x)(Ax \wedge Wxf) \wedge (\exists x)(Cx \wedge Wxf)$

**17.** $(\exists x)[(Ax \wedge Cx) \wedge Wxf]$

**18.** $\neg(\forall z)(Wzk \rightarrow Lkz)$

**19.** $\neg(\exists z)(Wzk \wedge Lkz)$
   $(\forall z)(Wzk \rightarrow \neg Lkz)$

**20.** $(\forall w)[Lkw \rightarrow (Uw \wedge Dw)]$
$\neg(\exists w)[Lkw \wedge \neg(Uw \wedge Dw)]$

**21.** $(\forall x)(Lxk \rightarrow Wxk)$
$\neg(\exists x)(Lxk \wedge \neg Wxk)$

**22.** $(\forall x)[(Ax \vee Cx) \rightarrow Lxk]$

**23.** $(\forall x)[(Ax \wedge Dx) \rightarrow Wxk] \wedge \neg(\exists x)[(Ax \wedge Lx) \wedge Wxk]$
$(\forall x)([(Ax \wedge Dx) \rightarrow Wxk] \wedge [(Ax \wedge Lx) \rightarrow \neg Wxk])$

**24.** $(\forall z)([(Az \wedge Uz) \wedge \neg Dz] \rightarrow [Wzk \vee Lfz])$

**25.** $(\forall x)[(Ax \wedge \neg Dx) \rightarrow \neg Wxk]$

**26.** $(\forall x)[([Ax \vee Cx] \wedge \neg Dx) \rightarrow \neg Wxk]$

**27.** $(\forall x)[([Ax \vee Cx] \wedge \neg[Dx \vee Ux]) \rightarrow \neg Wxk]$

**28.** $(\forall x)[([Ax \vee Cx] \wedge \neg[Dx \vee Ux]) \rightarrow (\neg Wxk \vee \neg Lkx)]$

**29.** $[(Cf \wedge Lf) \wedge Lff] \rightarrow (\exists w)[(Cw \wedge Lw) \wedge Lfw]$

**30.** $Wfk \rightarrow (\forall z)(Wzf \rightarrow Wzk)$

**31.** $(\forall x)(Wxk \rightarrow Lxk) \wedge (\forall x)(Lxk \rightarrow Wxk)$

**32.** $(\forall x)[(Wxk \rightarrow Lxk) \wedge (Lxk \rightarrow Wxk)]$

**33.** $(\forall x)(Wxk \leftrightarrow Lxk)$

Note: **31.**–**33.** are quantificationally equivalent.

# A.17 **Answers to 5.4.2**

**1.** $(\forall x)(\forall y)Lxy$

**2.** $(\exists x)(\exists y)Lxy$

**3.** $(\forall x)(\exists y)Lxy$

**4.** $(\forall x)(\exists y)Lyx$

**5.** $(\exists x)(\forall y)Lxy$

**6.** $(\exists x)(\forall y)Lyx$

**7.** $(\exists x)(\forall y)\neg Lyx$
$(\exists x)\neg(\exists y)Lyx$

**8.** $\neg(\exists x)(\forall y)\neg Lxy$
$\neg(\exists x)\neg(\exists y)Lxy$
$(\forall x)(\exists y)Lxy$

**9.** $(\exists x)\neg(\exists y)Lxy$
$(\exists x)(\forall y)\neg Lxy$

**10.** $\neg(\exists x)(\forall y)Lxy$
$(\forall x)\neg(\forall y)Lxy$

**11.** $(\exists x)\neg(\exists y)Lyx$
$(\exists x)(\forall y)\neg Lyx$

**12.** $\neg(\exists x)(\forall y)Lyx$
$(\forall x)(\exists y)\neg Lyx$

**13.** $(\exists x)(\exists y)\neg Lxy$
$(\exists x)\neg(\forall y)Lxy$

## A.18 **Answers to 5.4.4**

**1.** $(\exists x)(\exists y)(x \neq y \wedge (Cx \wedge Cy))$

**2.** $(\exists x)(\exists y)(x \neq y \wedge ((Cx \wedge Wxk) \wedge (Cy \wedge Wyk)))$

**3.** $(\exists x)(Cx \wedge (\forall y)(Cy \rightarrow x = y))$

**4.** $(\exists x)((Cx \wedge Wxk) \wedge (\forall y)((Cy \wedge Wyk) \rightarrow x = y))$

**5.** $(\exists x)(([Ux \wedge Cx] \wedge Wxk) \wedge (\forall y)(([Uy \wedge Cy] \wedge Wyk) \rightarrow x = y))$

**6.** $(\exists x)([(Ux \wedge Cx) \wedge (\forall y)((Uy \wedge Cy) \rightarrow x = y)] \wedge Wxk)$

**7.** $(\exists x)(\exists y)([x \neq y \wedge (Ax \wedge Ay)] \wedge (\forall z)(Az \rightarrow [z = x \vee z = y]))$

**8.** $(\exists x)(\exists y)([x \neq y \wedge ([Ax \wedge Wxk] \wedge [Ay \wedge Wyk])] \wedge (\forall z)([Az \wedge Wzk] \rightarrow [z = x \vee z = y]))$

**9.** $(\forall x)(x \neq f \rightarrow Wxk)$
$(\forall x)(x \neq f \rightarrow Wxk) \wedge \neg Wfk$

Note: these two answers are not equivalent, but represent two different ways of understanding the English. The first does not say whether or not Fred works for Kate, the second explicitly denies that he does.

**10.** $(\exists x)([Cx \wedge (\forall y)(Cy \rightarrow x = y)] \wedge Lx)$

**11.** $(\exists x)([(Cx \wedge Wxf) \wedge (\forall y)((Cy \wedge Wyf) \rightarrow x = y)] \wedge Lx)$

**12.** $(\exists x)([([Cx \wedge Lx] \wedge Wxf) \wedge (\forall y)(([Cy \wedge Ly] \wedge Wyf) \rightarrow x = y)] \wedge \neg Ux)$

**13.** $[(Df \wedge Af) \wedge Wfk] \wedge (\forall y)([(Dy \wedge Ay) \wedge Wyk] \rightarrow y = f)$

**14.** $(\exists x)([(Dx \wedge Ax) \wedge Wxk] \wedge (\forall y)([(Dy \wedge Ay) \wedge Wyk] \rightarrow y = x))$

**15.** $(\exists x)(Ax \wedge Wxk)$

**16.** $(\exists x)(\exists y)(x \neq y \wedge [(Ax \wedge Wxk) \wedge (Ay \wedge Wyk)])$

**17.** $(\exists x)([Ax \wedge Wxk] \wedge (\forall y)([Ay \wedge Wyk] \rightarrow x = y))$

**18.** $(\exists x)((Dx \wedge Ax) \wedge Wxk)$

**19.** $(\exists x)(\exists y)(x \neq y \wedge [((Dx \wedge Ax) \wedge Wxk) \wedge ((Dy \wedge Ay) \wedge Wyk)])$

**20.** $(\exists x)([(Dx \wedge Ax) \wedge Wxk] \wedge (\forall y)([(Dy \wedge Ay) \wedge Wyk] \rightarrow x = y))$

**21.** $(\exists x)[([Dx \wedge Ax] \wedge (\forall y)([Dy \wedge Ay] \rightarrow x = y)) \wedge Wxk]$

**22.** $(\exists x)[([Dx \wedge Ax] \wedge (\forall y)([Dy \wedge Ay] \rightarrow x = y)) \wedge Wxk]$

**23.** $(\exists x)[([(Dx \wedge Ax) \wedge Wxk] \wedge (\forall y)([(Dy \wedge Ay) \wedge Wyk] \rightarrow x = y)) \wedge Ux]$

**24.** $(\exists x)[([Dx \wedge Ax] \wedge (\forall y)([Dy \wedge Ay] \rightarrow x = y)) \wedge \neg(Wxk \vee Wxf)]$

## A.19 **Answers to 6.2.1**

Note: for answers that require models, there are an infinite number of correct answers. Here I give just one each, sometimes with a brief explanation. Further, for answers that require semantic meta-proofs, the proof I give is one of many correct answers.

1. To show the wff is not quantificationally true, we need an interpretation which makes it F by making the antecedent T and the consequent F. Since the antecedent is itself a conditional claim, the easiest way to make it T is to make its antecedent F. That means we need something in the **UD** that is not in the extension of 'K'. Next, to make the main consequent F, we need an object that is in the extension of 'K' but not in 'H'. Thus:

   **UD**: $\{a,b\}$
   K: $\{b\}$
   H: $\{\}$

2. To show the wff is not quantificationally false, we need an interpretation on which the wff is T. Hence one on which each conjunct is T. We simply need an object in the **UD** that is in 'B', an object in the **UD** that is in 'C', but no object that is in both. Thus:

   **UD**: $\{b,c\}$
   B: $\{b\}$
   C: $\{c\}$

**3.** To show the wff is not quantificationally true, we need an interpretation on which the wff is F. Again, we want the main antecedent T while the consequent is F. Thus:

> **UD:** {a,b}
> F: {a}
> G: {a}

**4.** To show the wff is not quantificationally true, we need an interpretation on which the wff is F. Again, we want the main antecedent T while the consequent is F. Thus:

> **UD:** {a,b}
> L: {(a,b),(b,a)}

**5.** To show the wff is quantificationally contingent, we need an interpretation on which the wff is T and an interpretation on which the wff is F. To make it T, it is easiest to make the antecedent F. To make it F, we must make the antecedent T and the consequent F. Thus:

> T:    **UD:** {a}         F:    **UD:** {a}
>       B: {}                     B: {(a,a)}

**6.** The wff is quantificationally true. Suppose the antecedent is T. Then either every object in the **UD** is in 'D' or every object in the **UD** is in 'E'. In each case, every object in the **UD** is either in 'D' or in 'E', making the consequent T as well. Since the consequent cannot be F when the antecedent is T, the wff is T on every interpretation—thus, it is quantificationally true.

**7.** The wff is quantificationally true. Suppose the antecedent is T. Then there is some object that is both in 'D' and in 'E'. Thus, there is both some object in 'D' and some object in 'E', making the consequent T as well. Since the consequent cannot be F when the antecedent is T, the wff is T on every interpretation—thus, it is quantificationally true.

# A.20 **Answers to 6.3.1**

**1.** To show the pair is not quantificationally equivalent, we need an interpretation on which they have different truth values. Thus:

> **UD:** {a,b}
> D: {a}
> E: {b}

Here '$(\forall x)Dx \lor (\forall x)Ex$' is F while '$(\forall x)(Dx \lor Ex)$' is T.

**2.** Again, we must produce different truth values. The previous interpretation suffices:

> **UD:** {a, b}
>   D: {a}
>   E: {b}

Here '$(\exists x)Dx \land (\exists x)Ex$' is T while '$(\exists x)(Dx \land Ex)$' is F.

**3.** Consider:

> **UD:** {a, b}
>   B: {b}
>   C: {}

Here '$(\exists x)Bx \rightarrow Ca$' is F. Object b satisfies 'B', making the antecedent T, while a does not satisfy 'C', making the consequent F. Thus the conditional is F. At the same time, '$(\exists x)(Bx \rightarrow Ca)$' is T. Object a does not satisfy 'B', but because of this a *does* satisfy 'Bx → Ca', thereby making the existential claim, '$(\exists x)(Bx \rightarrow Ca)$', T.

**4.** The previous interpretation suffices:

> **UD:** {a, b}
>   B: {b}
>   C: {}

Here '$(\forall x)Bx \rightarrow Ca$' is T, since the antecedent is false. At the same time, '$(\forall x)(Bx \rightarrow Ca)$' is F. Object b satisfies 'B' while a fails to satisfy 'C'. Thus, b fails to satisfy 'Bx → Ca'. As a result '$(\forall x)(Bx \rightarrow Ca)$' is F.

**5.** Consider:

> **UD:** {a, b, c}
>   D: {a}
>   E: {b}

Every member of the set is T.

**6.** Consider:

> **UD:** {a, b, c}
>   F: {a}
>   G: {a, b}
>   H: {a, b, c}

Every member of the set is T.

**7.** The previous interpretation suffices:

> **UD:** {a,b,c}
>  F: {a}
>  G: {a,b}
>  H: {a,b,c}

Every member of the set is T.

**8.** A slight change to the previous interpretation suffices:

> **UD:** {a,b,c}
>  F: {a}
>  G: {a,b}
>  H: {a,b}

Every member of the set is T.

**9.** Consider:

> **UD:** {a,b,c}
>  D: {a}
>  F: {b}
>  G: {a,b}

Every member of the set is T.

**10.** Consider:

> **UD:** {a}
>  F: {}
>  G: {a}

Every member of the set is T. Since every object in the **UD** (in this case, just a) fails to satisfy 'Fx', every object satisfies 'Fx → Gx'. So '(∀x)(Fx → Gx)' is T. Since no object satisfies 'F', '¬(∃x)Fx' is T. Object a satisfies 'H', so '(∃x)Gx' is T.

**11.** Consider:

> **UD:** {a}
>  F: {}
>  G: {}

Every member of the set is T. Since every object in the **UD** (in this case, just a) fails to satisfy 'Fx', every object satisfies 'Fx → Gx'. So '(∀x)(Fx → Gx)' is T. Since no object satisfies 'G', '¬(∃x)Gx' is T.

**12.** Consider:

> **UD**: $\{a,b\}$
> F: $\{a\}$
> G: $\{a\}$

Every member of the set is T, but the target is F.

**13.** Suppose, first, that the left hand wff (LHW) is T on some interpretation **M**. Then either i) some object **o** in the **UD** satisfies 'B' or ii) some object **m** in the **UD** satisfies 'C' (possibly both, and possibly **o** = **m**, though it doesn't matter). If i), then **o** satisfies 'Bx ∨ Cx', making the right hand wff (RHW) T. If ii), then **m** satisfies 'Bx ∨ Cx', making the right hand wff (RHW) T. So RHW is T on any such interpretation. Suppose, second, that RHW is T on some interpretation **M**. Then some object **o** satisfies 'Bx ∨ Cx', and does so by either i) satisfying 'B', making '(∃x)Bx' T; or ii) by satisfying 'C', making '(∃x)Cx' T (possibly both). In each case, at least one side of LHW is T, making LHW T. Being T on all and only the same interpretations, LHW and RHW are quantificationally equivalent.

**14.** Suppose that the left hand wff (LHW) is T on some interpretation **M**. This is equivalent to both conjuncts being T. So, every object **o** in the **UD** satisfies 'B' and a satisfies 'C'. But that is equivalent to saying that every object in the **UD** satisfies 'Bx ∧ Ca'. This is equivalent to the truth of '(∀x)(Bx ∧ Ca)'. We have proceeded by equivalences, so LHW is equivalent to RHW.

**15.** Suppose, first, that the left hand wff (LHW) is F on some interpretation **M**. Then its antecedent must be T and its consequent F. So there is some **o** in the **UD** that satisfies 'B', and a does not satisfy 'C'. Since **o** satisfies the antecedent 'Bx' and 'Ca' is F, **o** fails to satisfy 'Bx → Ca'. Thus the RHW is F. Suppose, second, that the LHW is T on some interpretation. Then either i) the antecedent is F or ii) the consequent is T. If i), then there is no object in the **UD** that satisfies 'B'. Thus, every object in the **UD** vacuously satisfies 'Bx → Ca' by failing to satisfy 'Bx'. So RHW is T. If ii), then every object in the **UD** vacuously satisfies 'Bx → Ca' because 'Ca' is T. So RHW is T. LHW and RHW have the same truth value on every interpretation, so they are quantificationally equivalent.

**16.** Suppose '(∃y)(∀x)Lxy' is T on some interpretation **M**. So there is at least one object **o** in the **UD**, such that every object **m** in the **UD** bears relation 'L' to **o**. But this means that every **m** in the **UD** bears the relation 'L' to at least that object **o**. So '(∀x)(∃y)Lxy' is T on **M** as well.

## A.21  **Answers to 6.4.1**

**1.** (Similar to question **13.** in Exercises 6.3.1.) Suppose, first, that the left hand wff (LHW) is T on some interpretation **M**. Then either i) some object **o** in the **UD** satisfies 𝔽x or ii) some object **m** in the **UD** satisfies 𝔾x (possibly both, and possibly

**o** = **m**, though it doesn't matter). If i), then **o** satisfies $\mathbb{F}x \vee \mathbb{G}x$, making the right hand wff (RHW) T. If ii), then **m** satisfies $\mathbb{F}x \vee \mathbb{G}x$, making the right hand wff (RHW) T. So RHW is T on any such interpretation. Suppose, second, that RHW is T on some interpretation **M**. Then some object **o** satisfies $\mathbb{F}x \vee \mathbb{G}x$, and does so by either i) satisfying $\mathbb{F}x$, making $(\exists x)\mathbb{F}x$ T; or ii) by satisfying $\mathbb{G}x$, making $(\exists x)\mathbb{G}x$ T (possibly both). In each case, at least one side of LHW is T, making LHW T. Being T on all and only the same interpretations, LHW and RHW are quantificationally equivalent.

2. (Similar to question **14.** in Exercises 6.3.1.) Suppose that LHW is T on some interpretation **M**. This is equivalent to both conjuncts being T. So, every object **o** in the **UD** satisfies $\mathbb{F}x$ and $\mathbb{P}$ is T. But that is equivalent to saying that every object in the **UD** satisfies $\mathbb{F}x \wedge \mathbb{P}$. This is equivalent to the truth of '$(\forall x)(Bx \wedge Ca)$'. We have proceeded by equivalences, so LHW is equivalent to RHW.

3. Suppose, first, LHW is T on some interpretation **M**. So both conjuncts are T. That is, there is some object **o** in the **UD** that satisfies $\mathbb{F}x$ and $\mathbb{P}$ is T (satisfied by every object in the **UD**). But this means there is at least one object **o** in the **UD** that satisfies $\mathbb{F}x \wedge \mathbb{P}$, so the RHW is T. Suppose, second, RHW is T on some interpretation **M**. So, there is at least one object **o** in the **UD** that satisfies $\mathbb{F}x \wedge \mathbb{P}$. So that object, **o** satisfies $\mathbb{F}x$, making $(\exists x)\mathbb{F}x$ T. Object **o** also satisfies $\mathbb{P}$. Since $\mathbb{P}$ contains no free variables, if one object satisfies it, then all do. So $\mathbb{P}$ is also T. So LHW is T.

4. Suppose, first, LHW is T on some interpretation **M**. So at least one disjunct is T. Either i) every object **o** in the **UD** satisfies $\mathbb{F}x$, or ii) $\mathbb{P}$ is T—satisfied by every object in the **UD**, since it contains no free variables. If i), then every object **o** satisfies $\mathbb{F}x \vee \mathbb{P}$ in virtue of every object satisfying the left disjunct. So RHW is T. If ii), then every object **o** satisfies $\mathbb{F}x \vee \mathbb{P}$ in virtue of every object satisfying the right disjunct. So in either case RHW is T. Suppose, second, RHW is T on some interpretation **M**. So, every object **o** in the **UD** satisfies $\mathbb{F}x \vee \mathbb{P}$. That is, for each **o** in the **UD**, either i) **o** satisfies $\mathbb{P}$, or ii) **o** satisfies $\mathbb{F}x$. If i), then every object in the **UD** satisfies $\mathbb{P}$ (some object satisfies a closed wff iff all do). This makes $\mathbb{P}$ T, making LHW T. If ii), but not i), then no object satisfies $\mathbb{P}$ (some object satisfies a closed wff iff all do), so every object must satisfy $\mathbb{F}x$. This makes $(\forall x)\mathbb{F}x$ T, making LHW T.

5. Suppose, first, LHW is T on some interpretation **M**. So either i) $\mathbb{P}$ is F or ii) $(\forall x)\mathbb{F}x$ is T. If i), then no object in the **UD** satisfies $\mathbb{P}$. But that means that every object satisfies $\mathbb{P} \rightarrow \mathbb{F}x$, making RHW T. If ii), then every object satisfies $\mathbb{F}x$, thereby satisfying $\mathbb{P} \rightarrow \mathbb{F}x$. This makes RHW T. Suppose, second, RHW is T on some interpretation **M**. So every object satisfies $\mathbb{P} \rightarrow \mathbb{F}x$. That is, for each **o** in the **UD**, either i) **o** fails to satisfy $\mathbb{P}$, or ii) **o** satisfies $\mathbb{F}x$. If i), then no object in the **UD** satisfies $\mathbb{P}$ (some object satisfies a closed wff iff all do). This makes $\mathbb{P}$ F, making LHW T. If ii), but not i), then every object satisfies $\mathbb{P}$ (some object satisfies a closed wff iff all do), so every object must satisfy $\mathbb{F}x$ (or else RHW would be F, contrary to supposition). This makes $(\forall x)\mathbb{F}x$ T, making LHW T.

6. Suppose, first, LHW is T on some interpretation **M**. So either i) $\mathbb{P}$ is F or ii) $(\exists x)\mathbb{F}x$ is T. If i), then no object in the **UD** satisfies $\mathbb{P}$. But that means that some object satisfies $\mathbb{P} \rightarrow \mathbb{F}x$, making RHW T. If ii), then some object satisfies $\mathbb{F}x$, thereby satisfying $\mathbb{P} \rightarrow \mathbb{F}x$. This makes RHW T. Suppose, second, RHW is T on some interpretation **M**. So some object satisfies $\mathbb{P} \rightarrow \mathbb{F}x$. That is, for some **o** in the **UD**, either i) **o** fails to satisfy $\mathbb{P}$, or ii) **o** satisfies $\mathbb{F}x$. If i), then no object in the **UD** satisfies $\mathbb{P}$ (some object satisfies a closed wff iff all do). This makes $\mathbb{P}$ F, making LHW T. If ii), then $(\exists x)\mathbb{F}x$ is T, making LHW T.

7. (Similar to question **15.** in Exercises 6.3.1.) Suppose, first, that LHW is T on some interpretation. Then either i) $(\exists x)\mathbb{F}x$ is F or ii) $\mathbb{P}$ is T. If i), then there is no object in the **UD** that satisfies $\mathbb{F}x$. So every object in the **UD** vacuously satisfies $\mathbb{F}x \rightarrow \mathbb{P}$. So RHW is T. If ii), then every object in the **UD** satisfies $\mathbb{P}$. But this means every object satisfies $\mathbb{F}x \rightarrow \mathbb{P}$. So RHW is T. Suppose, second, that RHW is T. Then every object **o** in the **UD** satisfies $\mathbb{F}x \rightarrow \mathbb{P}$. That is, for each **o** in the **UD**, either i) **o** fails to satisfy $\mathbb{F}x$, or ii) **o** satisfies $\mathbb{P}$. If ii), then every object in the **UD** satisfies $\mathbb{P}$ (some object satisfies a closed wff iff all do). This makes $\mathbb{P}$ T, making LHW T. If i), but not ii), then no object satisfies $\mathbb{P}$ (some object satisfies a closed wff iff all do), so every object must fail to satisfy $\mathbb{F}x$ (or else RHW would be F, contrary to supposition). This makes $(\exists x)\mathbb{F}x$ F, making LHW T.

# A.22 **Answers to 6.5.1**

A. We will take our **UD** for each of the following to be $\{a, b, c\}$.

1. **Reflexivity and Symmetry but not Transitivity.**

   R: $\{(a,a), (b,b), (c,c), (a,b), (b,a), (b,c), (c,b)\}$

   The first three pairs make this R reflexive. The remaining pairs yield symmetry. Transitivity fails because although we have $(a,b)$ and $(b,c)$, we do not have $(a,c)$. Similarly for $(c,b)$ and $(b,a)$, $(c,a)$ is absent, though only one failure of transitivity is needed.

2. **Reflexivity and Transitivity but not Symmetry.**

   R: $\{(a,a), (b,b), (c,c), (a,b), (b,c), (a,c)\}$

   The first three pairs make this R reflexive. The remaining pairs yeild transitivity. Symmetry fails because none of the non-identical pairs has its symmetric counterpart. E.g., we have $(a,b)$ but not $(b,a)$.

3. **Symmetry and Transitivity but not Reflexivity.**

   R: $\{(a,a), (b,b), (a,b), (b,a)\}$

   This R is not reflexive because $(c,c)$ is not in the relation. It is symmetric and transitive.

**4. Nonreflexivity, Nonsymmetry, and Nontransitivity.**

> R: $\{(a,b),(b,c)\}$

This R is nonreflexive because $(a,a)$ is not in the relation. It is nonsymmetric because $(a,b)$ is in the relation, but $(b,a)$ is not. Its is nontransitive because $(a,b)$ and $(b,c)$ are in the relation, but $(a,c)$ is not.

**B.** We will take our **UD** for each of the following to be $\{a,b,c,d\}$ and define a relation R for each answer:

**1.–4. Reflexivity, Symmetry, Transitivity, Seriality.**

> R: $\{(a,a),(b,b),(c,c),(d,d),(a,b),(b,a)\}$

This R suffices for the first four properties. The first four pairs ensure reflexivity and seriality. The remaining pairs ensure that this is not the identity relation. Moreover, they maintain symmetry and transitivity.

**5.–11. Irreflexivity, Asymmetry, Antisymmetry, Antitransitivity, Nonreflexivity, Nonsymmetry, Nontransitivity.**

> R: $\{(a,b),(b,c),(c,a)\}$

Here we achieve the remaining properties. The relation R is irreflexive because no object is related to itself. This also makes it nonreflexive. It is asymmetric because there are no symmetric pairs whatsoever. This also yields nonsymmetry and (vacuously) antisymmetry. The relation is antitransitive because in every case where we are set up for transitivity—Rab ∧ Rbc and Rca ∧ Rab—we do not have the transitive instance: there is no $(a,c)$ and no $(c,b)$. This also give us nontransitivity.

**C.** This is much more complicated than the previous set of questions. In addition to exhibiting the desired properties, we are trying to avoid every other property not implied by the one we are exhibiting. First, let's note what must be the case for each of these to *fail* (remember, we are not allowing the relation to be empty)...

**Reflexivity:**
> $(\forall x)\mathbb{R}xx$

For reflexivity to fail $(\exists x)\neg\mathbb{R}xx$ must be true—i.e., some pair $(a,a) \notin \mathbb{R}$.

**Symmetry:**
> $(\forall x)(\forall y)(\mathbb{R}xy \rightarrow \mathbb{R}yx)$

For symmetry to fail $(\exists x)(\exists y)(\mathbb{R}xy \wedge \neg\mathbb{R}yx)$ must be true—i.e., some pair $(a,b) \in \mathbb{R}$ but $(b,a) \notin \mathbb{R}$.

**Transitivity:**
> $(\forall x)(\forall y)(\forall z)((\mathbb{R}xy \wedge \mathbb{R}yz) \rightarrow \mathbb{R}xz)$

For transitivity to fail $(\exists x)(\exists y)(\exists z)((\mathbb{R}xy \wedge \mathbb{R}yz) \wedge \neg \mathbb{R}xz)$ must be true—i.e., there must be two linked pairs $(a,b),(b,c) \in \mathbb{R}$ but the jump pair $(a,c) \notin \mathbb{R}$.

**Seriality:**

$(\forall x)(\exists y)\mathbb{R}xy$

For seriality to fail $(\exists x)\neg(\exists y)\mathbb{R}xy$ must be true—i.e., there must be some $a \in \mathbf{UD}$ that does not appear as the first member of any pair in $\mathbb{R}$.

**Irreflexivity:**

$(\forall x)\neg \mathbb{R}xx$

For irreflexivity to fail $(\exists x)\mathbb{R}xx$ must be true—i.e., there must be some $(a,a) \in \mathbb{R}$.

**Asymmetry:**

$(\forall x)(\forall y)(\mathbb{R}xy \rightarrow \neg \mathbb{R}yx)$

For asymmetry to fail $(\exists x)(\exists y)(\mathbb{R}xy \wedge \mathbb{R}yx)$—i.e., there must be at least one pair of inverse pairs. Note that a single pair $(a,a)$ meets this condition.

**Antisymmetry:**

$(\forall x)(\forall y)((\mathbb{R}xy \wedge \mathbb{R}yx) \rightarrow x = y)$

For antisymmetry to fail $(\exists x)(\exists y)((\mathbb{R}xy \wedge \mathbb{R}yx) \wedge x \neq y)$—i.e., there must be at least one pair of inverse pairs, in which the objects are distinct.

**Antitransitivity:**

$(\forall x)(\forall y)(\forall z)((\mathbb{R}xy \wedge \mathbb{R}yz) \rightarrow \neg \mathbb{R}xz)$

For antitransitivity to fail $(\exists x)(\exists y)(\exists z)((\mathbb{R}xy \wedge \mathbb{R}yz) \wedge \mathbb{R}xz)$—i.e., there must be at least one pair of linked pairs with a jump pair $(a,b),(b,c),(a,c) \in \mathbb{R}$. Note that a single pair $(a,a)$ meets this condition.

We will take our **UD** for each of the following to be $\{a,b,c,d\}$ and define a different R for each answer:

1. **Reflexivity.**

    R: $\{(a,a),(b,b),(c,c),(d,d),(a,b),(b,a),(b,c)\}$

    The first four pairs ensure reflexivity. The presence of $(b,c)$ without $(c,b)$ negates symmetry. The presence of $(a,b)$ and $(b,c)$ without $(a,c)$ negates transitivity. Seriality is implied by reflexivity, so this R cannot avoid it. Irreflexivity fails precisely because this is reflexive. Asymmetry fails because we have $(a,a)$. The pairs $(a,b)$ and $(b,a)$ also show this. Antisymmetry fails because of $(a,b)$ and $(b,a)$ where we assume $a \neq b$. Antitransitivity fails because we have $(a,a)$. The pairs $(a,b),(b,a),(a,a)$ also show that antitransitivity fails.

2. **Symmetry.**

    R: $\{(a,a),(a,b),(b,a)\}$

The three pairs here give us a symmetric R. Reflexivity fails because we are missing, among others, $(b, b)$. Transitivity fails because we have $(b, a)$ and $(a, b)$ without $(b, b)$. Seriality fails because neither c nor d participate as the first member of any pair in R. Irreflexivity fails because of $(a, a)$. Symmetry implies the failure of asymmetry. The pairs $(a, a)$, $(a, b)$, and $(b, a)$ also show this. Antisymmetry fails because of $(a, b)$ and $(b, a)$ where we assume a $\neq$ b. Antitransitivity fails because we have $(a, a)$.

### 3. Transitivity.

$$R: \{(a, a), (a, b), (a, c), (b, a), (b, b), (b, c)\}$$

This R is transitive. Reflexivity fails because we are missing, among others, $(c, c)$. Symmetry fails because we have $(b, c)$ without $(c, b)$. Seriality fails because neither c nor d participate as the first member of any pair in R. Irreflexivity fails because of $(a, a)$. Asymmetry fails because of $(a, a)$. The pairs $(a, b)$ and $(b, a)$ also show this. Antisymmetry fails because of $(a, b)$ and $(b, a)$ where we assume a $\neq$ b. Antitransitivity fails because we have $(a, a)$. The pairs $(a, b), (b, c), (a, c)$ also show that antitransitivity fails.

### 4. Seriality.

$$R: \{(a, a), (a, b), (b, a), (c, a), (d, c)\}$$

This R is serial since every member of the **UD** appears as the first member of some pair. Reflexivity fails because we are missing, among others, $(c, c)$. Symmetry fails because we have $(c, a)$ without $(a, c)$. Transitivity fails because we have $(d, c)$ and $(c, a)$ without $(d, a)$. Irreflexivity fails because of $(a, a)$. Asymmetry fails because of $(a, a)$. Antisymmetry fails because of $(a, b)$ and $(b, a)$ where we assume a $\neq$ b. Antitransitivity fails because we have $(a, a)$. The pairs $(a, b), (b, a), (a, a)$ also show that antitransitivity fails.

### 5. Irreflexivity.

$$R: \{(a, b), (a, c), (b, a), (b, c)\}$$

This R is irreflexive since no object is paired with itself. Thus it also fails to be reflexive. Symmetry fails because we have $(a, c)$ without $(c, a)$. Transitivity fails because we have $(a, b)$ and $(b, a)$ without $(a, a)$. Seriality fails because neither c nor d participate as the first member of any pair in R. Asymmetry and antisymmetry fail because of $(a, b)$ and $(b, a)$ where we assume a $\neq$ b. Antitransitivity fails because we have $(a, b)$, $(b, c)$, and $(a, c)$.

### 6. Asymmetry:
$$R: \{(a, b), (b, c), (c, d), (b, d)\}$$

This R is asymmetric since the inverse of no pair appears. This also vacuously satisfies antisymmetry since no two pairs can satisfy the antecedent condition of antisymmetry. This also implies irreflexivity as well as the failure of reflexivity and symmetry. Transitivity fails because we have $(a, b)$ and $(b, c)$ without $(a, c)$. Seriality fails because d does not participate as the first member of any pair in R. Antitransitivity fails because we have $(b, c)$, $(c, d)$, and $(b, d)$.

7. **Antisymmetry.**

$$R: \{(a, a), (a, b), (b, c)\}$$

This R is antisymmetric since in any case in which we have a pair and its inverse, the members are identical—$(a, a)$. Reflexivity fails due to the lack of $(b, b)$, among others. Symmetry fails due to the lack of $(b, a)$. Transitivity fails because we have $(a, b)$ and $(b, c)$ without $(a, c)$. Seriality fails because d does not participate as the first member of any pair in R. Irreflexivity, asymmetry, and antitransitivity fail because of $(a, a)$.

8. **Antitransitivity.**

$$R: \{(a, b), (b, a), (b, c)\}$$

This R is antitransitive since, in every case where there are linked pairs, there is no jump pair. This implies irreflexivity, the failure of reflexivity, and the failure of transitivity. Symmetry fails due to the lack of $(c, b)$. Seriality fails because d does not participate as the first member of any pair in R. Asymmetry and antisymmetry fail because of $(a, b)$ and $(b, a)$ where we assume $a \neq b$.

D. Note: These implications can also be proved using the derivation rules of **PD** from Chapter 7

1. **Reflexivity implies Seriality.**

$$(\forall x)\mathbb{R}xx \vDash (\forall x)(\exists y)\mathbb{R}xy$$

Suppose $(\forall x)\mathbb{R}xx$. Since every object **o** in the **UD** bears $\mathbb{R}$ to itself, then it is obvious that for every object **o**, there is some object—namely **o** itself—to which it bears $\mathbb{R}$. Thus, $(\forall x)(\exists y)\mathbb{R}xy$.

2. **Asymmetry implies Irreflexivity.**

$$(\forall x)(\forall y)(\mathbb{R}xy \to \neg\mathbb{R}yx) \vDash (\forall x)\neg\mathbb{R}xx$$

Suppose LHW. Suppose also that $(\exists x)\mathbb{R}xx$ (equivalent to the negation of the RHW). So for some **o** in the **UD**, call it 'a', we have $\mathbb{R}aa$. If we instantiate both universals in the LHW with 'a', we get $\mathbb{R}aa \to \neg\mathbb{R}aa$. We now have $\mathbb{R}aa$ and $\neg\mathbb{R}aa$, a contradiction. So our second supposition must be false, yielding RHW.

3. **Asymmetry implies Antisymmetry.**

$$(\forall x)(\forall y)(\mathbb{R}xy \to \neg\mathbb{R}yx) \vDash (\forall x)(\forall y)((\mathbb{R}xy \land \mathbb{R}yx) \to x = y)$$

Suppose LHW. Suppose also that $(\exists x)(\exists y)((\mathbb{R}xy \wedge \mathbb{R}yx) \wedge x \neq y)$ (equivalent to the negation of RHW). So for some **o** and **m** in the **UD**, call them 'a' and 'b', we have $(\mathbb{R}ab \wedge \mathbb{R}ba) \wedge a \neq b$. If we instantiate LHW with 'a' and 'b', we have $\mathbb{R}ab \rightarrow \neg\mathbb{R}ba$. This and the first two conjuncts of the previous wff yield $\mathbb{R}ba$ and $\neg\mathbb{R}ba$, a contradiction. So our second supposition must be false, yielding RHW.

### 4. Antitransitivity implies Irreflexivity.

$$(\forall x)(\forall y)(\forall z)((\mathbb{R}xy \wedge \mathbb{R}yz) \rightarrow \neg\mathbb{R}xz) \vDash (\forall x)\neg\mathbb{R}xx$$

Suppose LHW. Suppose also that $(\exists x)\mathbb{R}xx$ (equivalent to the negation of the RHW). So for some **o** in the **UD**, call it 'a', we have $\mathbb{R}aa$. If we instantiate all three universals in the LHW with 'a', we get $(\mathbb{R}aa \wedge \mathbb{R}aa) \rightarrow \neg\mathbb{R}aa$. We now have $\mathbb{R}aa$ and $\neg\mathbb{R}aa$, a contradiction. So our second supposition must be false, yielding RHW.

### 5. Transitivity and Irreflexivity imply Asymmetry.

$$\{(\forall x)(\forall y)(\forall z)((\mathbb{R}xy \wedge \mathbb{R}yz) \rightarrow \mathbb{R}xz),$$
$$(\forall x)\neg\mathbb{R}xx\} \vDash (\forall x)(\forall y)(\mathbb{R}xy \rightarrow \neg\mathbb{R}yx)$$

Suppose both wffs in the set. Suppose also that $(\exists x)(\exists y)(\mathbb{R}xy \wedge \mathbb{R}yx)$ (equivalent to the negation of RHW). So for some **o** and **m** in the **UD**, call them 'a' and 'b', we have $\mathbb{R}ab \wedge \mathbb{R}ba$. If we instantiate the first wff in the set with 'a' for x and z, and 'b' for y, we have $(\mathbb{R}ab \wedge \mathbb{R}ba) \rightarrow \mathbb{R}aa$. This gives us $\mathbb{R}aa$. When we instantiate the second wff of the set with 'a' we have $\neg\mathbb{R}aa$. This is a contradiction, so our second supposition must be false, yielding RHW.

**E.**    **1.** Suppose $(\forall x)(\forall y)(\forall z)((x = y \wedge y = z) \rightarrow x = z)$ is F on some interpretation **M**. So $\neg(\forall x)(\forall y)(\forall z)((x = y \wedge y = z) \rightarrow x = z)$ is T. If we drive the negation across all three universal quantifiers and into the conditional, we have: $(\exists x)(\exists y)(\exists z)((x = y \wedge y = z) \wedge x \neq z)$. If we now instantiate each of the existentials we get: $(a = b \wedge b = c) \wedge a \neq c$. But the first two conjuncts imply $a = c$, so we have both $a = c$ and $a \neq c$, which is impossible, so $(\forall x)(\forall y)(\forall z)((x = y \wedge y = z) \rightarrow x = z)$ must be true on every interpretation **M**. So it is quantificationally true.

   **2.** Recall that each interpretation includes a set **A** of all variable assignments, **a**, in which every individual variable, $x_i$, of **P** is assigned an object **o** in the **UD**. When there are at least 2 objects in the **UD**, we will have an infinite number of variable assignments (see p. 210). But suppose the **UD** has only one member (which is the smallest **UD** we allow). Call that member $o_1$. In this case there will be only one variable assignment, call it $a_1$. In particular, $a_1$ assigns $o_1$ to each and every variable, $x_i$. For a variable assignment to differ from $a_1$, there would have to be some $o_n \neq o_1$ to assign to one or more of the $x_i$s. But then we would have at least 2 members in the **UD**.

**3.** Suppose R is transitive, irreflexive, and serial. The **UD** must have at least one object, call that $a_1$. Since R is serial, we must have $(\exists x)Ra_1x$, but since R is irreflexive, the value of x must be distinct from $a_1$, so the **UD** must have a second object. Call it $a_2$. So $Ra_1a_2$. Again, due to seriality, we must have $(\exists x)Ra_2x$. Because of irreflexivity, that x cannot be $a_2$. If we suppose that $Ra_2a_1$, then $Ra_1a_2$ and transitivity give us $Ra_2a_2$ violating irreflexivity. So we cannot have $Ra_2a_1$. So the **UD** must contain a third object, $a_3$, giving us $Ra_2a_3$. Seriality requires $(\exists x)Ra_3x$. Because of irreflexivity, that x cannot be $a_3$, and because of transitivity we cannot have $Ra_3a_2$ or $Ra_3a_1$. Either would give us $Ra_3a_3$ via transitivity. So the **UD** must contain a fourth object, $a_4$, giving us $Ra_3a_4$…

Technically, this is not a proof. We don't actually show that for any finite $n$ the **UD** must be larger than $n$. But you do see the idea. To convert it to a proof proper, we would have to use mathematical induction.

## A.23  **Answers to 7.1.5**

**1.**

| | | | |
|---|---|---|---|
| 1 | $(\forall z)(Kz \rightarrow Jz)$ | P | $\vdash Jn$ |
| 2 | Kn | P | |
| 3 | $Kn \rightarrow Jn$ | 1 $\forall$E | |
| 4 | Jn | 2, 3 $\rightarrow$E | |

**2.**

| | | | |
|---|---|---|---|
| 1 | $(\forall y)Gy$ | P | $\vdash (\exists x)Hx$ |
| 2 | $(\forall z)(Gz \rightarrow Hz)$ | P | |
| 3 | Gb | 1 $\forall$E | |
| 4 | $Gb \rightarrow Hb$ | 2 $\forall$E | |
| 5 | Hb | 3, 4 $\rightarrow$E | |
| 6 | $(\exists x)Hx$ | 5 $\exists$I | |

**3.**

| | | | |
|---|---|---|---|
| 1 | $(\forall w)(Dw \leftrightarrow Cw)$ | P | $\vdash (\forall x)Cx \rightarrow (\exists x)Dx$ |
| 2 | $(\forall x)Cx$ | A | |
| 3 | Cg | 2 $\forall$E | |
| 4 | $Dg \leftrightarrow Cg$ | 1 $\forall$E | |
| 5 | Dg | 3, 4 $\leftrightarrow$E | |
| 6 | $(\exists x)Dx$ | 5 $\exists$I | |
| 7 | $(\forall x)Cx \rightarrow (\exists x)Dx$ | 2–6 $\rightarrow$I | |

**4.**

| | | | |
|---|---|---|---|
| 1 | $(\forall x)(Mx \rightarrow Bx)$ | P | |
| 2 | $\neg Bc$ | P | $\vdash \neg Mc$ |
| 3 | $Mc \rightarrow Bc$ | 1 $\forall E$ | |
| 4 | $\quad$ Mc | A | |
| 5 | $\quad$ Bc | 3, 4 $\rightarrow$E | |
| 6 | $\quad \neg Bc$ | 2 R | |
| 7 | $\neg Mc$ | 4−6 $\neg$I | |

**5.**

| | | | |
|---|---|---|---|
| 1 | Fa | P | |
| 2 | $(\forall x)(Mx \leftrightarrow Gx)$ | P | |
| 3 | $(\forall x)(Fx \rightarrow Gx)$ | P | $\vdash (\exists z)(Mz \wedge Gz)$ |
| 4 | $Ma \leftrightarrow Ga$ | 2 $\forall E$ | |
| 5 | $Fa \rightarrow Ga$ | 3 $\forall E$ | |
| 6 | Ga | 1, 5 $\rightarrow$E | |
| 7 | Ma | 4, 6 $\leftrightarrow$E | |
| 8 | $Ma \wedge Ga$ | 6, 7 $\wedge$I | |
| 9 | $(\exists z)(Mz \wedge Gz)$ | 8 $\exists$I | |

**6.**

| | | | |
|---|---|---|---|
| 1 | $(\forall y)My$ | P | |
| 2 | $(\forall z)(Kz \leftrightarrow Mz)$ | P | $\vdash (\forall z)Kz$ |
| 3 | Md | 1 $\forall E$ | |
| 4 | $Kd \leftrightarrow Md$ | 2 $\forall E$ | |
| 5 | Kd | 3, 4 $\leftrightarrow$E | |
| 6 | $(\forall z)Kz$ | 5 $\forall$I | |

**7.**

| | | | |
|---|---|---|---|
| 1 | $(\exists x)Fx$ | P | |
| 2 | $(\forall x)(Fx \rightarrow Gx)$ | P | $\vdash (\exists x)Gx$ |
| 3 | $Ff \rightarrow Gf$ | 2 $\forall E$ | |
| 4 | $\quad$ Ff | A | |
| 5 | $\quad$ Gf | 3, 4 $\rightarrow$E | |
| 6 | $\quad (\exists x)Gx$ | 5 $\exists$I | |
| 7 | $(\exists x)Gx$ | 1, 4−6 $\exists$E | |

**8.**

| | | | |
|---|---|---|---|
| 1 | $(\forall x)(Ax \to Bx)$ | P | $\vdash (\exists x)Ax \to (\exists x)Bx$ |
| 2 | $(\exists x)Ax$ | A | |
| 3 | Aa | A | |
| 4 | $Aa \to Ba$ | $1 \forall E$ | |
| 5 | Ba | $3, 4 \to E$ | |
| 6 | $(\exists x)Bx$ | $5 \exists I$ | |
| 7 | $(\exists x)Bx$ | $2, 3-6 \exists E$ | |
| 8 | $(\exists x)Ax \to (\exists x)Bx$ | $2-7 \to I$ | |

**9.**

| | | | |
|---|---|---|---|
| 1 | $(\forall x)Cx$ | P | |
| 2 | $(\forall x)(Cx \to (Dx \land Ex))$ | P | $\vdash (\exists x)Dx \land (\forall y)Ey$ |
| 3 | Cf | $1 \forall E$ | |
| 4 | $Cf \to (Df \land Ef)$ | $2 \forall E$ | |
| 5 | $Df \land Ef$ | $3, 4 \to E$ | |
| 6 | Df | $5 \land E$ | |
| 7 | Ef | $5 \land E$ | |
| 8 | $(\exists x)Dx$ | $6 \exists I$ | |
| 9 | $(\forall y)Ey$ | $7 \forall I$ | |
| 10 | $(\exists x)Dx \land (\forall y)Ey$ | $8, 9 \land I$ | |

**10.**

| | | | |
|---|---|---|---|
| 1 | $(\exists x)Cx$ | P | |
| 2 | $(\forall x)(Cx \to (Dx \land Ex))$ | P | $\vdash (\exists x)Dx \land (\exists x)Ex$ |
| 3 | Ca | A | |
| 4 | $Ca \to (Da \land Ea)$ | $2 \forall E$ | |
| 5 | $Da \land Ea$ | $3, 4 \to E$ | |
| 6 | Da | $5 \land E$ | |
| 7 | Ea | $5 \land E$ | |
| 8 | $(\exists x)Dx$ | $6 \exists I$ | |
| 9 | $(\exists x)Ex$ | $7 \exists I$ | |
| 10 | $(\exists x)Dx \land (\exists x)Ex$ | $8, 9 \land I$ | |
| 11 | $(\exists x)Dx \land (\exists x)Ex$ | $1, 3-10 \exists E$ | |

# A.24  **Answers to 7.3.1**

**A.  1.**

| | | | |
|---|---|---|---|
| 1 | $(\forall x)Jx$ | P | $\vdash (\forall z)Jz$ |
| 2 | Ja | $1 \forall E$ | |
| 3 | $(\forall z)Jz$ | $2 \forall I$ | |

**2.**

| 1 | $(\forall x)(Dx \to Gx)$ | P |
| 2 | $(\forall w)(Gw \to Mw)$ | P |
| 3 | $(\forall y)Dy$ | P        ⊢ $(\forall z)Mz$ |
| 4 | $Da \to Ga$ | 1 $\forall$E |
| 5 | $Ga \to Ma$ | 2 $\forall$E |
| 6 | $Da$ | 3 $\forall$E |
| 7 | $Ga$ | 4, 6 $\to$E |
| 8 | $Ma$ | 5, 7 $\to$E |
| 9 | $(\forall z)Mz$ | 8 $\forall$I |

**3.**

| 1 | $(\exists z)(Az \land Kz)$ | P |
| 2 | $(\forall x)(Kx \to Cx)$ | P        ⊢ $(\exists x)(Cx \land Ax)$ |
| 3 | $\quad Ag \land Kg$ | A |
| 4 | $\quad Kg \to Cg$ | 2 $\forall$E |
| 5 | $\quad Kg$ | 3 $\land$E |
| 6 | $\quad Ag$ | 3 $\land$E |
| 7 | $\quad Cg$ | 4, 5 $\to$E |
| 8 | $\quad Cg \land Ag$ | 6, 7 $\land$I |
| 9 | $\quad (\exists x)(Cx \land Ax)$ | 8 $\exists$I |
| 10 | $(\exists x)(Cx \land Ax)$ | 1, 3–9 $\exists$E |

**4.**

| 1 | $(\forall z)(Gzz \to Hfz)$ | P |
| 2 | $(\exists x)((\forall y)Fxy \land Gxx)$ | P        ⊢ $(\exists x)((\forall y)Fxy \land Hfx)$ |
| 3 | $\quad (\forall y)Fky \land Gkk$ | A |
| 4 | $\quad Gkk \to Hfk$ | 1 $\forall$E |
| 5 | $\quad Gkk$ | 3 $\land$E |
| 6 | $\quad Hfk$ | 4, 5 $\to$E |
| 7 | $\quad (\forall y)Fky$ | 3 $\land$E |
| 8 | $\quad (\forall y)Fky \land Hfk$ | 7, 6 $\land$I |
| 9 | $\quad (\exists x)((\forall y)Fxy \land Hfx)$ | 8 $\exists$I |
| 10 | $(\exists x)((\forall y)Fxy \land Hfx)$ | 2, 3–9 $\exists$E |

**5.**

| | | |
|---|---|---|
| 1 | $(\forall z)(Gzz \rightarrow Hfz)$ | P |
| 2 | $(\exists x)((\forall y)Fxy \wedge Gxx)$ | P    $\vdash (\exists x)(\forall y)(Fxy \wedge Hfx)$ |
| 3 | $\quad (\forall y)Fky \wedge Gkk$ | A |
| 4 | $\quad Gkk \rightarrow Hfk$ | 1 $\forall$E |
| 5 | $\quad Gkk$ | 3 $\wedge$E |
| 6 | $\quad Hfk$ | 4, 5 $\rightarrow$E |
| 7 | $\quad (\forall y)Fky$ | 3 $\wedge$E |
| 8 | $\quad Fkb$ | 7 $\forall$E |
| 9 | $\quad Fkb \wedge Hfk$ | 6, 8 $\wedge$I |
| 10 | $\quad (\forall y)(Fky \wedge Hfk)$ | 9 $\forall$I |
| 11 | $\quad (\exists x)(\forall y)(Fxy \wedge Hfx)$ | 10 $\exists$I |
| 12 | $(\exists x)(\forall y)(Fxy \wedge Hfx)$ | 2, 3–11 $\exists$E |

**B.**    **1.**

| | | |
|---|---|---|
| 1 | $(\forall x)Mx$ | P |
| 2 | $(\forall y)\neg Dyy$ | P    $\vdash (\forall x)(\neg Dxx \wedge Mx)$ |
| 3 | $Mg$ | 1 $\forall$E |
| 4 | $\neg Dgg$ | 2 $\forall$E |
| 5 | $\neg Dgg \wedge Mg$ | 3, 4 $\wedge$I |
| 6 | $(\forall x)(\neg Dxx \wedge Mx)$ | 5 $\forall$I |

**2.**

| | | |
|---|---|---|
| 1 | $(\forall x)Mx$ | P |
| 2 | $(\forall y)\neg Dyy$ | P    $\vdash (\exists x)(\forall y)(\neg Dxx \wedge My)$ |
| 3 | $Mg$ | 1 $\forall$E |
| 4 | $\neg Drr$ | 2 $\forall$E |
| 5 | $\neg Drr \wedge Mg$ | 3, 4 $\wedge$I |
| 6 | $(\forall y)(\neg Drr \wedge My)$ | 5 $\forall$I |
| 7 | $(\exists x)(\forall y)(\neg Dxx \wedge My)$ | 6 $\forall$I |

**3.**

| | | |
|---|---|---|
| 1 | $(\forall x)(Mx \rightarrow Wx)$ | P |
| 2 | $(\exists y)\neg Wy$ | P    $\vdash (\exists z)\neg Mz$ |
| 3 | $\quad \neg Wa$ | A |
| 4 | $\quad Ma \rightarrow Wa$ | 1 $\forall$E |
| 5 | $\quad\quad Ma$ | A |
| 6 | $\quad\quad Wa$ | 4, 5 $\rightarrow$E |
| 7 | $\quad\quad \neg Wa$ | 3 R |
| 8 | $\quad \neg Ma$ | 5–7 $\neg$I |
| 9 | $\quad (\exists z)\neg Mz$ | 8 $\exists$I |
| 10 | $(\exists z)\neg Mz$ | 2, 3–9 $\exists$E |

**4.**

| | | | |
|---|---|---|---|
| 1 | (∃y)(∀x)Lxy | P | ⊢ (∀x)(∃y)Lxy |
| 2 | (∀x)Lxa | A | |
| 3 | Lba | 2 ∀E | |
| 4 | (∃y)Lby | 3 ∃I | |
| 5 | (∀x)(∃y)Lxy | 4 ∀I | |
| 6 | (∀x)(∃y)Lxy | 1, 2–5 ∃E | |

**5.**

| | | | |
|---|---|---|---|
| 1 | (∃y)(∀x)Lyx | P | ⊢ (∀x)(∃y)Lyx |
| 2 | (∀x)Lax | A | |
| 3 | Lab | 2 ∀E | |
| 4 | (∃y)Lyb | 3 ∃I | |
| 5 | (∀x)(∃y)Lyx | 4 ∀I | |
| 6 | (∀x)(∃y)Lyx | 1, 2–5 ∃E | |

**C.    1.**                                          ⊢ (∀x)(Jx → Kx) → ((∀x)Jx → (∀x)Kx)

| | | |
|---|---|---|
| 1 | (∀x)(Jx → Kx) | A |
| 2 | (∀x)Jx | A |
| 3 | Jb | 2 ∀E |
| 4 | Jb → Kb | 1 ∀E |
| 5 | Kb | 3, 4 →E |
| 6 | (∀x)Kx | 5 ∀I |
| 7 | (∀x)Jx → (∀x)Kx | 2–6 →I |
| 8 | (∀x)(Jx → Kx) → ((∀x)Jx → (∀x)Kx) | 1–7 →I |

**2.**                                          ⊢ (∀x)(Fx → Gx) → ((∃x)Fx → (∃x)Gx)

| | | |
|---|---|---|
| 1 | (∀x)(Fx → Gx) | A |
| 2 | (∃x)Fx | A |
| 3 | Fa | A |
| 4 | Fa → Ga | 1 ∀E |
| 5 | Ga | 3, 4 →E |
| 6 | (∃x)Gx | 5 ∃I |
| 7 | (∃x)Gx | 2, 3–6 ∃E |
| 8 | (∃x)Fx → (∃x)Gx | 2–7 →I |
| 9 | (∀x)(Fx → Gx) → ((∃x)Fx → (∃x)Gx) | 1–8 →I |

**3.**                                                        $\vdash (\exists x)(Fx \wedge Gx) \rightarrow ((\exists x)Fx \wedge (\exists x)Gx)$

| | | | |
|---|---|---|---|
| 1 | | $(\exists x)(Fx \wedge Gx)$ | A |
| 2 | | $Fa \wedge Ga$ | A |
| 3 | | $Fa$ | 2 $\wedge$E |
| 4 | | $Ga$ | 2 $\wedge$E |
| 5 | | $(\exists x)Fx$ | 3 $\exists$I |
| 6 | | $(\exists x)Gx$ | 4 $\exists$I |
| 7 | | $(\exists x)Fx \wedge (\exists x)Gx$ | 5, 6 $\wedge$I |
| 8 | | $(\exists x)Fx \wedge (\exists x)Gx$ | 1, 2–7 $\exists$E |
| 9 | $(\exists x)(Fx \wedge Gx) \rightarrow ((\exists x)Fx \wedge (\exists x)Gx)$ | | 1–8 $\rightarrow$I |

**4.**                                                                $\vdash (\forall x)(\exists y)(Fy \rightarrow Fx)$

| | | |
|---|---|---|
| 1 | $Fa$ | A |
| 2 | $Fa$ | 1 R |
| 3 | $Fa \rightarrow Fa$ | 1–2 $\rightarrow$I |
| 4 | $(\exists y)(Fy \rightarrow Fa)$ | 3 $\exists$I |
| 5 | $(\forall x)(\exists y)(Fy \rightarrow Fx)$ | 4 $\forall$I |

**D.   1.**

| | | | |
|---|---|---|---|
| 1 | $(\forall x)Fx \wedge (\forall x)Gx$ | P | $\vdash (\forall x)(Fx \wedge Gx)$ |
| 2 | $(\forall x)Fx$ | 1 $\wedge$E | |
| 3 | $(\forall x)Gx$ | 1 $\wedge$E | |
| 4 | $Fa$ | 2 $\forall$E | |
| 5 | $Ga$ | 3 $\forall$E | |
| 6 | $Fa \wedge Ga$ | 4, 5 $\wedge$I | |
| 7 | $(\forall x)(Fx \wedge Gx)$ | 6 $\forall$I | |

| | | | |
|---|---|---|---|
| 1 | $(\forall x)(Fx \wedge Gx)$ | P | $\vdash (\forall x)Fx \wedge (\forall x)Gx$ |
| 2 | $Fa \wedge Ga$ | 1 $\forall$E | |
| 3 | $Fa$ | 2 $\wedge$E | |
| 4 | $Ga$ | 2 $\wedge$E | |
| 5 | $(\forall x)Fx$ | 3 $\forall$I | |
| 6 | $(\forall x)Gx$ | 4 $\forall$I | |
| 7 | $(\forall x)Fx \wedge (\forall x)Gx$ | 5, 6 $\wedge$I | |

**2.**

| | 1 | $(\exists x)Fx \lor (\exists x)Gx$ | P | $\vdash (\exists x)(Fx \lor Gx)$ |
|---|---|---|---|---|
| | 2 | $(\exists x)Fx$ | A | |
| | 3 | Fa | A | |
| | 4 | $Fa \lor Ga$ | 3 $\lor$I | |
| | 5 | $(\exists x)(Fx \lor Gx)$ | 4 $\exists$I | |
| | 6 | $(\exists x)(Fx \lor Gx)$ | 2, 3–5 $\exists$E | |
| | 7 | $(\exists x)Gx$ | A | |
| | 8 | Ga | A | |
| | 9 | $Fa \lor Ga$ | 8 $\lor$I | |
| | 10 | $(\exists x)(Fx \lor Gx)$ | 9 $\exists$I | |
| | 11 | $(\exists x)(Fx \lor Gx)$ | 7, 8–10 $\exists$E | |
| | 12 | $(\exists x)(Fx \lor Gx)$ | 1, 2–6, 7–11 $\lor$E | |

| 1 | $(\exists x)(Fx \lor Gx)$ | P | $\vdash (\exists x)Fx \lor (\exists x)Gx$ |
|---|---|---|---|
| 2 | $Fa \lor Ga$ | A | |
| 3 | Fa | A | |
| 4 | $(\exists x)Fx$ | 3 $\exists$I | |
| 5 | $(\exists x)Fx \lor (\exists x)Gx$ | 4 $\lor$I | |
| 6 | Ga | A | |
| 7 | $(\exists x)Gx$ | 6 $\exists$I | |
| 8 | $(\exists x)Fx \lor (\exists x)Gx$ | 7 $\lor$I | |
| 9 | $(\exists x)Fx \lor (\exists x)Gx$ | 2, 3–5, 6–8 $\lor$E | |
| 10 | $(\exists x)Fx \lor (\exists x)Gx$ | 1, 2–9 $\exists$E | |

**3.**

| | 1 | $\neg(\exists x)Bx$ | P | $\vdash (\forall x)\neg Bx$ |
|---|---|---|---|---|
| | 2 | Bd | A | |
| | 3 | $(\exists x)Bx$ | 2 $\exists$I | |
| | 4 | $\neg(\exists x)Bx$ | 1 R | |
| | 5 | $\neg Bd$ | 2–4 $\neg$I | |
| | 6 | $(\forall x)\neg Bx$ | 5 $\forall$I | |

| 1 | $(\forall x)\neg Bx$ | P | $\vdash \neg(\exists x)Bx$ |
|---|---|---|---|
| 2 | $(\exists x)Bx$ | A | |
| 3 | $Ba$ | A | |
| 4 | $(\exists x)Bx$ | A | |
| 5 | $Ba$ | 3 R | |
| 6 | $\neg Ba$ | 1 $\forall$E | |
| 7 | $\neg(\exists x)Bx$ | 4–6 $\neg$I | |
| 8 | $\neg(\exists x)Bx$ | 2, 3–7 $\exists$E | |
| 9 | $(\exists x)Bx$ | 2 R | |
| 10 | $\neg(\exists x)Bx$ | 2–9 $\neg$I | |

## A.25  **Answers to 7.4.2**

**A.**  **1.**

| 1 | $(\forall x)Fx$ | P | $\vdash \neg(\forall x)\neg Fx$ |
|---|---|---|---|
| 2 | $Fa$ | 1 $\forall$E | |
| 3 | $(\exists x)Fx$ | 2 $\exists$I | |
| 4 | $\neg\neg(\exists x)Fx$ | 3 DN | |
| 5 | $\neg(\forall x)\neg Fx$ | 4 QN | |

**2.**

| 1 | $\neg(\exists x)Fx$ | P | $\vdash \neg(\forall x)Fx$ |
|---|---|---|---|
| 2 | $(\forall x)\neg Fx$ | 1 QN | |
| 3 | $\neg Fs$ | 2 $\forall$E | |
| 4 | $(\exists x)\neg Fx$ | 3 $\exists$I | |
| 5 | $\neg(\forall x)Fx$ | 4 QN | |

**3.**

| 1 | $\neg(\exists x)(Fx \wedge Gx)$ | P | $\vdash (\forall x)(Fx \rightarrow \neg Gx)$ |
|---|---|---|---|
| 2 | $(\forall x)\neg(Fx \wedge Gx)$ | 1 QN | |
| 3 | $(\forall x)(\neg Fx \vee \neg Gx)$ | 2 DeM | |
| 4 | $(\forall x)(Fx \rightarrow \neg Gx)$ | 3 Impl | |

**4.**

| 1 | $(\exists y)\neg Ey$ | P | $\vdash \neg(\forall x)(Nx \wedge Ex)$ |
|---|---|---|---|
| 2 | $\neg Ea$ | A | |
| 3 | $\neg Na \vee \neg Ea$ | 2 $\vee$I | |
| 4 | $\neg(Na \wedge Ea)$ | 3 DeM | |
| 5 | $(\exists x)\neg(Nx \wedge Ex)$ | 4 $\exists$I | |
| 6 | $(\exists x)\neg(Nx \wedge Ex)$ | 1, 2–5 $\exists$E | |
| 7 | $\neg(\forall x)(Nx \wedge Ex)$ | 6 QN | |

**5.**

| | | | |
|---|---|---|---|
| 1 | $(\forall x)Fx \wedge P$ | P | $\vdash (\forall x)(Fx \wedge P)$ |
| 2 | P | 1 $\wedge$E | |
| 3 | $(\forall x)Fx$ | 1 $\wedge$E | |
| 4 | Fa | 3 $\forall$E | |
| 5 | Fa $\wedge$ P | 2, 4 $\wedge$I | |
| 6 | $(\forall x)(Fx \wedge P)$ | 5 $\forall$I | |

**6.**

| | | | |
|---|---|---|---|
| 1 | $(\forall x)(Fx \wedge P)$ | P | $\vdash (\forall x)Fx \wedge P$ |
| 2 | Fa $\wedge$ P | 1 $\forall$E | |
| 3 | Fa | 2 $\wedge$E | |
| 4 | P | 2 $\wedge$E | |
| 5 | $(\forall x)Fx$ | 3 $\forall$I | |
| 6 | $(\forall x)Fx \wedge P$ | 4, 5 $\wedge$I | |

**7.**

| | | | |
|---|---|---|---|
| 1 | $(\exists x)(Fx \rightarrow P)$ | P | $\vdash (\forall x)Fx \rightarrow P$ |
| 2 | $(\forall x)Fx$ | A | |
| 3 | Fa $\rightarrow$ P | A | |
| 4 | Fa | 2 $\forall$E | |
| 5 | P | 3, 4 $\rightarrow$E | |
| 6 | P | 1, 3–5 $\exists$E | |
| 7 | $(\forall x)Fx \rightarrow P$ | 2–6 $\rightarrow$I | |

**8.**

| | | | |
|---|---|---|---|
| 1 | $(\forall x)Fx \rightarrow P$ | P | $\vdash (\exists x)(Fx \rightarrow P)$ |
| 2 | $\neg(\exists x)(Fx \rightarrow P)$ | A | |
| 3 | $(\forall x)\neg(Fx \rightarrow P)$ | 2 QN | |
| 4 | $(\forall x)\neg(\neg Fx \vee P)$ | 3 Impl | |
| 5 | $(\forall x)(\neg\neg Fx \wedge \neg P)$ | 4 DeM | |
| 6 | $\neg\neg Fa \wedge \neg P$ | 5 $\forall$E | |
| 7 | Fa $\wedge \neg P$ | 6 DN | |
| 8 | Fa | 7 $\wedge$E | |
| 9 | $(\forall x)Fx$ | 8 $\forall$I | |
| 10 | P | 1, 9 $\rightarrow$E | |
| 11 | $\neg P$ | 7 $\wedge$E | |
| 12 | $(\exists x)(Fx \rightarrow P)$ | 2–11 $\neg$E | |

**B.**  **1.**

| | | | |
|---|---|---|---|
| 1 | $(\exists y)Ky \rightarrow (\forall x)Lx$ | P | |
| 2 | $(\exists x)\neg Lx$ | P | $\vdash (\forall y)\neg Ky$ |
| 3 | $\neg(\forall x)Lx$ | 2 QN | |
| 4 | $\neg(\exists y)Ky$ | 1, 3 MT | |
| 5 | $(\forall y)\neg Ky$ | 4 QN | |

**2.**

| | | |
|---|---|---|
| 1 | ¬(∀x)Kx | P |
| 2 | (∀y)(Dy → Ky) | P          ⊢ ¬(∀x)Dx |
| 3 | (∃x)¬Kx | 1 QN |
| 4 | ¬Kc | A |
| 5 | Dc → Kc | 2 ∀E |
| 6 | ¬Dc | 4, 5 MT |
| 7 | (∃x)¬Dx | 6 ∃I |
| 8 | (∃x)¬Dx | 3, 4–7 ∃E |
| 9 | ¬(∀x)Dx | 8 QN |

**3.**

| | | |
|---|---|---|
| 1 | (∀x)(Dx → Hx) | P |
| 2 | ¬(∃x)Hx | P          ⊢ ¬(∃x)Dx |
| 3 | (∀x)¬Hx | 2 QN |
| 4 | Dc → Hc | 1 ∀E |
| 5 | ¬Hc | 3 ∀E |
| 6 | ¬Dc | 4, 5 MT |
| 7 | (∀x)¬Dx | 6 ∀I |
| 8 | ¬(∃x)Dx | 7 QN |

**4.**

| | | |
|---|---|---|
| 1 | (∀x)(Tx → (∃y)(¬Sx ∧ Ryx)) | P |
| 2 | (∃y)¬Uyy | P |
| 3 | (∀z)(¬Sz → Uzz) | P          ⊢ (∃w)¬Tw |
| 4 | ¬Ubb | A |
| 5 | ¬Sb → Ubb | 3 ∀E |
| 6 | ¬¬Sb | 4, 5 MT |
| 7 | ¬¬Sb ∨ ¬Rcb | 6 ∨I |
| 8 | ¬(¬Sb ∧ Rcb) | 7 DeM |
| 9 | (∀y)¬(¬Sb ∧ Ryb) | 8 ∀I |
| 10 | ¬(∃y)(¬Sb ∧ Ryb) | 9 QN |
| 11 | Tb → (∃y)(¬Sb ∧ Ryb) | 1 ∀E |
| 12 | ¬Tb | 10, 11 MT |
| 13 | (∃w)¬Tw | 12 ∃I |
| 14 | (∃w)¬Tw | 2, 4–13 ∃E |

**C.    1.**                                             $\vdash \neg((\forall x)Nx \land (\forall z)\neg Nz)$

| 1 | $(\forall x)Nx \land (\forall z)\neg Nz$ | A |
|---|---|---|
| 2 | $(\forall x)Nx$ | 1 $\land$E |
| 3 | $(\forall z)\neg Nz$ | 1 $\land$E |
| 4 | Nb | 2 $\forall$E |
| 5 | $\neg Nb$ | 3 $\forall$E |
| 6 | $\neg((\forall x)Nx \land (\forall z)\neg Nz)$ | 1$-$5 $\neg$I |

**2.**                                  $\vdash (\forall x)(Fx \rightarrow Gx) \leftrightarrow \neg(\exists x)(Fx \land \neg Gx)$

| 1 | $(\forall x)(Fx \rightarrow Gx)$ | A |
|---|---|---|
| 2 | $(\forall x)(\neg Fx \lor Gx)$ | 1 Impl |
| 3 | $(\forall x)(\neg Fx \lor \neg\neg Gx)$ | 2 DN |
| 4 | $(\forall x)\neg(Fx \land \neg Gx)$ | 3 DeM |
| 5 | $\neg(\exists x)(Fx \land \neg Gx)$ | 4 QN |
| 6 | $\neg(\exists x)(Fx \land \neg Gx)$ | A |
| 7 | $(\forall x)\neg(Fx \land \neg Gx)$ | 6 QN |
| 8 | $(\forall x)(\neg Fx \lor \neg\neg Gx)$ | 7 DeM |
| 9 | $(\forall x)(\neg Fx \lor Gx)$ | 8 DN |
| 10 | $(\forall x)(Fx \rightarrow Gx)$ | 9 Impl |
| 11 | $(\forall x)(Fx \rightarrow Gx) \leftrightarrow \neg(\exists x)(Fx \land \neg Gx)$ | 1$-$5, 6$-$10 $\leftrightarrow$I |

**3.**                                        $\vdash \neg(\exists x)(\forall y)(Rxy \leftrightarrow \neg Ryy)$

| 1 | $(\forall y)(Ray \leftrightarrow \neg Ryy)$ | A |
|---|---|---|
| 2 | $Raa \leftrightarrow \neg Raa$ | 1 $\forall$E |
| 3 | $(Raa \land \neg Raa) \lor (\neg Raa \land \neg\neg Raa)$ | 2 Equiv |
| 4 | $(Raa \land \neg Raa) \lor (\neg Raa \land Raa)$ | 3 DN |
| 5 | $(Raa \land \neg Raa) \lor (Raa \land \neg Raa)$ | 4 Com |
| 6 | $Raa \land \neg Raa$ | 5 Idem |
| 7 | $Raa$ | 6 $\land$E |
| 8 | $\neg Raa$ | 6 $\land$E |
| 9 | $\neg(\forall y)(Ray \leftrightarrow \neg Ryy)$ | 1$-$8 $\neg$I |
| 10 | $(\forall x)\neg(\forall y)(Rxy \leftrightarrow \neg Ryy)$ | 9 $\forall$I |
| 11 | $\neg(\exists x)(\forall y)(Rxy \leftrightarrow \neg Ryy)$ | 10 QN |

**4.**                                         $\vdash (\forall x)((\forall y)(Rxy \leftrightarrow (Fy \land \neg Ryy)) \rightarrow \neg Fx)$

| 1 | $(\forall y)(Ray \leftrightarrow (Fy \land \neg Ryy))$ | A |
|---|---|---|
| 2 | Fa | A |
| 3 | $Raa \leftrightarrow (Fa \land \neg Raa)$ | 1 $\forall$E |
| 4 | Raa | A |
| 5 | $Fa \land \neg Raa$ | 3, 4 $\leftrightarrow$E |
| 6 | $\neg Raa$ | 5 $\land$E |
| 7 | Raa | 4R |
| 8 | $\neg Raa$ | 4–7 $\neg$I |
| 9 | $Fa \land \neg Raa$ | 2, 8 $\land$I |
| 10 | Raa | 3, 9 $\leftrightarrow$E |
| 11 | $\neg Fa$ | 2–10 $\neg$I |
| 12 | $(\forall y)(Ray \leftrightarrow (Fy \land \neg Ryy)) \rightarrow \neg Fa$ | 1–11 $\rightarrow$I |
| 13 | $(\forall x)((\forall y)(Rxy \leftrightarrow (Fy \land \neg Ryy)) \rightarrow \neg Fx)$ | 12 $\forall$I |

**D.**  **1.**

| 1 | $(\exists x)Fx \land P$ | P | $\vdash (\exists x)(Fx \land P)$ |
|---|---|---|---|
| 2 | $(\exists x)Fx$ | 1 $\land$E | |
| 3 | Fa | A | |
| 4 | P | 1 $\land$E | |
| 5 | $Fa \land P$ | 3, 4 $\land$I | |
| 6 | $(\exists x)(Fx \land P)$ | 5 $\exists$I | |
| 7 | $(\exists x)(Fx \land P)$ | 2, 3–6 $\exists$E | |

| 1 | $(\exists x)(Fx \land P)$ | P | $\vdash (\exists x)Fx \land P$ |
|---|---|---|---|
| 2 | $Fa \land P$ | A | |
| 3 | Fa | 2 $\land$E | |
| 4 | $(\exists x)Fx$ | 3 $\exists$I | |
| 5 | P | 2 $\land$E | |
| 6 | $(\exists x)Fx \land P$ | 4, 5 $\land$I | |
| 7 | $(\exists x)Fx \land P$ | 1, 2–6 $\exists$E | |

**2.**

| 1 | $(\forall x)Fx \lor P$ | P | $\vdash (\forall x)(Fx \lor P)$ |
|---|---|---|---|
| 2 | $(\forall x)Fx$ | A | |
| 3 | Fa | 2 $\forall$E | |
| 4 | $Fa \lor P$ | 3 $\lor$I | |
| 5 | P | A | |
| 6 | $Fa \lor P$ | 5 $\lor$I | |
| 7 | $Fa \lor P$ | 1, 2–4, 5–6 $\lor$E | |
| 8 | $(\forall x)(Fx \lor P)$ | 7 $\forall$I | |

| 1 | $(\forall x)(Fx \vee P)$ | P | $\vdash (\forall x)Fx \vee P$ |
|---|---|---|---|
| 2 | $\neg((\forall x)Fx \vee P)$ | A | |
| 3 | $\neg(\forall x)Fx \wedge \neg P$ | 2 DeM | |
| 4 | $\neg P$ | 3 $\wedge$E | |
| 5 | $Fa \vee P$ | 1 $\forall$E | |
| 6 | $Fa$ | 4, 5 DS | |
| 7 | $(\forall x)Fx$ | 6 $\forall$I | |
| 8 | $\neg(\forall x)Fx$ | 3 $\wedge$E | |
| 9 | $(\forall x)Fx \vee P$ | 2–8 $\neg$E | |

**3.**

| 1 | $(\exists x)Fx \vee P$ | P | $\vdash (\exists x)(Fx \vee P)$ |
|---|---|---|---|
| 2 | $(\exists x)Fx$ | A | |
| 3 | $Fa$ | A | |
| 4 | $Fa \vee P$ | 3 $\vee$I | |
| 5 | $(\exists x)(Fx \vee P)$ | 4 $\exists$I | |
| 6 | $(\exists x)(Fx \vee P)$ | 2, 3–5 $\exists$E | |
| 7 | $P$ | A | |
| 8 | $Fa \vee P$ | 7 $\vee$I | |
| 9 | $(\exists x)(Fx \vee P)$ | 8 $\exists$I | |
| 10 | $(\exists x)(Fx \vee P)$ | 1, 2–6, 7–9 $\vee$E | |

| 1 | $(\exists x)(Fx \vee P)$ | P | $\vdash (\exists x)Fx \vee P$ |
|---|---|---|---|
| 2 | $Fa \vee P$ | A | |
| 3 | $Fa$ | A | |
| 4 | $(\exists x)Fx$ | 3 $\exists$I | |
| 5 | $(\exists x)Fx \vee P$ | 4 $\vee$I | |
| 6 | $P$ | A | |
| 7 | $(\exists x)Fx \vee P$ | 6 $\vee$I | |
| 8 | $(\exists x)Fx \vee P$ | 2, 3–5, 6–7 $\vee$E | |
| 9 | $(\exists x)Fx \vee P$ | 1, 2–8 $\exists$E | |

**4.**

| 1 | $P \to (\forall x)Fx$ | P | $\vdash (\forall x)(P \to Fx)$ |
|---|---|---|---|
| 2 | $P$ | A | |
| 3 | $(\forall x)Fx$ | 1, 2 $\to$E | |
| 4 | $Fa$ | 3 $\forall$E | |
| 5 | $P \to Fa$ | 2–4 $\to$I | |
| 6 | $(\forall x)(P \to Fx)$ | 5 $\forall$I | |

| 1 | $(\forall x)(P \rightarrow Fx)$ | P | $\vdash P \rightarrow (\forall x)Fx$ |
|---|---|---|---|
| 2 | P | A | |
| 3 | $P \rightarrow Fa$ | 1 $\forall$E | |
| 4 | Fa | 2, 3 $\rightarrow$E | |
| 5 | $(\forall x)Fx$ | 4 $\forall$I | |
| 6 | $P \rightarrow (\forall x)Fx$ | 2–5 $\rightarrow$I | |

**5.**

| 1 | $(\exists x)(P \rightarrow Fx)$ | P | $\vdash P \rightarrow (\exists x)Fx$ |
|---|---|---|---|
| 2 | P | A | |
| 3 | $P \rightarrow Fa$ | A | |
| 4 | Fa | 2, 3 $\rightarrow$E | |
| 5 | $(\exists x)Fx$ | 4 $\exists$I | |
| 6 | $(\exists x)Fx$ | 1, 3–5 $\exists$E | |
| 7 | $P \rightarrow (\exists x)Fx$ | 2–6 $\rightarrow$I | |

| 1 | $P \rightarrow (\exists x)Fx$ | P | $\vdash (\exists x)(P \rightarrow Fx)$ |
|---|---|---|---|
| 2 | $\neg(\exists x)(P \rightarrow Fx)$ | A | |
| 3 | $(\forall x)\neg(P \rightarrow Fx)$ | 2 QN | |
| 4 | $(\forall x)\neg(\neg P \vee Fx)$ | 3 Impl | |
| 5 | $(\forall x)(\neg\neg P \wedge \neg Fx)$ | 4 DeM | |
| 6 | $\neg\neg P \wedge \neg Fa$ | 5 $\forall$E | |
| 7 | $\neg\neg P$ | 6 $\wedge$E | |
| 8 | P | 7 DN | |
| 9 | $\neg Fa$ | 6 $\wedge$E | |
| 10 | $(\exists x)Fx$ | 1, 8 $\rightarrow$E | |
| 11 | $(\forall x)\neg Fx$ | 9 $\forall$I | |
| 12 | $\neg(\exists x)Fx$ | 11 QN | |
| 13 | $(\exists x)(P \rightarrow Fx)$ | 2–12 $\neg$E | |

**6.**

| 1 | $(\forall x)(Fx \rightarrow P)$ | P | $\vdash (\exists x)Fx \rightarrow P$ |
|---|---|---|---|
| 2 | $(\exists x)Fx$ | A | |
| 3 | Fa | A | |
| 4 | $Fa \rightarrow P$ | 1 $\forall$E | |
| 5 | P | 3, 4 $\rightarrow$E | |
| 6 | P | 2, 3–5 $\exists$E | |
| 7 | $(\exists x)Fx \rightarrow P$ | 2–6 $\rightarrow$I | |

|   |   |   |   |
|---|---|---|---|
| 1 | $(\exists x)Fx \rightarrow P$ | P | $\vdash (\forall x)(Fx \rightarrow P)$ |
| 2 | Fa | A |   |
| 3 | $(\exists x)Fx$ | $2\ \exists I$ |   |
| 4 | P | $1,\ 3 \rightarrow E$ |   |
| 5 | $Fa \rightarrow P$ | $2\text{--}4 \rightarrow I$ |   |
| 6 | $(\forall x)(Fx \rightarrow P)$ | $5\ \forall I$ |   |

**E.  1.**

|   |   |   |   |
|---|---|---|---|
| 1 | $(\exists x)(Fx \wedge Gx)$ | P |   |
| 2 | $(\forall x)(\neg Dx \rightarrow \neg Fx)$ | P |   |
| 3 | $\neg(\exists x)Dx$ | P | $\vdash \mathbb{P},\ \neg\mathbb{P}$ |
| 4 | $\neg Da \rightarrow \neg Fa$ | $2\ \forall E$ |   |
| 5 | $(\forall x)\neg Dx$ | $3\ QN$ |   |
| 6 | $\neg Da$ | $5\ \forall E$ |   |
| 7 | $\neg Fa$ | $4,\ 6 \rightarrow E$ |   |
| 8 | $\neg Fa \vee \neg Ga$ | $7\ \vee I$ |   |
| 9 | $\neg(Fa \wedge Ga)$ | $8\ DeM$ |   |
| 10 | $(\forall x)\neg(Fx \wedge Gx)$ | $9\ \forall I$ |   |
| 11 | $\neg(\exists x)(Fx \wedge Gx)$ | $10\ QN$ |   |
| 12 | $(\exists x)(Fx \wedge Gx)$ | $1R$ |   |

**2.**

|   |   |   |   |
|---|---|---|---|
| 1 | $(\exists y)(\forall x)Lxy$ | P |   |
| 2 | $\neg(\exists x)Lxx$ | P | $\vdash \mathbb{P},\ \neg\mathbb{P}$ |
| 3 | $(\forall x)\neg Lxx$ | $2\ QN$ |   |
| 4 | $\neg Laa$ | $3\ \forall E$ |   |
| 5 | $(\exists x)\neg Lxa$ | $4\ \exists I$ |   |
| 6 | $(\forall y)(\exists x)\neg Lxy$ | $5\ \forall I$ |   |
| 7 | $(\forall y)\neg(\forall x)Lxy$ | $6\ QN$ |   |
| 8 | $\neg(\exists y)(\forall x)Lxy$ | $7\ QN$ |   |
| 9 | $(\exists y)(\forall x)Lxy$ | $1R$ |   |

| 3. | 1 | $(\forall x)(\forall y)(\forall z)((Fxy \wedge Fyz) \rightarrow Fxz)$ | P | |
|---|---|---|---|---|
| | 2 | $(\exists x)(\exists y)(Fxy \wedge Fyx)$ | P | |
| | 3 | $\neg(\exists x)Fxx$ | P | $\vdash \mathbb{P}, \neg\mathbb{P}$ |
| | 4 | $(\forall y)(\forall z)((Fay \wedge Fyz) \rightarrow Faz)$ | 1 $\forall$E | |
| | 5 | $(\forall z)((Fab \wedge Fbz) \rightarrow Faz)$ | 4 $\forall$E | |
| | 6 | $(Fab \wedge Fba) \rightarrow Faa$ | 5 $\forall$E | |
| | 7 | $(\forall x)\neg Fxx$ | 3 QN | |
| | 8 | $\neg Faa$ | 7 $\forall$E | |
| | 9 | $\neg(Fab \wedge Fba)$ | 6, 8 MT | |
| | 10 | $(\forall y)\neg(Fay \wedge Fya)$ | 9 $\forall$I | |
| | 11 | $(\forall x)(\forall y)\neg(Fxy \wedge Fyx)$ | 10 $\forall$I | |
| | 12 | $(\forall x)\neg(\exists y)(Fxy \wedge Fyx)$ | 11 QN | |
| | 13 | $\neg(\exists x)(\exists y)(Fxy \wedge Fyx)$ | 12 QN | |
| | 14 | $(\exists x)(\exists y)(Fxy \wedge Fyx)$ | 2 R | |

# A.26 **Answers to 8.4.1**

Note: many of these proofs can be done in **PDE**, here I have treated them less formally.

1. Suppose $A = \{a\}$, $B = \{A\}$, and $C = \{B\}$. Then we have $A \in B$, and $B \in C$, but $A \notin C$.

2. While, as argued in the text, for all $\Gamma$, $\varnothing \subseteq \Gamma$, the situation is different for proper subset. It is not the case that for all $\Gamma$, $\varnothing \subset \Gamma$. In particular, $\varnothing \not\subset \varnothing$. Suppose the opposite: $\varnothing \subset \varnothing$. Then by the definition of proper subset, we have $(\forall x)((x \in \varnothing \rightarrow x \in \varnothing) \wedge \varnothing \neq \varnothing)$. While the left conjunct is vacuously true, the right conjunct is always false, making the whole universal claim false. So it must be that $\varnothing \not\subset \varnothing$.

3. Generalize the argument above: Suppose the opposite: $\Gamma \subset \Gamma$. By the definition of proper subset, we have $(\forall x)((x \in \Gamma \rightarrow x \in \Gamma) \wedge \Gamma \neq \Gamma)$. While the left conjunct is vacuously true, the right conjunct is always false, making the whole universal claim false. So it must be that $\Gamma \not\subset \Gamma$.

4. Suppose the antecedent, $A \subseteq B \wedge B \subseteq C$. By the definition of subset, we have for all $x$, $x \in A \rightarrow x \in B \wedge x \in B \rightarrow x \in C$. Hypothetical Syllogism yields $x \in A \rightarrow x \in C$, that is: $A \subseteq C$.

5. Suppose the antecedent, $A \subseteq B \wedge B \subseteq A$. By the definition of subset, we have for all $x$, $x \in A \rightarrow x \in B \wedge x \in B \rightarrow x \in A$. So we have $x \in A \leftrightarrow x \in B$. Thus, by the Axiom of Extensionality, $A = B$.

6. By definition: $A \cup \varnothing = \{x \mid x \in A \vee x \in \varnothing\}$. Since $\varnothing$ has no members, this is equivalent to $\{x \mid x \in A\}$, which is A. So, $A \cup \varnothing = A$.

7. This follows from the commutativity of disjunction:

$$A \cup B = \{x \mid x \in A \vee x \in B\}$$
$$= \{x \mid x \in B \vee x \in A\}$$
$$= B \cup A$$

**8.** This follows from the associativity of disjunction:

$$A \cup (B \cup C) = \{x \mid x \in A \vee (x \in B \vee x \in C)\}$$
$$= \{x \mid (x \in A \vee x \in B) \vee x \in C\}$$
$$= (A \cup B) \cup C$$

**9.** This follows from the idempotence of disjunction:

$$A \cup A = \{x \mid x \in A \vee x \in A\}$$
$$= \{x \mid x \in A\}$$
$$= A$$

**10.** By definition: $A \cap \varnothing = \{x \mid x \in A \wedge x \in \varnothing\}$. Since $\varnothing$ has no members, $A \cap \varnothing$ has no members, so it is $\varnothing$. So, $A \cup \varnothing = A$.

**11.** This follows from the commutativity of conjunction, similar to 7.

**12.** This follows from the associativity of conjunction, similar to 8.

**13.** This follows from the idempotence of conjunction, similar to 9.

**14.** Suppose $A \subseteq B$. Then for every x, $x \in A \rightarrow x \in B$. Now suppose that $A \cup B \neq B$. So either i) there is some x, $x \in A \cup B \wedge x \notin B$, which implies there is some x, $x \in A \wedge x \notin B$, which contradicts the initial assumption; or ii) there is some x, $x \notin A \cup B \wedge x \in B$, which implies that there is some x, $x \notin B \wedge x \in B$, which is a contradiction. So $A \cup B = B$.

Suppose $A \cup B = B$. Now suppose $A \not\subseteq B$. So there must be some x, $x \in A \wedge x \notin B$. But this implies that there is some x, $x \in A \cup B \wedge x \notin B$, which contradicts the initial assumption. So $A \subseteq B$.

**15.** Suppose $A \subseteq B$. Then for every x, $x \in A \rightarrow x \in B$. Now suppose that $A \cap B \neq A$. So either i) there is some x, $x \in A \cap B \wedge x \notin A$, which implies there is some x, $x \in A \wedge x \notin A$, which is a contradiction; or ii) there is some x, $x \notin A \cap B \wedge x \in A$, which implies there is some x, $(x \notin A \vee x \notin B) \wedge x \in A$. But this implies that there is some x, $x \in A \wedge x \notin B$, which contradicts the initial assumption. So $A \cap B = A$.

Suppose $A \cap B = A$. So for every x, $x \in A \cap B \leftrightarrow x \in A$. This implies that for every x, $x \in A \rightarrow x \in B$. I.e., $A \subseteq B$.

**16.** This follows from the distributive property of conjunction and disjunction:

$$A \cap (B \cup C) = \{x \mid x \in A \wedge (x \in B \vee x \in C)\}$$
$$= \{x \mid (x \in A \wedge x \in B) \vee (x \in A \wedge x \in C)\}$$
$$= (A \cap B) \cup (A \cap C)$$

**17.** This follows from the distributive property of conjunction and disjunction, similar to 16.

**18.** Suppose $A = \{a, b, c\}$ and $B = \{b, c, d\}$. Then $A - B = \{a\}$ and $B - A = \{d\}$.

**19.** By definition: $A - A = \{x \mid x \in A \wedge x \notin A\} = \varnothing$

**20.** By definition:[1] $\overline{\overline{A}} = \{x \notin \overline{A}\} = \{x \in A\} = A$

**21.** By definition: $\overline{\varnothing} = \{x \in K \mid x \notin \varnothing\} = \{x \in K\} = K$

**22.** By definition: $A \cap \overline{A} = \{x \in A \wedge x \notin A\} = \varnothing$

**23.** By definition: $A \cup \overline{A} = \{x \in K \mid x \in A \vee x \notin A\} = K$

**24.** This follows from the transposition equivalence:

$$A \subseteq B \leftrightarrow (\forall x)(x \in A \rightarrow x \in B)$$
$$\leftrightarrow (\forall x)(x \notin B \rightarrow x \notin A)$$
$$\leftrightarrow (\forall x)(x \in \overline{B} \rightarrow x \in \overline{A})$$
$$\leftrightarrow \overline{B} \subseteq \overline{A}$$

**25.** This follows from the De Morgan equivalences for negation, conjunction, and disjunction:

$$\overline{A \cap B} = \{x \mid x \notin A \cap B\}$$
$$= \{x \mid \neg(x \in A \wedge x \in B)\}$$
$$= \{x \mid x \notin A \vee x \notin B\}$$
$$= \overline{A} \cup \overline{B}$$

**26.** This follows from the De Morgan equivalences for negation, conjunction, and disjunction, similar to 25.

**27.** By definition:

$$A - B = \{x \mid x \in A \wedge x \notin B\}$$
$$= \{x \mid x \in A \wedge x \in \overline{B}\}$$
$$= A \cap \overline{B}$$

**28.** This follows from the definition of subset and difference, implication, De Morgan's, and quantifier negation:

$$A \subseteq B \leftrightarrow (\forall x)(x \in A \rightarrow x \in B)$$
$$\leftrightarrow (\forall x)(x \notin A \vee x \in B)$$
$$\leftrightarrow (\forall x)\neg(x \in A \wedge x \notin B)$$
$$\leftrightarrow \neg(\exists x)(x \in A \wedge x \notin B)$$
$$\leftrightarrow \neg(\exists x)(x \in A - B)$$
$$\leftrightarrow A - B = \varnothing$$

---

1. Recall that when dealing with the temporarily absolute complement, K is a superset of all the sets of current interest. We invoke it only when needed. See, p. 278.

**29.** By definition, De Morgan, distribution:

$$
\begin{aligned}
A - (A - B) &= \{x \mid x \in A \wedge x \notin A - B\} \\
&= \{x \mid x \in A \wedge \neg(x \in A \wedge x \notin B)\} \\
&= \{x \mid x \in A \wedge (x \notin A \vee x \in B)\} \\
&= \{x \mid (x \in A \wedge x \notin A) \vee (x \in A \wedge x \in B)\} \\
&= \{x \mid x \in A \wedge x \in B\} \\
&= A \cap B
\end{aligned}
$$

**30.** By definition, De Morgan, and this equivalence, $\mathbb{P} \wedge \mathbb{Q} \Leftrightarrow \mathbb{P} \wedge (\neg \mathbb{P} \vee \mathbb{Q})$:

$$
\begin{aligned}
A \cap (B - C) &= \{x \mid x \in A \wedge (x \in B \wedge x \notin C)\} \\
&= \{x \mid (x \in A \wedge x \in B) \wedge x \notin C\} \\
&= \{x \mid (x \in A \wedge x \in B) \wedge (x \notin A \vee x \notin C)\} \\
&= \{x \mid (x \in A \wedge x \in B) \wedge \neg(x \in A \wedge x \in C)\} \\
&= (A \cap B) - (A \cap C)
\end{aligned}
$$

**31.** By definition, and $(\forall x) x \notin \varnothing$:

$$
\begin{aligned}
A \ominus \varnothing &= (A - \varnothing) \cup (\varnothing - A) \\
&= \{x \mid (x \in A \wedge x \notin \varnothing) \vee (x \in \varnothing \wedge x \notin A)\} \\
&= \{x \mid (x \in A \wedge x \notin \varnothing)\} \\
&= \{x \mid x \in A\} \\
&= A
\end{aligned}
$$

**32.** By definition, idempotence, and contradiction:

$$
\begin{aligned}
A \ominus A &= (A - A) \cup (A - A) \\
&= \{x \mid (x \in A \wedge x \notin A) \vee (x \in A \wedge x \notin A)\} \\
&= \{x \mid x \in A \wedge x \notin A\} \\
&= \varnothing
\end{aligned}
$$

**33.** By definition, the equivalence $(\mathbb{P} \wedge \neg \mathbb{Q}) \vee (\neg \mathbb{P} \wedge \mathbb{Q}) \Leftrightarrow (\mathbb{P} \vee \mathbb{Q}) \wedge (\neg \mathbb{P} \vee \neg \mathbb{Q})$:

$$
\begin{aligned}
A \ominus B &= (A - B) \cup (B - A) \\
&= \{x \mid (x \in A \wedge x \notin B) \vee (x \in B \wedge x \notin A)\} \\
&= \{x \mid (x \in A \vee x \in B) \wedge (x \notin A \vee x \notin B)\} \\
&= \{x \mid (x \in A \vee x \in B) \wedge \neg(x \in A \wedge x \in B)\} \\
&= (A \cup B) - (A \cap B)
\end{aligned}
$$

**34.** By definition, and commutativity of union (7. above):

$$
\begin{aligned}
A \ominus B &= (A - B) \cup (B - A) \\
&= (B - A) \cup (A - B) \\
&= B \ominus A
\end{aligned}
$$

## A.27 **Answers to 8.6.1**

1. Here is one: $f(x) = \frac{x+1}{2}$

$$\mathbb{O} = \{1, \quad 3, \quad 5, \quad 7, \quad 9,\ldots\}$$
$$\downarrow \quad \downarrow \quad \downarrow \quad \downarrow \quad \downarrow \cdots$$
$$\mathbb{Z}_{>0} = \{1, \quad 2, \quad 3, \quad 4, \quad 5,\ldots\}$$

2. Here is one: $f(x) = \frac{x+1}{2} - 1$

$$\mathbb{O} = \{1, \quad 3, \quad 5, \quad 7, \quad 9,\ldots\}$$
$$\downarrow \quad \downarrow \quad \downarrow \quad \downarrow \quad \downarrow \cdots$$
$$\mathbb{N} = \{0, \quad 1, \quad 2, \quad 3, \quad 4,\ldots\}$$

3. Here is one: $f(x) = x + 1$

$$\mathbb{O} = \{1, \quad 3, \quad 5, \quad 7, \quad 9,\ldots\}$$
$$\downarrow \quad \downarrow \quad \downarrow \quad \downarrow \quad \downarrow \cdots$$
$$\mathbb{E} = \{2, \quad 4, \quad 6, \quad 8, \quad 10,\ldots\}$$

4. No. Suppose there were a barber who lives in town—and therefore is a townsperson—who shaves all and only those townspeople who do not shave themselves. Who shaves the barber? If he shaves himself, then he does not shave himself. If he does not shave himself, then he does. Thus he shaves himself if and only if he does not shave himself—a contradiction! He cannot exist as described. $(\exists x)(\forall y)(Tx \wedge (Sxy \leftrightarrow (Ty \wedge \neg Syy)))$ is quantificationally false. I.e., $\neg(\exists x)(\forall y)(Tx \wedge (Sxy \leftrightarrow (Ty \wedge \neg Syy)))$ is quantificationally true.

   This, of course, is Russell's paradox in different guise. See next question for more.

5. Yes, but he must not live in town. Suppose there is a barber who shaves all and only those townspeople who do not shave themselves. He must not live in town. Suppose he does live in town. Then he is a townsperson, and by the argument above, he cannot exist. Thus he must not live in town. $(\exists x)(\forall y)(Sxy \leftrightarrow (Ty \wedge \neg Syy))$ is quantificationally satisfiable, as long as the x in question does does not satisfy T. Thus, $(\forall x)((\forall y)(Sxy \leftrightarrow (Ty \wedge \neg Syy)) \rightarrow \neg Tx)$ is quantificationally true.

   This answer is the solution to Russell's Paradox offered by the Axiom Schema of Separation. Here, the barber can exist, but he must not live in town and he must not shave himself. Above (Sec. 8.3), the set B must not be in set A and must not be in itself.

## A.28 **Answers to 9.2.2**

**A.**    1. Suppose $\Diamond \mathbb{P}$ is T. So $\mathbb{P}$ is T in some world. Thus, $\neg \mathbb{P}$ is F in some world. So $\Box \neg \mathbb{P}$ is F. Thus $\neg \Box \neg \mathbb{P}$ is T. Each step is an equivalence, so we have proved the equivalence.

2. Suppose that $\Diamond\neg\mathbb{P}$ is T. So there is some world at which $\neg\mathbb{P}$ is T; i.e., some world at which $\mathbb{P}$ is F. This is equivalent to $\Box\mathbb{P}$ is F, which is to say, $\neg\Box\mathbb{P}$ is T.

3. Suppose $\neg\Diamond\mathbb{P}$ is T. Thus, there is no world at which $\mathbb{P}$ is T. So at every world $\neg\mathbb{P}$ is T. This equates to $\Box\neg\mathbb{P}$.

4. Suppose $\Diamond\mathbb{P}\wedge\Diamond\neg\mathbb{P}$ is T. So there is some world, $w_1$, at which $\mathbb{P}$ is T, and there is some world, $w_2$, at which $\mathbb{P}$ is F. The existence of $w_2$ means that $\Box\mathbb{P}$ is F, and the existence of $w_1$ means that $\Box\neg\mathbb{P}$ is F. This is equivalent to $\neg(\Box\mathbb{P}\vee\Box\neg\mathbb{P})$.

5. Since the interpretation of $\Box\mathbb{P}$ involves truth in all worlds there is implicit *universal* quantification. Since the interpretation of $\Diamond\mathbb{P}$ involves truth in at least one world there is implicit *existential* quantification. $(\forall x)\mathbb{F}x$ is interdefinable as $\neg(\exists x)\neg\mathbb{F}x$, so, equally, $\Box\mathbb{P}$ is interdefinable as $\neg\Diamond\neg\mathbb{P}$.

**B.**

1. $\Box E$

2. $\neg\Box F$

3. $\neg\neg\Diamond F$
   $\Diamond F$

4. $\neg\Diamond A$
   $\Box\neg A$

5. $\Diamond M$

6. $\Diamond\neg M$

7. $\Diamond F\wedge\Diamond\neg F$

8. $\Box(E\wedge F)$

9. $\Box E\wedge\Box F$

10. $\Box(A\vee M)$

11. $\Box A\vee\Box M$

12. $\Diamond(M\wedge E)$

13. $\Diamond M\wedge\Diamond E$

14. $\Diamond A\vee\Diamond F$

15. $\Diamond(A\vee F)$

16. $\Box(M\vee\neg M)$

17. $\Box M\vee\Box\neg M$

18. $\Box(\Diamond A\vee\neg\Diamond A)$

19. $\neg\Box\neg\Diamond A$
    $\Diamond\Diamond A$

20. $\Box F\rightarrow\Box E$

21. $\Box(F\rightarrow E)$

22. $\neg\Diamond A\rightarrow\neg\Box\neg M$
    $\neg\Diamond A\rightarrow\Diamond M$

23. $(\Diamond F\wedge\Diamond\neg F)\leftrightarrow\neg(\Diamond E\wedge\Diamond\neg E)$
    $(\Diamond F\wedge\Diamond\neg F)\leftrightarrow(\Box\neg E\vee\Box E)$

24. $\Box((E\wedge F)\rightarrow\neg\Diamond A)$

**25.** $\Box(E \wedge F) \rightarrow \neg \Diamond A$

# A.29 **Answers to 9.3.2**

**A.** Using the definition of truth at a world in $S^\Box$, prove each of the following modal equivalences:

**1.** $\Box P \Leftrightarrow \Box \Box P$

Suppose $\Box P$ is T at some $w \in \mathbf{W}$, this means that $P$ is T at every $x \in \mathbf{W}$. Thus, no matter what $x \in \mathbf{W}$ we consider, $\Box P$ is T, which means $\Box \Box P$ is T at w. Now suppose $\Box \Box P$ is T at some $w \in \mathbf{W}$, thus $\Box P$ is T at every $x \in \mathbf{W}$, in particular w. So the equivalence holds for any world w.

**2.** $\Diamond P \Leftrightarrow \Box \Diamond P$

Suppose $\Diamond P$ is T at some $w \in \mathbf{W}$, this means that $P$ is T at some $x \in \mathbf{W}$, making $\Diamond P$ T at every $y \in \mathbf{W}$. So, $\Box \Diamond P$ is T at w. Now suppose $\Box \Diamond P$ is T at w, thus $\Diamond P$ is T at every $x \in \mathbf{W}$, in particular w. So the equivalence holds for any world w.

**3.** $\Box P \Leftrightarrow \Diamond \Box P$

Suppose $\Box P$ is T at some $w \in \mathbf{W}$, this means that $\Diamond \Box P$ is T at w. Now suppose $\Diamond \Box P$ is T at some $w \in \mathbf{W}$, thus $\Box P$ is T at some $x \in \mathbf{W}$. Thus, $P$ is T at every $y \in \mathbf{W}$, making $\Box P$ T at w. So the equivalence holds for any world w.

**4.** $\Diamond P \Leftrightarrow \Diamond \Diamond P$

Suppose $\Diamond P$ is T at some $w \in \mathbf{W}$, this means that $\Diamond \Diamond P$ is T at w. Now suppose $\Diamond \Diamond P$ is T at some $w \in \mathbf{W}$, thus $\Diamond P$ is T at some $x \in \mathbf{W}$. Thus, $P$ is T at some $y \in \mathbf{W}$, making $\Diamond P$ T at w. So the equivalence holds for any world w.

**B.**  **1.**

| $\bullet w_1$ | $\bullet w_2$ |
|---|---|
| A | $\neg$A |
| B | B |

**2.**

| $\bullet w_1$ | $\bullet w_2$ |
|---|---|
| G | $\neg$G |
| $\neg$H | H |

**3.**

| $\bullet w_1$ | $\bullet w_2$ |
|---|---|
| A | $\neg$A |
| B | B |

**4.**

| $\bullet w_1$ | $\bullet w_2$ |
|---|---|
| H | $\neg$H |
| J | $\neg$J |

# A.30 **Answers to 9.3.4**

**A.**    **1.** A → □A

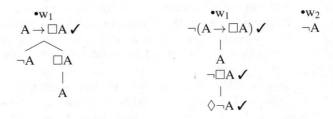

Since tree structures for it and its negation are open, 'A → □A' is Modally Satisfiable, Modally Falsifiable, and Modally Contingent.

**2.** A → ◊A

•w₁
¬(A → ◊A) ✔
|
A
¬◊A ✔
|
□¬A ✔
|
¬A
✗

Since the tree structure for its negation is closed, '¬(A → ◊A)' is Modally True.

**3.** $\Diamond(A \wedge B) \wedge (\neg \Diamond A \vee \neg \Diamond B)$

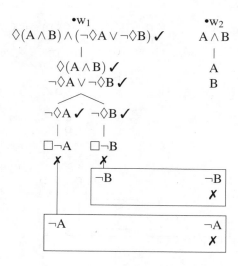

Since the tree at $w_1$ closes, the whole tree structure is closed. So, '$\Diamond(A \wedge B) \wedge (\neg \Diamond A \vee \neg \Diamond B)$' is Modally False.

**B.** **1.** A    $\Box$A

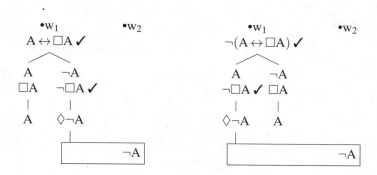

Since both tree structures are open, 'A' and '$\Box$A' are neither Modally Equivalent nor Modally Contradictory.

**2.** $\Box(C \wedge D)$     $\Box C \wedge \Box D$

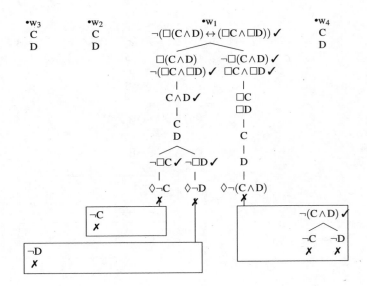

Since the tree structure closes, '$\Box(C \wedge D)$' and '$\Box C \wedge \Box D$' are Modally Equivalent.

**3.** $\Box(G \wedge H)$     $\Diamond(\neg G \vee \neg H)$

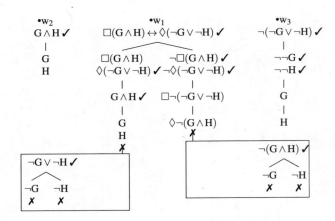

Since the tree structure closes, '$\Box(G \wedge H)$' and '$\Diamond(\neg G \vee \neg H)$' are Modally Contradictory.

**C.    1.** $\{\Box(A \rightarrow B), \Box(A \land C), \Diamond \neg B\}$

```
              •w₁                    •w₂
           □(A→B)                    ¬B
           □(A∧C)                   A∧C ✓
           ◇¬B ✓                   A→B ✓
                                     |
                                     A
                                     C
                                    ╱╲
                                  ¬A   B
                                  ✗    ✗
```

Since the tree structure is closed, the set is Modally Inconsistent.

**2.** $\{\Diamond(D \land E), \Diamond \neg E, \neg \Box D, \Box(D \lor E)\}$

```
      •w₁              •w₂           •w₃          •w₃
   ◇(D∧E) ✓          D∧E ✓          ¬E           ¬D
    ◇¬E ✓             |              |            |
    ¬□D ✓             D            D∨E ✓        D∨E ✓
   □(D∨E)             E             ╱╲           ╱╲
     |                |            D   E         D   E
   ◇¬D ✓            D∨E ✓          ✗             ✗
     |               ╱╲
   D∨E ✓            D   E
    ╱╲
   D   E
```

Since the tree structure is open, the set is Modally Consistent.

**D.    1.** $A \vDash \Box \Diamond A$

```
              •w₁                    •w₂
               A                    □¬A
            ¬□◇A ✓                    |
               |                     ¬A
             ◇¬◇A ✓
               |
             ◇□¬A ✓
               |
              ¬A
               ✗
```

Since the tree structure is closed, the entailment holds.

**2.** $B \rightarrow \Diamond B \nvDash \Diamond B \rightarrow \Box B$

$$
\begin{array}{lll}
\bullet w_1 & \bullet w_2 & \bullet w_3 \\
B \rightarrow \Diamond B \checkmark & \neg B & B \\
\neg(\Diamond B \rightarrow \Box B) \checkmark & & \\
\mid & & \\
\Diamond B \checkmark & & \\
\neg \Box B \checkmark & & \\
\mid & & \\
\Diamond \neg B \checkmark & & \\
\end{array}
$$

$\neg B \qquad \Diamond B$

Since the tree structure is open, the entailment fails. Note that we don't decompose the '$\Diamond B$' in the right branch of $w_1$, because it is accounted for in $w_3$.

**3.** $\Diamond D \wedge \Diamond E \nvDash \Diamond(D \wedge E)$

$$
\begin{array}{lll}
\bullet w_1 & \bullet w_2 & \bullet w_2 \\
\Diamond D \wedge \Diamond E \checkmark & D & E \\
\neg\Diamond(D \wedge E) \checkmark & \neg(D \wedge E) \checkmark & \neg(D \wedge E) \checkmark \\
\mid & & \\
\Box\neg(D \wedge E) & \neg D \quad E & \neg D \quad E \\
\mid & \times & \times \\
\Diamond D \checkmark & & \\
\Diamond E \checkmark & & \\
\mid & & \\
\neg(D \wedge E) & & \\
\end{array}
$$

$\neg D \qquad \neg E$

Since the tree structure is open, the entailment fails.

**4.** $\Diamond(D \wedge E) \vDash \Diamond D \wedge \Diamond E$

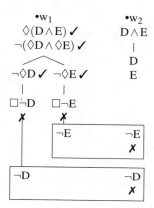

Since the tree at $w_1$ closes, the whole tree structure is closed. So the entailment holds.

**5.** $\Box A \vDash \Box(B \to A)$

| $\bullet w_1$ | $\bullet w_2$ |
|---|---|
| $\Box A$ | $\neg(B \to A)$ ✓ |
| $\neg\Box(B \to A)$ ✓ | $\mid$ |
| $\mid$ | $B$ |
| $\Diamond\neg(B \to A)$ ✓ | $\neg A$ |
| $\mid$ | $\mid$ |
| $A$ | $A$ |
|  | ✗ |

Since the tree structure is closed, the entailment holds.

**6.** $A \to \Diamond A \nvDash \Diamond A \to \Box A$

| $\bullet w_1$ | $\bullet w_2$ | $\bullet w_3$ | $\bullet w_4$ |
|---|---|---|---|
| $A \to \Diamond A$ ✓ | $A$ | $\neg A$ |  |
| $\neg(\Diamond A \to \Box A)$ ✓ |  |  |  |
| $\mid$ |  |  |  |
| $\Diamond A$ ✓ |  |  |  |
| $\neg\Box A$ ✓ |  |  |  |
| $\mid$ |  |  |  |
| $\Diamond\neg A$ ✓ |  |  |  |

$$A \quad\quad \Diamond A$$

| | | | $A$ |
|---|---|---|---|

Since the tree structure is open, the entailment fails.

**7.** ◊◊H ⊭ ◊H

| •w₁ | •w₂ | •w₃ |
|---|---|---|
| ◊◊H ✓ | ◊H ✓ | H |
| ¬◊H ✓ | \| | \| |
| \| | ¬H | ¬H |
| □¬H | | ✗ |
| \| | | |
| ¬H | | |

Since the tree structure is open, the entailment fails.

**8.** {□(A → B), □B → □C, ¬□C} ⊨ ◊¬A

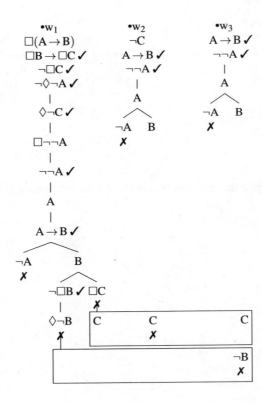

Since the tree structure is closed, the entailment holds.

## A.31 **Answers to 9.4.8**

**A.**    **1.**   
| | | |
|---|---|---|
| 1 | $\neg\Diamond P$ | P |
| 2 | $\quad\neg\Box\neg P$ | A |
| 3 | $\quad\Diamond P$ | 2 Dual |
| 4 | $\quad\neg\Diamond P$ | 1 R |
| 5 | $\Box\neg P$ | 2–4 $\neg$I |

**2.**  
| | | |
|---|---|---|
| 1 | $\Box\neg P$ | P |
| 2 | $\neg\neg\Box\neg P$ | 1 DN |
| 3 | $\neg\Diamond P$ | 2 Dual |

**3.**  
| | | |
|---|---|---|
| 1 | $\neg\Box P$ | P |
| 2 | $\quad\neg\Diamond\neg P$ | A |
| 3 | $\quad\Box P$ | 2 Dual |
| 4 | $\quad\neg\Box P$ | 1 R |
| 5 | $\Diamond\neg P$ | 2–4 $\neg$I |

**4.**  
| | | |
|---|---|---|
| 1 | $\Diamond\neg P$ | P |
| 2 | $\neg\neg\Diamond\neg P$ | 1 DN |
| 3 | $\neg\Box P$ | 2 Dual |

**5.**  
| | | |
|---|---|---|
| 1 | $\Box(P\rightarrow Q)$ | P |
| 2 | $\Box(\neg Q\rightarrow\neg P)$ | 1 Trans |
| 3 | $\Box\neg Q\rightarrow\Box\neg P$ | 2 K |
| 4 | $\neg\Diamond Q\rightarrow\Box\neg P$ | 3 ME |
| 5 | $\neg\Diamond Q\rightarrow\neg\Diamond P$ | 4 ME |
| 6 | $\Diamond P\rightarrow\Diamond Q$ | 5 Trans |

**6.**

| | | |
|---|---|---|
| 1 | $\Box(P \wedge Q)$ | P |
| 2 | $(P \wedge Q) \rightarrow P$ | $\vdash$ |
| 3 | $(P \wedge Q) \rightarrow Q$ | $\vdash$ |
| 4 | $\Box((P \wedge Q) \rightarrow P)$ | 2 N |
| 5 | $\Box((P \wedge Q) \rightarrow Q)$ | 3 N |
| 6 | $\Box(P \wedge Q) \rightarrow \Box P$ | 4 K |
| 7 | $\Box(P \wedge Q) \rightarrow \Box Q$ | 5 K |
| 8 | $\Box P$ | 1, 6 $\rightarrow$E |
| 9 | $\Box Q$ | 1, 7 $\rightarrow$E |
| 10 | $\Box P \wedge \Box Q$ | 8, 9 $\wedge$I |

**7.**

| | | |
|---|---|---|
| 1 | $\Box P \wedge \Box Q$ | P |
| 2 | $P \rightarrow (Q \rightarrow (P \wedge Q))$ | $\vdash$ |
| 3 | $\Box(P \rightarrow (Q \rightarrow (P \wedge Q)))$ | 2 N |
| 4 | $\Box P \rightarrow \Box(Q \rightarrow (P \wedge Q))$ | 3 K |
| 5 | $\Box P$ | 1 $\wedge$E |
| 6 | $\Box(Q \rightarrow (P \wedge Q))$ | 4, 5 $\rightarrow$E |
| 7 | $\Box Q \rightarrow \Box(P \wedge Q)$ | 6 K |
| 8 | $\Box Q$ | 1 $\wedge$E |
| 9 | $\Box(P \wedge Q)$ | 7, 8 $\rightarrow$E |

**8.**

| | | |
|---|---|---|
| 1 | $\Diamond(P \vee Q)$ | P |
| 2 | $\neg\Box\neg(P \vee Q)$ | 1 Dual |
| 3 | $\neg\Box(\neg P \wedge \neg Q)$ | 2 DeM |
| 4 | $\neg(\Box\neg P \wedge \Box\neg Q)$ | 3 Dist$\Box\Diamond$ |
| 5 | $\neg(\neg\Diamond P \wedge \Box\neg Q)$ | 4 ME |
| 6 | $\neg(\neg\Diamond P \wedge \neg\Diamond Q)$ | 5 ME |
| 7 | $\Diamond P \vee \Diamond Q$ | 6 DeM |

**9.**

| | | |
|---|---|---|
| 1 | $\Diamond P \vee \Diamond Q$ | P |
| 2 | $\neg\Diamond(P \vee Q)$ | A |
| 3 | $\Box\neg(P \vee Q)$ | 2 ME |
| 4 | $\Box(\neg P \wedge \neg Q)$ | 3 DeM |
| 5 | $\Box\neg P \wedge \Box\neg Q$ | 4 Dist$\Box\Diamond$ |
| 6 | $\neg\Diamond P \wedge \Box\neg Q$ | 5 ME |
| 7 | $\neg\Diamond P \wedge \neg\Diamond Q$ | 6 ME |
| 8 | $\neg(\Diamond P \vee \Diamond Q)$ | 7 DeM |
| 9 | $\Diamond P \vee \Diamond Q$ | 1 R |
| 10 | $\Diamond(P \vee Q)$ | 2–9 $\neg$E |

**B.**  **1.**  1 | P                          P

2 |   | ¬◇P                A

3 |   | □¬P                2 ME

4 |   | ¬P                  3 T

5 |   | P                    1 R

6 | ◇P                       2–5 ¬E


**2.**  1 | □P                         P

2 | P                          1 T

3 | ◇P                        2 T◇


**3.**  1 | ¬P                         P

2 | ◇¬P                      1 T◇

3 | ¬□P                       2 ME


**4.**  1 | P                          P

2 | ◇P                        1 T◇

3 | ◇◇P                      2 T◇

4 | ◇◇◇P                    3 T◇


**C.**  **1.**  1 | ◇□P                       P

2 |   | ¬P                   A

3 |   | □◇¬P              2 B

4 |   | □¬□P              3 ME

5 |   | ¬◇□P              4 ME

6 |   | ◇□P               1 R

7 | P                          2–6 ¬E


**D.**  **1.**  1 | ◇◇P                      P

2 |   | ¬◇P                A

3 |   | □¬P                2 ME

4 |   | □□¬P              3 S4

5 |   | □¬◇P              4 ME

6 |   | ¬◇◇P              5 ME

7 |   | ◇◇P               1 R

8 | ◇P                        2–7 ¬E

**2.**

| | | |
|---|---|---|
| 1 | $\square\square\square\square P$ | P |
| 2 | $\square\square\square P$ | 1 T |
| 3 | $\square\square P$ | 2 T |
| 4 | $\square P$ | 3 T |

**3.**

| | | |
|---|---|---|
| 1 | $\square P$ | P |
| 2 | $\square\square P$ | 1 S4 |
| 3 | $\square\square\square P$ | 2 S4 |
| 4 | $\square\square\square\square P$ | 3 S4 |

**4.**

| | | |
|---|---|---|
| 1 | $\Diamond\Diamond\Diamond\Diamond P$ | P |
| 2 | $\Diamond\Diamond\Diamond P$ | 1 S4$\Diamond$ |
| 3 | $\Diamond\Diamond P$ | 2 S4$\Diamond$ |
| 4 | $\Diamond P$ | 3 S4$\Diamond$ |

**5.**

| | | |
|---|---|---|
| 1 | $\Diamond P$ | P |
| 2 | $\Diamond\Diamond P$ | 1 T$\Diamond$ |
| 3 | $\Diamond\Diamond\Diamond P$ | 2 T$\Diamond$ |
| 4 | $\Diamond\Diamond\Diamond\Diamond P$ | 3 T$\Diamond$ |

**E.** **1.**

| | | |
|---|---|---|
| 1 | P | P |
| 2 | $\Diamond P$ | 1 T$\Diamond$ |
| 3 | $\square\Diamond P$ | 2 S5 |

**2.**

| | | |
|---|---|---|
| 1 | $\Diamond\square P$ | P |
| 2 | $\neg\square P$ | A |
| 3 | $\Diamond\neg P$ | 2 ME |
| 4 | $\square\Diamond\neg P$ | 3 S5 |
| 5 | $\square\neg\square P$ | 4 ME |
| 6 | $\neg\Diamond\square P$ | 5 ME |
| 7 | $\Diamond\square P$ | 1 R |
| 8 | $\square P$ | 2–7 ¬E |

**3.**  1 | $\Box P$ | P
     2 | $\Box\Diamond\Box P$ | 1 B
     3 | $\Diamond\Box P \to \Box P$ | $\vdash$ (S5$\Diamond\Box$)
     4 | $\Box(\Diamond\Box P \to \Box P)$ | 3 N
     5 | $\Box\Diamond\Box P \to \Box\Box P$ | 4 K
     6 | $\Box\Box P$ | 2, 5 $\to$E

**4.**  1 | $\Box\Diamond\Box P$ | P
     2 | $\Diamond\Box P$ | 1 T
     3 | $\Box P$ | 2 S5$\Diamond\Box$

**5.**  1 | $\Diamond\Box\Diamond P$ | P
     2 | $\Box\Diamond P$ | 1 S5$\Diamond\Box$
     3 | $\Diamond P$ | 2 T

# B  Glossary

**Abduction:**

*Abduction* or *abductive reasoning*, also known as *inference to the best explanation*, is a category of reasoning subject to inductive criteria in which the conclusion is supposed to explain the truth of the premises. (17)

**Absolute Complement:**

Given K, a superset of all the sets of current interest, the temporarily absolute complement of a set A, $\overline{A}$ is the set of all elements of K, not in A:

$$\overline{A} = \{x \in K \mid x \notin A\} \quad (278)$$

**Alethic Modality:**

The *alethic modalities*—necessity, possibility, contingency, impossibility—qualify the truth value of a statement. (300)

**Antisymmetry:**

$$(\forall x)(\forall y)((\mathbb{R}xy \wedge \mathbb{R}yx) \to x = y) \quad (229)$$

**Antitransitivity:**

$$(\forall x)(\forall y)(\forall z)((\mathbb{R}xy \wedge \mathbb{R}yz) \to \neg \mathbb{R}xz) \quad (230)$$

**Argument, Premise, Conclusion:**

An *argument* is a (finite) set of statements, some of which—the *premises*—are supposed to support, or give reasons for, the remaining statement—the *conclusion*. (6)

**Argument Form and Instance:**

An *argument form* (or schema) is the framework of an argument that results when certain portions of the component statements are replaced by blanks, schematic letters, or other symbols. An *argument instance* is what results when the blanks in a form are appropriately filled in. (7)

**Argument of P:**

An *argument* of *P* is a finite set of two or more wffs of *P*, one of which is the conclusion, while the others are premises. (221)

**Argument of S:**

An *argument* of *S* is a finite set of two or more wffs of *S*, one of which is the

425

conclusion, while the others are premises. (81)

**Asymmetry:**

$$(\forall x)(\forall y)(\mathbb{R}xy \rightarrow \neg\mathbb{R}yx) \quad (229)$$

**Atomic Formula, Molecular Formula:**

Any wff that qualifies simply in virtue of clause (1) of the definition of a wff (that is, any wff that just is some statement letter or some predicate letter with the appropriate number of terms), is called an *atomic* wff. By analogy, all other wffs are *molecular*. (50, 157)

**Axiom Schema of Abstraction:**

For any clearly stated condition, Sx, there exists a set B whose elements are exactly those objects which satisfy Sx:

$$(\exists B)(\forall x)(x \in B \leftrightarrow Sx)$$
$$B = \{x \mid Sx\} \quad (270)$$

**Axiom Schema of Separation:**

For every set A and every condition Sx there is a set B whose elements are exactly those members of A for which Sx holds (we require that y not be free in S).

$$(\exists B)(\forall x)(x \in B \leftrightarrow (x \in A \wedge Sx))$$
$$B = \{x \in A \mid Sx\} \quad (274)$$

**Axiom of Extensionality:**

If sets A and B have exactly the same members, then they are identical:

$$(\forall x)((x \in A \leftrightarrow x \in B) \rightarrow A = B) \quad (271)$$

**Axiom of Pairing:**

For any two elements, x and y, there exists a set, A, consisting of just those elements.

$$(\exists A)(\forall z)(z \in A \leftrightarrow z = x \vee z = y)$$
$$\{x, y\} = \{z \mid z = x \vee z = y\} \quad (281)$$

**Bijective:**

Function $f: A \rightarrow B$ is *bijective*, or one-one and onto, iff it is both injective and surjective. Such a function is called a bijection. (286)

**Bound Variable, Free Variable:**

An occurrence of a variable x in a wff $\mathbb{P}$ is *bound* iff it is within the scope of an x-quantifier. An occurrence of a variable is *free* iff it is not bound. (158)

**Cardinality:**

The number of elements in a set (considered without regard to any order relations on the elements) is called the cardinality of a set. We write the cardinality of a set A as:

$$|A| \quad (286)$$

**Cartesian Product:**

The *Cartesian product* of two sets A and B is the set of all ordered pairs $(x, y)$ with $x \in A$ and $y \in B$:

$$A \times B = \{(x, y) \mid x \in A \wedge y \in B\} \quad (282)$$

**Closed Branch:**

A branch on a truth tree (or modal tree) is a *closed branch* iff the branch contains both some atomic wff $\mathbb{P}$ and its negation $\neg \mathbb{P}$. (94, 321)

**Closed Tree:**

A truth tree (or modal tree) is a *closed tree* iff every branch is closed. (94, 321)

**Closed Tree Structure:**

A modal tree structure is a *closed structure* iff there is some world at which the tree is closed. (321)

**Closed Wff, Sentence of *P*:**

A wff of *P* is *closed* iff it contains no free occurrences of variables. We also call such wffs *sentences* of *P*. (159)

**Codomain:**

The *codomain* of a function is the set of possible images. The codomain is a superset of the range, where the range is the set of actual images. We can indicate the domain, A, and codomain, B, of a function, f, as follows:

$$f: A \to B$$

Here, f is a function from A to B. (285)

**Cogency:**

An argument is *cogent* if and only if it is inductively strong AND all the premises are true. (14)

**Compound Sentence:**

A *compound sentence* is a sentence that either contains one or more simple sentences and at least one compounding phrase, or contains a compound subject or a compound predicate. (31)

**Conclusion:**

See **Argument, Premise, Conclusion**. (6)

**Conclusion Indicators:**
> therefore, hence, thus, so, we may infer, consequently, it follows that (6)

**Constant:**
> See **Individual Terms**. (154)

**Contingency:**
> A statement, $\mathbb{P}$, is *contingent*, or is a *contingency* iff it is neither necessary nor impossible, that is, if it is possibly true and possibly false. (300)

**Counterexample:**
> A *counterexample* to an argument (form) is an argument instance of exactly the same form having all true premises and a false conclusion. Production of a counterexample shows that the argument form and all instances thereof are invalid. (Failure to produce a counterexample shows nothing, however.) (9)

**Deductive Validity, Invalidity:**
> An argument (form) is *deductively valid* if and only if it is NOT possible for ALL the premises to be true AND the conclusion false. An argument (form) is *deductively invalid* if and only if it is not valid. (8)

**Definite Description:**
> A *definite description* is a phrase that is supposed to designate an object via a unique description of it—i.e., a description that, if satisfied, is satisfied by one and only one object. (192)

**Derivable in *SD*, *PD*, $\Gamma \vdash \mathbb{P}$:**
> A wff $\mathbb{P}$ of $P$ is *derivable in SD, or PD*, from a set $\Gamma$ of wffs of $P$ iff there is a derivation in *SD*, or *PD* the primary assumptions of which are members of $\Gamma$ and $\mathbb{P}$ depends on only those assumptions. (125, 257)

**Derivation in *SD*, *PD*:**
> A *derivation in SD, or PD* is a finite sequence of wffs of $P$ such that each wff is either an assumption with scope indicated or justified by one of the rules of *SD*, or *PD*. (125, 257)

**Difference:**
> The difference (or relative complement) between sets A and B is the set containing every member of A that is not in B:

$$A - B = \{x \mid x \in A \wedge x \notin B\} \quad (278)$$

**Domain, Range, Field:**
> For each (2-place) relation R, we can distinguish the set of elements in the first position, called the *domain* of the relation, the set of elements in the second position, called the *range* of the relation, and the union of these two sets, called the *field* of

the relation:

$$\mathscr{D}(R) = \{x \mid (\exists y)Rxy\}$$
$$\mathscr{R}(R) = \{y \mid (\exists x)Rxy\}$$
$$\mathscr{F}(R) = \mathscr{D}(R) \cup \mathscr{R}(R) \quad (283)$$

**Equivalent in *SD*, *PD*:**

Two wffs $\mathbb{P}$ and $\mathbb{Q}$ are *equivalent in SD, or PD* iff they are interderivable in *SD*, or *PD*; i.e., iff both $\mathbb{P} \vdash \mathbb{Q}$ and $\mathbb{Q} \vdash \mathbb{P}$. (128, 258)

**Expression of *S*, *P*, $S^\square$:**

An *expression of S, P, or $S^\square$* is any finite sequence of the symbols of *S*, *P*, or $S^\square$. (46, 155, 305)

**Field:**

See **Domain, Range, Field**. (283)

**Free Variable:**

See **Bound Variable**. (158)

**Function:**

A relation F is a *function* iff it is many-one; that is, iff:

$$(\forall x)(\forall y)(\forall z)((Fzx \wedge Fzy) \rightarrow x = y)$$

In this case we write:

$$f(x) = y$$

Here, x is the argument and y is the image of x under f. (285)

**Identity:**

By *identity* we mean numerical identity—that what appear to be distinct objects are actually *one and the same thing*. Object x is *identical* to object y iff x is y. The two place predicate '① = ②' is always interpreted as ① is identical to ②. (188, 189)

**Impossibility:**

A statement, $\mathbb{P}$, is *impossible*, or is an *impossibility* iff it could not be true. (300)

**Inconsistent in *SD*, *PD*:**

A set $\Gamma$ of wffs is *inconsistent in SD, or PD* iff, for some wff $\mathbb{P}$, both $\Gamma \vdash \mathbb{P}$ and $\Gamma \vdash \neg \mathbb{P}$. (128, 259)

**Individual Terms:**

**Individual Constants:**

$a, b, \ldots, v, a_1, b_1, \ldots, v_1, a_2, \ldots$

**Individual Variables:**

$$w, x, y, z, w_1, x_1, y_1, z_1, w_2, \ldots \quad (154)$$

**Inductive Strength:**

An argument is *inductively strong* to the degree to which the premises provide evidence to make the truth of the conclusion plausible or probable. If an argument is not strong, it is *weak*. (14)

**Injective, One-One:**

Function $f: A \to B$ is *injective*, or one-one, iff every element of B is the value of at most one element of A,

$$(\forall x)(\forall y)(((x \in A \land y \in A) \land f(x) = f(y)) \to x = y)$$

Such a function is called an injection. (286)

**Interpretation of *P*:**

An *interpretation M of P* consists of: a non-empty universe of discourse, **UD**; a function, **E**, that assigns extensions to predicate letters and objects to constants; the set **A** of all variable assignments, **a**; and a denotation function $\mathbf{d_a^M}$ that assigns objects from the **UD** to individual terms of *P*. See full definition in text. (208)

**Interpretation of *P* (Informal):**

An *interpretation of P* consists of 3 components: (i) a non-empty universe of discourse, (ii) an assignment of truth values or natural language statements to statement letters and of natural language predicates or extensions to predicate letters, and (iii) an assignment of objects from the universe of discourse to constants. See full definition in text. (146)

**Interpretation of *S*:**

An *interpretation of S* assigns semantic value to the statement letters of *S*, either via a translation key or a truth value assignment. See also **Truth Value Assignment**. (52)

**Interpretation of $S^\square$:**

An *interpretation M of $S^\square$* consists of: a non-empty set, **W**, of worlds, w; and a function, **E**, which for each $w \in W$, assigns T or F to each statement letter of $S^\square$. (307)

**Interpretation of $S^\square$ (Informal):**

$\square \mathbb{P}$ is T iff $\mathbb{P}$ is T in every world.
$\lozenge \mathbb{P}$ is T iff $\mathbb{P}$ is T in at least one world. (304)

**Intersection:**

The intersection of a pair of sets A and B, $A \cap B$, the set consisting of all and only the elements in both sets:

$$A \cap B = \{x \mid x \in A \land x \in B\} \quad (276)$$

**Intersection, Generalized:**
> For every set of sets A, there is some set C whose elements are all and only the elements in every $B \in A$:

$$(\exists C)(\forall x)(x \in C \leftrightarrow (\forall B)(B \in A \rightarrow x \in B))$$
$$\cap A = \{x \mid \text{for every } B \in A, x \in B\} \quad (276)$$

**Invalid, Valid:**
> See **Deductive Validity**. (8)

**Invalidity, Validity:**
> See **Deductive Validity**. (8)

**Irreflexivity:**
> $(\forall x)\neg \mathbb{R}xx$     (229)

**Logic:**
> *Logic* is the study of (i) criteria for distinguishing successful from unsuccessful argument, (ii) methods for applying those criteria, and (iii) related properties of statements such as implication, equivalence, logical truth, consistency, etc. (5)

**Logically Consistent:**
> A set of statements is *logically consistent* if and only if it is possible for all the statements to be true. (22)

**Logically Contingent:**
> A statement is *logically contingent* if and only if it is neither logically true nor logically false; i.e., it is both possible for the statement to be true, and possible for the statement to be false. (21)

**Logically Contradictory:**
> A pair of statements is *logically contradictory* if and only if it is not possible for the statements to have the same truth values. (22)

**Logically Entails, Logically Follows:**
> A set of statements *logically entails* a target statement if and only if it is NOT possible for every member of the set to be true AND the target statement false. We also say that the target statement *logically follows* from the set. (23)

**Logically Equivalent:**
> A pair of statements is *logically equivalent* if and only if it is not possible for the statements to have different truth values. (22)

**Logically False:**
> A statement is *logically false* if and only if it is not possible for the statement to be true. Such statements are sometimes called self-contradictions. (21)

**Logically Inconsistent:**
 A set of statements is *logically inconsistent* if and only if it is not possible for all
 the statements to be true. (23)

**Logically True:**
 A statement is *logically true* if and only if it is not possible for the statement to be
 false. Such statements are sometimes called tautologies. (21)

**Main Connective, Main Operator, Well-Formed Components:**
 Atomic wffs have no main connective (operator). The *main connective or operator*
 of a molecular wff $\mathbb{R}$ is the connective (operator) appearing in the clause of the
 definition of a wff cited last in showing $\mathbb{R}$ to be a wff. The *immediate well-formed
 components* of a molecular wff are the values of $\mathbb{P}$ and $\mathbb{Q}$ (in some cases simply $\mathbb{P}$)
 in the last-cited clause of the definition of a wff. The *well-formed components* of a
 wff are the wff itself, its immediate well-formed components, and the well-formed
 components of its immediate well-formed components. The *atomic components* of
 a wff are the well-formed components that are atomic wffs. (50, 158)

**Many-Many:**
 A relation R is *many-many* when it is neither one-many nor many-one.

 $\neg(\forall x)(\forall y)(\forall z)((Rxz \land Ryz) \to x = y) \land \neg(\forall x)(\forall y)(\forall z)((Rzx \land Rzy) \to x = y)$
 $(\exists x)(\exists y)(\exists z)((Rxz \land Ryz) \land x \neq y) \land (\exists x)(\exists y)(\exists z)((Rzx \land Rzy) \land x \neq y)$
 (284)

**Many-One:**
 A relation R is *many-one* when, if an object z bears R to x, then z bears R to no
 other object y. The relation son to biological father is many-one.

 $$(\forall x)(\forall y)(\forall z)((Rzx \land Rzy) \to x = y) \quad (284)$$

**Metalanguage:**
 When one is talking about a language, the *metalanguage* is the language in which
 one is talking about the object language. (42)

**Metalogic:**
 *Metalogic* is the study of the properties of logical systems. In particular, it is systematic reasoning in a metalanguage about the properties of object language systems
 of logic. (5, 85)

**Metavariables:**
 *Metavariables* are variables of the metalanguage that range over (take as possible
 values) expressions of the object language. We use Blackboard Bold:

 $$\mathbb{A}, \mathbb{B}, \mathbb{C}, \ldots, \mathbb{Z}, \mathbb{A}_1, \ldots$$
 $$\mathbb{A}_k^n, \mathbb{B}_k^n, \ldots, \mathbb{Z}_k^n$$
 $$\mathbb{a}, \mathbb{b}, \mathbb{c}, \ldots, \mathbb{z}, \mathbb{a}_1, \ldots \quad (44, 155)$$

**Modal Tree Structure:**

A *modal tree structure* for a set of wffs is a structure that lists all the wffs, and all the well-formed components that must be true in attempting to show the set modally consistent, typically including trees at various worlds. (321)

**Modality Operators:**

□◇    (305)

**Modally Consistent Set of Wffs:**

A set $\Gamma$ of wffs of $S^{\square}$ is *modally consistent* iff on at least one interpretation there is a world in which all members of $\Gamma$ are true. $\Gamma$ is *modally inconsistent* iff it is not modally consistent. (310)

**Modal Tree Test:** A set $\Gamma$ is modally consistent iff it has an open tree structure. (321)

**Modal Tree Test:** A set $\Gamma$ is modally inconsistent iff it has a closed tree structure. (321)

**Modally Contingent Wff:**

A wff $\mathbb{P}$ of $S^{\square}$ is *modally contingent* iff $\mathbb{P}$ is neither modally true nor modally false; i.e., iff on some interpretation it is false in some world and on some interpretation it is true in some world. (309)

**Modal Tree Test:** A wff $\mathbb{P}$ is modally contingent iff $\{\mathbb{P}\}$ has an open tree structure and $\{\neg\mathbb{P}\}$ has an open tree structure. (321)

**Modally Contradictory Pair of Wffs:**

Wffs $\mathbb{P}$ and $\mathbb{Q}$ of $S^{\square}$ are *modally contradictory* iff there is no interpretation containing a world in which they have the same truth value. (310)

**Modal Tree Test:** Wffs $\mathbb{P}$ and $\mathbb{Q}$ are modally contradictory iff $\{\mathbb{P} \leftrightarrow \mathbb{Q}\}$ has a closed tree structure. (322)

**Modally Entails:**

A set $\Gamma$ of wffs of $S^{\square}$ *modally entails* a wff $\mathbb{P}$, $\Gamma \vDash \mathbb{P}$, iff there is no interpretation containing a world in which all the members of $\Gamma$ are true and $\mathbb{P}$ is false. (310)

**Modal Tree Test:** A set of wffs, $\Gamma$, modally entails a target wff, $\mathbb{P}$, iff $\Gamma \cup \{\neg\mathbb{P}\}$ has a closed tree structure. (322)

**Modally Equivalent Pair of Wffs:**

Wffs $\mathbb{P}$ and $\mathbb{Q}$ of $S^{\square}$ are *modally equivalent* iff there is no interpretation containing a world in which they differ in truth value. (309)

**Modal Tree Test:** Wffs $\mathbb{P}$ and $\mathbb{Q}$ are modally equivalent iff $\{\neg(\mathbb{P} \leftrightarrow \mathbb{Q})\}$ has a closed tree structure. (322)

**Modally False Wff:**

A wff $\mathbb{P}$ of $S^{\square}$ is *modally false* iff on every interpretation $\mathbb{P}$ is false in every world. (309)

**Modal Tree Test:** A wff $\mathbb{P}$ is modally false iff $\{\mathbb{P}\}$ has a closed tree structure. (321)

**Modally Falsifiable Wff:**

A wff $\mathbb{P}$ of $S^\square$ is *modally falsifiable* iff $\mathbb{P}$ is not modally true; i.e., on some interpretation it is false in some world. (309)

**Modal Tree Test:** A wff $\mathbb{P}$ is modally falsifiable iff $\{\neg\mathbb{P}\}$ has an open tree structure. (322)

**Modally Satisfiable Wff:**

A wff $\mathbb{P}$ of $S^\square$ is *modally satisfiable* iff $\mathbb{P}$ is not modally false; i.e., on some interpretation it is true in some world. (309)

**Modal Tree Test:** A wff $\mathbb{P}$ is modally satisfiable iff $\{\mathbb{P}\}$ has an open tree structure. (322)

**Modally True Wff:**

A wff $\mathbb{P}$ of $S^\square$ is *modally true* iff on every interpretation $\mathbb{P}$ is true in every world. (309)

**Modal Tree Test:** A wff $\mathbb{P}$ is modally true iff $\{\neg\mathbb{P}\}$ has a closed tree structure. (321)

**Modally Valid Argument:**

An argument of $S^\square$ is *modally valid* iff there is no interpretation containing a world in which all the premises are true and the conclusion is false. An argument of $S^\square$ is *modally invalid* iff it is not modally valid. (310)

**Modal Tree Test:** An argument is modally valid iff the set consisting of all and only the premises and the negation of the conclusion has a closed tree structure. (322)

**Model:**

A tva is a *model* for (or *models*) a wff $\mathbb{P}$ (or set of wffs $\Gamma$) iff the wff $\mathbb{P}$ (or all wffs in $\Gamma$) are true on that tva. (72)

**Molecular Formula:**

See **Atomic Formula**. (50)

**Necessity:**

A statement, $\mathbb{P}$, is *necessary*, *necessarily true*, or is a *necessity* iff it could not be false. (300)

**Non-Necessity:**

A statement, $\mathbb{P}$, is *non-necessary*, or is *possibly false* iff it is not necessary, that is, if its negation is possibly true. (300)

**Nonreflexivity:**

$\neg(\forall x)Rxx$

$(\exists x)\neg Rxx$    (229)

**Nonsymmetry:**
$$\neg(\forall x)(\forall y)(\mathbb{R}xy \to \mathbb{R}yx)$$
$$(\exists x)(\exists y)(\mathbb{R}xy \wedge \neg\mathbb{R}yx) \qquad (229)$$

**Nontransitivity:**
$$\neg(\forall x)(\forall y)(\forall z)((\mathbb{R}xy \wedge \mathbb{R}yz) \to \mathbb{R}xz)$$
$$(\exists x)(\exists y)(\exists z)((\mathbb{R}xy \wedge \mathbb{R}yz) \wedge \neg\mathbb{R}xz) \qquad (229)$$

**Object Language:**
When one is talking about a language, the *object language* is the language being talked about. (42)

**One-Many:**
A relation R is *one-many* when, if an object x bears R to an object z, then no other object y bears R to z. The relation biological father to son is one-many.

$$(\forall x)(\forall y)(\forall z)((Rxz \wedge Ryz) \to x = y) \qquad (284)$$

**One-One:**
A relation R is *one-one* when it is both one-many and many-one.

$$(\forall x)(\forall y)(\forall z)((Rxz \wedge Ryz) \to x = y) \wedge (\forall x)(\forall y)(\forall z)((Rzx \wedge Rzy) \to x = y)$$
$$(\forall x)(\forall y)(\forall z)(((Rxz \wedge Ryz) \vee (Rzx \wedge Rzy)) \to x = y) \qquad (284)$$

**Open Branch:**
A branch on a truth tree (or modal tree) is an *open branch* iff it is not closed. (94, 321)

**Open Tree:**
A truth tree (or modal tree) is an *open tree* iff at least one branch is open. (94, 321)

**Open Tree Structure:**
A modal tree structure is an *open structure* iff there is an open tree at every world. (321)

**Open Wff:**
A wff of *P* is *open* iff it contains at least one free occurrence of a variable. Otherwise it is a closed wff. (158) See also **Closed Wff, Sentence of *P*.**

**Operator:**
The truth-functional connectives and quantifiers are *operators*. (157)

**Ordered Pair:**
For any two elements, x and y, there exists a set, $(x, y)$, such that

$$(x, y) = \{\{x\}, \{x, y\}\} \qquad (281)$$

**Possibility:**
    A statement, $\mathbb{P}$, is *possible*, *possibly true*, or is a *possibility* iff it is not impossible. (300)

**Powerset Axiom:**
    For each set A, there is a set C containing all subsets B of A:

$$(\exists C)(\forall B)(B \in C \leftrightarrow B \subseteq A)$$
$$\mathscr{P}(A) = \{B \mid B \subseteq A\} \qquad (282)$$

**Predicate:**
    A *predicate* is a series of words with one or more blanks that yields a sentence when all its blanks are filled with singular terms. Conversely, we could think of a predicate as what remains after removing one or more singular terms from a sentence. (145)

**Predicate Letters:**
$$A_1^0, B_1^0, \ldots, Z_1^0, A_2^0, B_2^0, \ldots, Z_2^0, A_3^0, \ldots$$
$$A_1^1, B_1^1, \ldots, Z_1^1, A_2^1, B_2^1, \ldots, Z_2^1, A_3^1, \ldots$$
$$A_1^2, B_1^2, \ldots, Z_1^2, A_2^2, B_2^2, \ldots, Z_2^2, A_3^2, \ldots$$
$$A_1^3, B_1^3, \ldots, Z_1^3, A_2^3, B_2^3, \ldots, Z_2^3, A_3^3, \ldots$$
$$\vdots \qquad \qquad \ddots$$

That is, any uppercase letter with zero or positive integer superscript, $n$, indicating the number of places, and positive integer subscript, $k$, to give us an infinite (denumerable) supply. (154)

**Premise:**
    See **Argument, Premise, Conclusion**. (6)

**Premise Indicators:**
    as, since, for, because, given that, for the reason that, inasmuch as (6)

**Principle of Substitutivity:**
    Where $\mathbb{a}$ and $\mathbb{b}$ are singular terms of $\boldsymbol{P}$,

$$\text{if } \mathbb{a} = \mathbb{b}, \text{ then } \mathbb{F}\mathbb{b} \Leftrightarrow \mathbb{F}(\mathbb{a}/\mathbb{b}) \qquad (228)$$

**Proper Subset, Proper Inclusion:**
    Set A is a *proper subset of* or is *properly included in* B iff every member of A is also a member of B, but $A \neq B$. We use the '$\subset$' symbol:

$$A \subset B \leftrightarrow (A \subseteq B \wedge A \neq B) \qquad (275)$$

**Punctuation Marks:**
    ( )    (46, 154)

**Quantificational Logic:**
    *Quantificational Logic* is the logic of sentences involving quantifiers, predicates,

and names. It investigates the properties that arguments, sentences, and sets of sentences have in virtue of their quantificational structure. (144)

**Quantificationally Consistent Set of Wffs:**

A set $\Gamma$ of wffs of **P** is *quantificationally consistent* iff there is at least one interpretation on which all members of $\Gamma$ are true. We also say that $\Gamma$ has a *model* or is *modeled*. $\Gamma$ is *quantificationally inconsistent* iff it is not quantificationally consistent. (220)

**Quantificationally Contingent Wff:**

A wff $\mathbb{P}$ of **P** is *quantificationally contingent* iff $\mathbb{P}$ is neither quantificationally true nor quantificationally false; i.e., iff it is false on at least one interpretation and true on at least one interpretation (at least one interpretation is model, but not all are). (216)

**Quantificationally Contradictory Pair of Wffs:**

Wffs $\mathbb{P}$ and $\mathbb{Q}$ of **P** are *quantificationally contradictory* iff there is no interpretation on which they have the same truth value (no model of $\mathbb{P}$ is a model of $\mathbb{Q}$, and no model of $\mathbb{Q}$ is a model of $\mathbb{P}$). (219)

**Quantificationally Entails:**

A set $\Gamma$ of wffs of **P** *quantificationally entails* a wff $\mathbb{P}$ iff there is no interpretation on which all the members of $\Gamma$ are true and $\mathbb{P}$ is false. (Every model of $\Gamma$ is a model of $\mathbb{P}$.) (221)

**Quantificationally Equivalent Pair of Wffs:**

Wffs $\mathbb{P}$ and $\mathbb{Q}$ of **P** are *quantificationally equivalent* iff there is no interpretation on which they differ in truth value (every model of $\mathbb{P}$ is a model of $\mathbb{Q}$, and every model of $\mathbb{Q}$ is a model of $\mathbb{P}$). (219)

**Quantificationally False Wff:**

A wff $\mathbb{P}$ of **P** is *quantificationally false* iff $\mathbb{P}$ is false on every interpretation (no interpretation is a model). (216)

**Quantificationally Falsifiable Wff:**

A wff $\mathbb{P}$ of **P** is *quantificationally falsifiable* iff $\mathbb{P}$ is not quantificationally true; i.e., it is false on at least one interpretation (at least one interpretation is not a model). (217)

**Quantificationally Satisfiable Wff:**

A wff $\mathbb{P}$ of **P** is *quantificationally satisfiable* iff $\mathbb{P}$ is not quantificationally false; i.e., it is true on at least one interpretation. We also say that the wff has a *model* or is *modeled*. (216)

**Quantificationally True Wff:**

A wff $\mathbb{P}$ of **P** is *quantificationally true* iff $\mathbb{P}$ is true on every interpretation (every interpretation is a model). (216)

**Quantificationally Valid Argument:**
>An argument of $P$ is *quantificationally valid* iff there is no interpretation on which all the premises are true and the conclusion is false. An argument of $P$ is *quantificationally invalid* iff it is not quantificationally valid. (221)

**Quantifier of $P$:**
>Where x ranges over individual variables, expressions of the form $(\forall x)$ are called *universal quantifiers*, while expressions of the form $(\exists x)$ are called *existential quantifiers*. We may also refer to a quantifier by the particular variable it contains— e.g., '$(\forall y_3)$' is a universal $y_3$-quantifier, while '$(\exists x)$' is an existential x-quantifier. (155)

**Quantifier Symbols:**
>$\forall$  $\exists$     (154)

**Range:**
>See **Domain, Range, Field**. (283)

**Recursive Definition:**
>*Recursive definitions* consist of three parts: the basis, the recursive clause(s), and the extremal clause. (47)

**Reflexivity:**
>$(\forall x)\mathbb{R}xx$
>$(\forall x)((x,x) \in R)$     (226, 283)

**Relation:**
>A (2-place) *relation* R is a set of ordered pairs:

$$R \text{ is a relation } \leftrightarrow (\forall z)(z \in R \rightarrow (\exists x)(\exists y)(z = (x,y)))     \quad (283)$$

**Satisfaction and Truth in $P$ (Informal):**
>Given an interpretation, where $\mathbb{F}x$ is a wff with only instances of the variable x free, and a is a constant:

>(1) The denotation of a *satisfies* $\mathbb{F}x$ (or $\mathbb{F}x$ is *true of* the object named by a) iff $\mathbb{F}a$ is T

>(2) A universally quantified wff $(\forall x)\mathbb{F}x$ is true iff the condition expressed by the immediate subcomponent $\mathbb{F}x$ is satisfied by *every* object in the **UD**

>(3) An existentially quantified wff $(\exists x)\mathbb{F}x$ is true iff the condition expressed by the immediate subcomponent $\mathbb{F}x$ is satisfied by *at least one* object in the **UD** (201)

**Satisfaction in $P$:**
>Given an interpretation **M**, satisfaction is a relation between a variable assignment, **a**, and $\mathbb{P}$, a wff of the language $P$. See full definition in text. (212)

**Scope:**

   The *scope* of a connective or operator is that portion of the wff containing its immediate sentential component(s). (50, 158)

**Semantics:**

   *Semantics* is the study of language with regard to meaningful interpretations or valuations of the components. (45)

**Seriality:**

   $(\forall x)(\exists y)\mathbb{R}xy$    (229)

**Simple Sentence:**

   A *simple sentence* is a sentence that contains one subject and one predicate. (31)

**Singular Term:**

   A *singular term* is a word or phrase that designates or is supposed to designate some individual object. Natural language singular terms are either proper nouns or definite descriptions (a phrase that is supposed to designate an object via a unique description of it). (144)

**Soundness:**

   An argument is *sound* if and only if it is deductively valid AND all its premises are true. (8)

**Statement:**

   A *statement* is a declarative sentence; a sentence that attempts to state a fact—as opposed to a question, a command, an exclamation. (5)

**Statement Letters:**

   $A, B, C, \ldots, Z, A_1, B_1, C_1, \ldots, Z_1, A_2, \ldots$    (46)

**Strength:**

   See **Inductive Strength**.

**Subset, Inclusion:**

   Set A is a *subset of* or is *included in* B iff every member of A is also a member of B. We use the '$\subseteq$' symbol: (275)

   $$A \subseteq B \leftrightarrow (\forall x)(x \in A \rightarrow x \in B)$$

**Substitution Instance:**

   Let $\mathbb{Q}(a/x)$ indicate the wff that is just like $\mathbb{Q}$ except for having the constant a in every position where the variable x appears in $\mathbb{Q}$. Where $\mathbb{P}$ is a closed wff of the form $(\forall x)\mathbb{Q}$ or $(\exists x)\mathbb{Q}$, then $\mathbb{Q}(a/x)$ is a *substitution instance* of $\mathbb{P}$, with a as the *instantiating constant*. (159, 238)

**Surjective, Onto:**

   Function $f: A \rightarrow B$ is *surjective*, or *onto*, iff every element in B is the value of at least one element of A,

$$(\forall y)(y \in B \rightarrow (\exists x)(x \in A \wedge f(x) = y))$$

Such a function is called a surjection. (286)

**Symbols of *P*:**
See **Predicate Letters** (154), **Individual Terms, Individual Constants, Individual Variables** (154), **Truth-Functional Connectives** (154), **Quantifier Symbols** (154), **Punctuation Marks** (154).

**Symbols of *S*:**
See **Statement Letters** (46), **Truth-Functional Connectives** (46), **Punctuation Marks** (46).

**Symmetric Difference:**
The symmetric difference (or symmetric complement) of sets A and B is the union of the differences:

$$A \ominus B = (A - B) \cup (B - A)$$

(279)

**Symmetry:**
$(\forall x)(\forall y)(\mathbb{R}xy \rightarrow \mathbb{R}yx)$
$(\forall x)(\forall x)((x, y) \in R \rightarrow (y, x) \in R)$    (227, 283)

**Syntax:**
*Syntax* is the study of the signs of a language with regard only to their formal properties. (45)

**Theorem of *SD, PD*:**
A wff $\mathbb{P}$ is a *theorem of SD, or PD* iff $\mathbb{P}$ is derivable from the empty set; i.e., iff $\vdash \mathbb{P}$. (127, 258)

**Transitivity:**
$(\forall x)(\forall y)(\forall z)((\mathbb{R}xy \wedge \mathbb{R}yz) \rightarrow \mathbb{R}xz)$
$(\forall x)(\forall y)(\forall z)(((x, y) \in R \wedge (y, z) \in R) \rightarrow (x, z) \in R)$    (227, 284)

**Truth Table:**
A *truth table* for a wff or set of wffs is a structure that lists the wff or wffs, all relevant truth value assignments, and the truth value of each wff on each truth value assignment. (75)

**Truth Tree:**
A *truth tree* for a set of wffs is a structure that lists all the wffs, and all the well-formed components that must be true in attempting to show the set truth-functionally consistent. (94)

**Truth Value:**

The *truth value* of a statement is just its truth or falsehood. At this point we make the assumption that every statement is either true (has the truth value true) or false (has the truth value false) but not both. The truth value of a given statement is fixed whether or not we *know* what that truth value is. (5)

**Truth Value Assignment:**

A *truth value assignment* (or tva) is an assignment of the value true (abbreviated by a 'T') or the value false ('F'), but not both, to each of the atomic wffs of *S*. A truth value assignment is a function from the set of atomic wffs into the set {T, F}. Also called an interpretation. (71)

**Truth at a World in $S^\square$:**

Truth in $S^\square$ is determined by a valuation function **V**, relative to an interpretation **M**. See full definition in text. (307)

**Truth in *P*:**

A closed wff, or sentence, $\mathbb{P}$ of *P* is T *on interpretation* **M** iff every variable assignment **a** satisfies $\mathbb{P}$ on **M**. A closed wff of $\mathbb{P}$ is F on **M** iff no **a** satisfies $\mathbb{P}$ on **M**. (213)

**Truth-Functional Compound:**

A sentence is a *truth-functional compound* iff the truth value of the compound sentence is completely and uniquely determined by (is a function of) the truth values of the simple component sentences. Otherwise, the compound sentence is *non-truth-functional*. (32)

**Truth-Functional Connectives:**

$\neg \wedge \vee \rightarrow \leftrightarrow$   (46, 154)

**Truth-Functional Logic:**

*Truth-functional logic* is the logic of truth-functional combinations of simple sentences. It investigates the properties that arguments, sentences, and sets of sentences have in virtue of their truth-functional structure. (33)

**Truth-Functionally Consistent Set of Wffs:**

A set $\Gamma$ of wffs of *S* is *truth-functionally consistent* iff there is at least one truth value assignment on which all members of $\Gamma$ are true. We also say that $\Gamma$ has a *model* or is *modeled*. $\Gamma$ is *truth-functionally inconsistent* iff it is not truth-functionally consistent. (79)

**Truth Table Test:** A set $\Gamma$ of wffs is T-Fly consistent iff, in the set's joint truth table, there is at least one row in which all the members of $\Gamma$ have a T under their main connectives. $\Gamma$ is T-Fly inconsistent iff there is no such row. (79)

**Truth Tree Test:** A set $\Gamma$ is T-Fly consistent iff it has an open tree. $\Gamma$ is T-Fly inconsistent iff it has a closed tree. (95)

**Truth-Functionally Contingent Wff:**

A wff $\mathbb{P}$ of $S$ is *truth-functionally contingent* iff $\mathbb{P}$ is neither truth-functionally true nor truth-functionally false; i.e., iff it is false on at least one truth value assignment and true on at least one truth value assignment. (75)

**Truth Table Test:** $\mathbb{P}$ is T-Fly contingent iff, in its truth table, at least one F and at least one T appear in the column under the main connective. (75)

**Truth Tree Test:** A wff $\mathbb{P}$ is T-Fly contingent iff $\{\mathbb{P}\}$ has an open tree and $\{\neg\mathbb{P}\}$ has an open tree. (95)

**Truth-Functionally Contradictory Pair of Wffs:**

Wffs $\mathbb{P}$ and $\mathbb{Q}$ of $S$ are *truth-functionally contradictory* iff there is no truth value assignment on which they have the same truth value. (78)

**Truth Table Test:** Wffs $\mathbb{P}$ and $\mathbb{Q}$ are T-Fly contradictory iff, in their joint truth table, there is no row in which the values under their main connectives are the same. (78)

**Truth Tree Test:** Wffs $\mathbb{P}$ and $\mathbb{Q}$ are T-Fly contradictory iff $\{\mathbb{P} \leftrightarrow \mathbb{Q}\}$ has a closed tree. (97)

**Truth-Functionally Entails:**

A set $\Gamma$ of wffs of $S$ *truth-functionally entails* a wff $\mathbb{P}$ iff there is no truth value assignment on which all the members of $\Gamma$ are true and $\mathbb{P}$ is false. (Every model of $\Gamma$ is a model of $\mathbb{P}$.) (80)

**Truth Table Test:** A set $\Gamma$ T-Fly entails a wff $\mathbb{P}$ iff there is no row of their joint truth table on which all the members of $\Gamma$ have a T under their main connectives and $\mathbb{P}$ has an F under its main connective. (80)

**Truth Tree Test:** A set of wffs, $\Gamma$, T-Fly entails a target wff, $\mathbb{P}$, iff $\Gamma$ together with $\neg\mathbb{P}$ has a closed tree. (99)

**Truth-Functionally Equivalent Pair of Wffs:**

Wffs $\mathbb{P}$ and $\mathbb{Q}$ of $S$ are *truth-functionally equivalent* iff there is no truth value assignment on which they differ in truth value (every model of $\mathbb{P}$ is a model of $\mathbb{Q}$, and every model of $\mathbb{Q}$ is a model of $\mathbb{P}$). (77)

**Truth Table Test:** Wffs $\mathbb{P}$ and $\mathbb{Q}$ are T-Fly equivalent iff, in their joint truth table, there is no row in which the values under their main connectives differ. (77)

**Truth Tree Test:** Wffs $\mathbb{P}$ and $\mathbb{Q}$ are T-Fly equivalent iff $\{\neg(\mathbb{P} \leftrightarrow \mathbb{Q})\}$ has a closed tree. (97)

**Truth-Functionally False Wff:**

A wff $\mathbb{P}$ of $S$ is *truth-functionally false* iff $\mathbb{P}$ is false on every truth value assignment (no tva is a model). (75)

**Truth Table Test:** $\mathbb{P}$ is T-Fly false iff, in its truth table, only Fs appear in the column under the main connective. (75)

**Truth Tree Test:** A wff $\mathbb{P}$ is T-Fly false iff $\{\mathbb{P}\}$ has a closed tree. (95)

**Truth-Functionally Falsifiable Wff:**

A wff $\mathbb{P}$ of $S$ is *truth-functionally falsifiable* iff $\mathbb{P}$ is not truth-functionally true; i.e., it is false on at least one truth value assignment. (76)

**Truth Table Test:** $\mathbb{P}$ is T-Fly falsifiable iff, in its truth table, at least one F appears in the column under the main connective. (76)

**Truth Tree Test:** A wff $\mathbb{P}$ is T-Fly falsifiable iff $\{\neg\mathbb{P}\}$ has an open tree. (95)

**Truth-Functionally Satisfiable Wff:**

A wff $\mathbb{P}$ of $S$ is *truth-functionally satisfiable* iff $\mathbb{P}$ is not truth-functionally false; i.e., it is true on at least one truth value assignment. We also say that the wff has a *model* or is *modeled*. (76)

**Truth Table Test:** $\mathbb{P}$ is T-Fly satisfiable (has a model) iff, in its truth table, at least one T appears in the column under the main connective. (76)

**Truth Tree Test:** A wff $\mathbb{P}$ is T-Fly satisfiable iff $\{\mathbb{P}\}$ has an open tree. (95)

**Truth-Functionally True Wff:**

A wff $\mathbb{P}$ of $S$ is *truth-functionally true* iff $\mathbb{P}$ is true on every truth value assignment (every tva is a model). (75)

**Truth Table Test:** $\mathbb{P}$ is T-Fly true iff, in its truth table, only Ts appear in the column under the main connective. (75)

**Truth Tree Test:** A wff $\mathbb{P}$ is T-Fly true iff $\{\neg\mathbb{P}\}$ has a closed tree. (95)

**Truth-Functionally Valid Argument:**

An argument of $S$ is *truth-functionally valid* iff there is no truth value assignment on which all the premises are true and the conclusion is false. An argument of $S$ is *truth-functionally invalid* iff it is not truth-functionally valid. (81)

**Truth Table Test:** An argument is T-Fly valid iff there is no row of their joint truth table on which all the premises have a T under their main connectives and the conclusion has an F under its main connective. If there is such a row, the argument is T-Fly invalid. (81)

**Truth Tree Test:** An argument is T-Fly valid iff the set consisting of all and only the premises and the negation of the conclusion has a closed tree. If the tree is open, the argument is T-Fly invalid. (99)

**Union:**

The union of a pair of sets A and B, $A \cup B$, the set consisting of all and only the elements in either set:

$$A \cup B = \{x \mid x \in A \vee x \in B\} \quad (277)$$

**Union Axiom:**

For every set of sets A, there is some set C whose elements are all and only the elements of at least one $B \in A$:

$$(\exists C)(\forall x)(x \in C \leftrightarrow (\exists B)(B \in A \land x \in B))$$
$$\bigcup A = \{x \mid \text{for some } B \in A, x \in B\} \quad (278)$$

**Valid, Invalid:**
>   See **Deductive Validity**. (8)

**Valid in *PD*:**
>   An argument of *P* is *valid in PD* iff the conclusion is derivable from the set consisting of only the premises, otherwise it is invalid in *PD*. (258)

**Valid in *SD*:**
>   An argument of *S* is *valid in SD* iff the conclusion is derivable from the set consisting of only the premises, otherwise it is invalid in *SD*. (127)

**Validity, Invalidity:**
>   See **Deductive Validity**. (154)

**Variable:**
>   See **Individual Terms**. (154)

**Well-Formed Formula, Wff:**
>   A *well-formed formula* or *wff* is a grammatical formula of the language *S*, *P*, or $S^{\square}$. Wffs of *P* may be open or closed. See full definitions in text. (47, 155, 305)

# C Truth Tables, Tree Rules, and Derivation Rules

## C.1 Characteristic Truth Tables

| $\mathbb{P}\,\mathbb{Q}$ | $\neg\mathbb{P}$ | $\mathbb{P}\wedge\mathbb{Q}$ | $\mathbb{P}\vee\mathbb{Q}$ | $\mathbb{P}\to\mathbb{Q}$ | $\mathbb{P}\leftrightarrow\mathbb{Q}$ |
|---|---|---|---|---|---|
| T T | F | T | T | T | T |
| T F |   | F | T | F | F |
| F T | T | F | T | T | F |
| F F |   | F | F | T | T |

| $\mathbb{P}$ | $\neg\mathbb{P}$ |
|---|---|
| T | F |
| F | T |

| $\mathbb{P}\,\mathbb{Q}$ | $\mathbb{P}\wedge\mathbb{Q}$ |
|---|---|
| T T | T |
| T F | F |
| F T | F |
| F F | F |

| $\mathbb{P}\,\mathbb{Q}$ | $\mathbb{P}\vee\mathbb{Q}$ |
|---|---|
| T T | T |
| T F | T |
| F T | T |
| F F | F |

| $\mathbb{P}\,\mathbb{Q}$ | $\mathbb{P}\to\mathbb{Q}$ |
|---|---|
| T T | T |
| T F | F |
| F T | T |
| F F | T |

| $\mathbb{P}\,\mathbb{Q}$ | $\mathbb{P}\leftrightarrow\mathbb{Q}$ |
|---|---|
| T T | T |
| T F | F |
| F T | F |
| F F | T |

## C.2  **Truth Tree Rules for** $S$

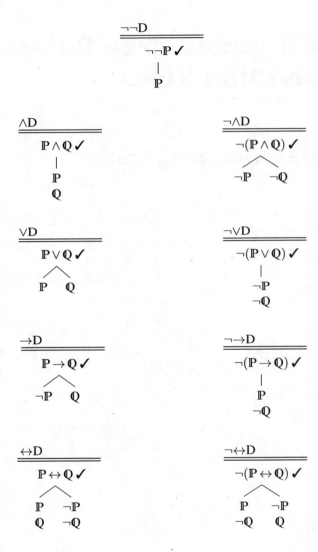

# C.3  **Modal Tree Rules for** $S^\square$

The tree rules for $S$, plus

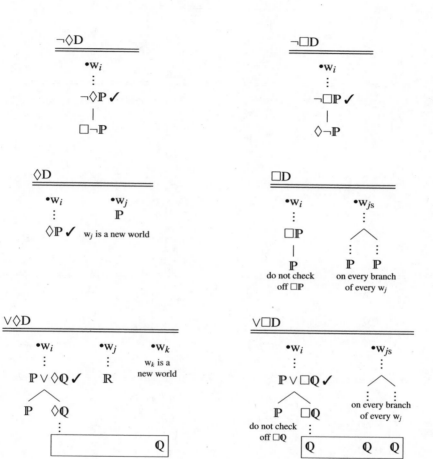

## C.4 **The Derivation System** *SD*

R
═══════════════════════════

$i$ | P

▷ | P                                          $i$ R

---

∧I
═══════════════════════════

$i$ | P
$j$ | Q

▷ | P ∧ Q                   $i, j$ ∧I

---

∧E
═══════════════════════════

$i$ | P ∧ Q

▷ | P                                          $i$ ∧E
-or-
▷ | Q                                          $i$ ∧E

---

→I
═══════════════════════════

$i$ |  | P                   A

$j$ |  | Q
▷ | P → Q                   $i$–$j$ →I

---

→E
═══════════════════════════

$i$ | P → Q
$j$ | P

▷ | Q                        $i, j$ →E

---

¬I
═══════════════════════════

$i$ |  | P                   A

  |  | Q
$j$ |  | ¬Q
▷ | ¬P                       $i$–$j$ ¬I

---

¬E
═══════════════════════════

$i$ |  | ¬P                  A

  |  | Q
$j$ |  | ¬Q
▷ | P                        $i$–$j$ ¬E

---

∨I
═══════════════════════════

$i$ | P

▷ | P ∨ Q                    $i$ ∨I
-or-
▷ | Q ∨ P                    $i$ ∨I

---

∨E
═══════════════════════════

$i$ | P ∨ Q
$j$ |  | P                   A

$k$ |  | R
$l$ |  | Q                   A

$m$ |  | R
▷ | R                        $i, j$–$k, l$–$m$ ∨E

---

↔I
═══════════════════════════

$i$ |  | P                   A

$j$ |  | Q
$k$ |  | Q                   A

$l$ |  | P
▷ | P ↔ Q                    $i$–$j$, $k$–$l$ ↔I

---

↔E
═══════════════════════════

$i$ | P ↔ Q
$j$ | P

▷ | Q                        $i, j$ ↔E
-or-
$i$ | P ↔ Q
$j$ | Q

▷ | P                        $i, j$ ↔E

# C.5  **The Derivation System** *SDE*

The rules of *SD*, plus

## Inference Rules

**MT**

| | |
|---|---|
| $i$ | $\mathbb{P} \to \mathbb{Q}$ |
| $j$ | $\neg\mathbb{Q}$ |
| $\triangleright$ | $\neg\mathbb{P}$    $i, j$ MT |

**HS**

| | |
|---|---|
| $i$ | $\mathbb{P} \to \mathbb{Q}$ |
| $j$ | $\mathbb{Q} \to \mathbb{R}$ |
| $\triangleright$ | $\mathbb{P} \to \mathbb{R}$    $i, j$ HS |

**DS**

| | | | |
|---|---|---|---|
| $i$ | $\mathbb{P} \vee \mathbb{Q}$ | $i$ | $\mathbb{P} \vee \mathbb{Q}$ |
| $j$ | $\neg\mathbb{P}$ | $j$ | $\neg\mathbb{Q}$ |
| $\triangleright$ | $\mathbb{Q}$    $i, j$ DS | $\triangleright$ | $\mathbb{P}$    $i, j$ DS |

*-or-*

## Replacement Rules

**Com**

$\mathbb{P} \wedge \mathbb{Q} \quad \triangleleft\triangleright \quad \mathbb{Q} \wedge \mathbb{P}$

$\mathbb{P} \vee \mathbb{Q} \quad \triangleleft\triangleright \quad \mathbb{Q} \vee \mathbb{P}$

**Assoc**

$\mathbb{P} \wedge (\mathbb{Q} \wedge \mathbb{R}) \quad \triangleleft\triangleright \quad (\mathbb{P} \wedge \mathbb{Q}) \wedge \mathbb{R}$

$\mathbb{P} \vee (\mathbb{Q} \vee \mathbb{R}) \quad \triangleleft\triangleright \quad (\mathbb{P} \vee \mathbb{Q}) \vee \mathbb{R}$

**Impl**

$\mathbb{P} \to \mathbb{Q} \quad \triangleleft\triangleright \quad \neg\mathbb{P} \vee \mathbb{Q}$

**DN**

$\mathbb{P} \quad \triangleleft\triangleright \quad \neg\neg\mathbb{P}$

**DeM**

$\neg(\mathbb{P} \wedge \mathbb{Q}) \quad \triangleleft\triangleright \quad \neg\mathbb{P} \vee \neg\mathbb{Q}$

$\neg(\mathbb{P} \vee \mathbb{Q}) \quad \triangleleft\triangleright \quad \neg\mathbb{P} \wedge \neg\mathbb{Q}$

**Idem**

$\mathbb{P} \quad \triangleleft\triangleright \quad \mathbb{P} \wedge \mathbb{P}$

$\mathbb{P} \quad \triangleleft\triangleright \quad \mathbb{P} \vee \mathbb{P}$

**Trans**

$\mathbb{P} \to \mathbb{Q} \quad \triangleleft\triangleright \quad \neg\mathbb{Q} \to \neg\mathbb{P}$

**Exp**

$\mathbb{P} \to (\mathbb{Q} \to \mathbb{R}) \quad \triangleleft\triangleright \quad (\mathbb{P} \wedge \mathbb{Q}) \to \mathbb{R}$

**Dist**

$\mathbb{P} \wedge (\mathbb{Q} \vee \mathbb{R}) \quad \triangleleft\triangleright \quad (\mathbb{P} \wedge \mathbb{Q}) \vee (\mathbb{P} \wedge \mathbb{R})$

$\mathbb{P} \vee (\mathbb{Q} \wedge \mathbb{R}) \quad \triangleleft\triangleright \quad (\mathbb{P} \vee \mathbb{Q}) \wedge (\mathbb{P} \vee \mathbb{R})$

**Equiv**

$\mathbb{P} \leftrightarrow \mathbb{Q} \quad \triangleleft\triangleright \quad (\mathbb{P} \to \mathbb{Q}) \wedge (\mathbb{Q} \to \mathbb{P})$

$\mathbb{P} \leftrightarrow \mathbb{Q} \quad \triangleleft\triangleright \quad (\mathbb{P} \wedge \mathbb{Q}) \vee (\neg\mathbb{P} \wedge \neg\mathbb{Q})$

## C.6 **The Derivation System** *PD*

The rules of *SDE*, plus

∀I

$$i \mid \mathbb{P}(a/x)$$

$$\triangleright \mid (\forall x)\mathbb{P} \qquad i \, \forall I$$

Provided:
- (i)  a does not occur in an undischarged assumption.
- (ii)  a does not occur in $(\forall x)\mathbb{P}$.

∀E

$$i \mid (\forall x)\mathbb{P}$$

$$\triangleright \mid \mathbb{P}(a/x) \qquad i \, \forall E$$

∃I

$$i \mid \mathbb{P}(a/x)$$

$$\triangleright \mid (\exists x)\mathbb{P} \qquad i \, \exists I$$

∃E

$$i \mid (\exists x)\mathbb{P}$$

$$j \mid \mid \mathbb{P}(a/x) \qquad A$$

$$k \mid \mid \mathbb{Q}$$

$$\triangleright \mid \mathbb{Q} \qquad i, j{-}k \, \exists E$$

Provided:
- (i)  a does not occur in an undischarged assumption.
- (ii)  a does not occur in $(\exists x)\mathbb{P}$.
- (iii)  a does not occur in $\mathbb{Q}$.

## C.7 **The Derivation System** *PDE*

The rules of *PD*, plus

QN

$$\neg(\forall x)\mathbb{P} \;\; \triangleleft\triangleright \;\; (\exists x)\neg\mathbb{P}$$
$$\neg(\exists x)\mathbb{P} \;\; \triangleleft\triangleright \;\; (\forall x)\neg\mathbb{P}$$

## C.8 **Modal Derivation Systems**

The rules of *SDE*, plus

### System *K*

Primitive Rules:

Derived Rules:

---

Dual

$\Diamond \mathbb{P}$ ◁▷ $\neg \Box \neg \mathbb{P}$
$\Box \mathbb{P}$ ◁▷ $\neg \Diamond \neg \mathbb{P}$

---

K◊

| | |
|---|---|
| $i$ | $\Box (\mathbb{P} \to \mathbb{Q})$ |
| ▷ | $\Diamond \mathbb{P} \to \Diamond \mathbb{Q}$        $i$ K◊ |

---

N

| | | |
|---|---|---|
| $i$ | $\mathbb{P}$ | ⊢ |
| ▷ | $\Box \mathbb{P}$ | $i$ N |

---

Dist

$\Box (\mathbb{P} \land \mathbb{Q})$ ◁▷ $\Box \mathbb{P} \land \Box \mathbb{Q}$
$\Diamond (\mathbb{P} \lor \mathbb{Q})$ ◁▷ $\Diamond \mathbb{P} \lor \Diamond \mathbb{Q}$

---

K

| | | |
|---|---|---|
| $i$ | $\Box (\mathbb{P} \to \mathbb{Q})$ | |
| ▷ | $\Box \mathbb{P} \to \Box \mathbb{Q}$ | $i$ K |

---

ME

$\neg \Diamond \mathbb{P}$ ◁▷ $\Box \neg \mathbb{P}$
$\neg \Box \mathbb{P}$ ◁▷ $\Diamond \neg \mathbb{P}$

---

### System *D*

*K* plus D

Primitive Rule:

---

D

| | | |
|---|---|---|
| $i$ | $\Box \mathbb{P}$ | |
| ▷ | $\Diamond \mathbb{P}$ | $i$ D |

## System *T*

*K* plus T

Primitive Rule:                                    Derived Rules: D, and …

T                                                  T◇

i │ □P                                             i │ P

▷ │ P              i T                              ▷ │ ◇P              i T◇

## System *B*

*T* plus B

Primitive Rule:                                    Derived Rule:

B                                                  B◇□

i │ P                                              i │ ◇□P

▷ │ □◇P            i B                              ▷ │ P              i B◇□

## System *S4*

*T* plus S4

Primitive Rule:                                    Derived Rule:

S4                                                 S4◇

i │ □P                                             i │ ◇◇P

▷ │ □□P            i S4                             ▷ │ ◇P              i S4◇

## System *S5*

*T* plus S5

Primitive Rule:                    Derived Rules: B, S4, and …

S5
─────────

| $i$ | $\Diamond \mathbb{P}$ |
|---|---|
| ▷ | $\Box \Diamond \mathbb{P}$       *i* S5 |

S5◇□
─────────

| $i$ | $\Diamond \Box \mathbb{P}$ |
|---|---|
| ▷ | $\Box \mathbb{P}$       *i* S5◇□ |

## Relations Between Modal Systems

S → S′ indicates that S′ is a proper extension of S. That is, everything derivable in S is derivable in S′, but some things derivable in S′ are not derivable in S. Thus, S′ is stronger than S.

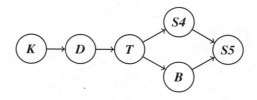

# From the Publisher

A name never says it all, but the word "Broadview" expresses a good deal of the philosophy behind our company. We are open to a broad range of academic approaches and political viewpoints. We pay attention to the broad impact book publishing and book printing has in the wider world; we began using recycled stock more than a decade ago, and for some years now we have used 100% recycled paper for most titles. Our publishing program is internationally oriented and broad-ranging. Our individual titles often appeal to a broad readership too; many are of interest as much to general readers as to academics and students.

Founded in 1985, Broadview remains a fully independent company owned by its shareholders—not an imprint or subsidiary of a larger multinational.

For the most accurate information on our books (including information on pricing, editions, and formats) please visit our website at www.broadviewpress.com. Our print books and ebooks are also available for sale on our site.

On the Broadview website we also offer several goods that are not books—among them the Broadview coffee mug, the Broadview beer stein (inscribed with a line from Geoffrey Chaucer's *Canterbury Tales*), the Broadview fridge magnets (your choice of philosophical or literary), and a range of T-shirts (made from combinations of hemp, bamboo, and/or high-quality pima cotton, with no child labor, sweatshop labor, or environmental degradation involved in their manufacture).

All these goods are available through the "merchandise" section of the Broadview website. When you buy Broadview goods you can support other goods too.

broadview press
www.broadviewpress.com

The interior of this book is printed on 100% recycled paper.

PERMANENT   100%   BIO GAS® ENERGY   Ancient Forest Friendly™